Maritime Autonomous Vehicles

Other related titles:

You may also like

- SBRA537 | Weimin Huang and Eric W. Gill | Ocean Remote Sensing Technologies: High frequency, marine and GNSS-based radar | 2021
- SBRA525 | Frank Ehlers | Autonomous Underwater Vehicles: Design and practice | 2020
- PBTR011 | Sanjay Sharma and Bidyadhar Subudhi | Navigation and Control of Autonomous Marine Vehicles | 2019

We also publish a wide range of books on the following topics:
Computing and Networks
Control, Robotics and Sensors
Electrical Regulations
Electromagnetics and Radar
Energy Engineering
Healthcare Technologies
History and Management of Technology
IET Codes and Guidance
Materials, Circuits and Devices
Model Forms
Nanomaterials and Nanotechnologies
Optics, Photonics and Lasers
Production, Design and Manufacturing
Security
Telecommunications
Transportation

All books are available in print via https://shop.theiet.org or as eBooks via our Digital Library https://digital-library.theiet.org.

Radar, Sonar and Navigation 570

Maritime Autonomous Vehicles

Methods and measurements

Edited by
Frank Ehlers

Institution of Engineering and Technology

About the IET

This book is published by the Institution of Engineering and Technology (The IET).

We inspire, inform and influence the global engineering community to engineer a better world. As a diverse home across engineering and technology, we share knowledge that helps make better sense of the world, to accelerate innovation and solve the global challenges that matter.

The IET is a not-for-profit organisation. The surplus we make from our books is used to support activities and products for the engineering community and promote the positive role of science, engineering and technology in the world. This includes education resources and outreach, scholarships and awards, events and courses, publications, professional development and mentoring, and advocacy to governments.

To discover more about the IET, please visit https://www.theiet.org/.

About IET books

The IET publishes books across many engineering and technology disciplines. Our authors and editors offer fresh perspectives from universities and industry. Within our subject areas, we have several book series steered by editorial boards made up of leading subject experts.

We peer review each book at the proposal stage to ensure the quality and relevance of our publications.

Get involved

If you are interested in becoming an author, editor, series advisor, or peer reviewer, please visit https://www.theiet.org/publishing/publishing-with-iet-books/ or contact author_support@theiet.org.

Discovering our electronic content

All of our books are available online via the IET's Digital Library. Our Digital Library is the home of technical documents, eBooks, conference publications, real-life case studies and journal articles. To find out more, please visit https://digital-library.theiet.org.

In collaboration with the United Nations and the International Publishers Association, the IET is a Signatory member of the SDG Publishers Compact. The Compact aims to accelerate progress to achieve the Sustainable Development Goals (SDGs) by 2030. Signatories aspire to develop sustainable practices and act as champions of the SDGs during the Decade of Action (2020-2030), publishing books and journals that will help inform, develop and inspire action in that direction.

In line with our sustainable goals, our UK printing partner has FSC accreditation, which helps reduce our environmental impact on the planet. We use a print-on-demand model to further reduce our carbon footprint.

British Library Cataloguing in Publication Data

A catalogue record for this product is available from the British Library

ISBN 978-1-83953-918-3 (hardback)
ISBN 978-1-83953-919-0 (PDF)

Typeset in India by MPS Limited

Cover Image: Drone shot of coral reefs in a tropical ocean By Tdub303 via Getty Images

Contents

Foreword

This book is a natural follow-on of the Edited IET-Book 'Autonomous Underwater Vehicles (AUVs) - Design and practice' (published 2020), because it extends the specific underwater-related topics for AUVs to maritime topics. By that extension, it is abstracting design processes and best practices to methods, and it is specifying their interfaces and standards for real-world measurements. That is, this book is embedding 'design and practice' in the digitalization process, which, in many domains and governed by model-based engineering paradigms and digital twin concepts, is rapidly changing our society, economy and security.

The two exciting fields of work, robotics and maritime research, which are to be combined when maritime unmanned vehicles enter the scene, meet again autonomy, as a disruptive technology, whereby now (five years later than the AUVs-Design and practice Book) also the methods of how to implement effective, robust and efficient systems, are in the reach of a 'learning at machine speed'.

Written by a team of top-class international contributors, this book provides an entrance into the hot topic of Maritime Autonomous Vehicles–Methods and measurements. The reader is invited to take a guided tour from the state-of-the-art in maritime robotics to highlighted topics in coordination, communication, control and biomimetic design, via real-world application of Artificial Intelligence tool chains for underwater datasets, to a sophisticated reference/simulation/surrogate model world capable of addressing the necessity of maintaining holistically the interrelation of the many details real maritime environments need to have to be accurately described. I would like to thank all authors for their excellent work.

Many thanks to Nicki Dennis who supported this editorial work. Furthermore, the process continued and concluded with the excellent cooperation with Olivia Watkins, Paul Deards and Brittany Insull from IET Books. Thank you for your help, support, guidelines and for letting this book become a reality.

Despite the fact that a lot of literature on Maritime Autonomous Vehicles is already available, compiling a book from the viewpoint of embedding design and practice into the digitalization process has been an effort worthwhile. Generating a digital framework that enables automated decision-making between relevant and irrelevant information is a prerequisite to understand the interconnectivity between Methods and Measurements. As understanding seems to be the most sustainable result of scientific and engineering work, I hope that the reader, after returning from the guided tour presented in this book, gets motivated to invest in generating such digital frameworks, which are, by design, great collaborative tools for scientists and engineers.

Frank Ehlers,
02.06.2025

About the editor

Frank Ehlers is a principal scientist at the Bundeswehr Technical Center for Ships and Naval Weapons, Maritime Technology and Research (WTD 71) in Germany. He has worked as a scientist, project manager, program manager or principal scientist in German and international research institutes dealing with maritime technology for over 30 years. He has conducted both application-oriented research and more fundamental research in the fields of signal processing, detection, classification, data fusion, coordinated distributed sensor systems and autonomy for maritime scenarios. He is also the editor of *Autonomous Underwater Vehicles – Design and practice* (2020).

List of authors

Akbulut, Batuhan	Bahcesehir University, Turkey
Baaj, Ahmad	Bahcesehir University, Turkey
Bechlioulis, Charalampos P.	University of Patras, Greece
Garafano, Vittorio	Delft University, The Netherlands
Gavriilidis, Konstantinos	University of Pisa, Italy
Gur, Berke	Marmara University, Turkey
Hastie, Helen	Heriot-Watt University, UK
Heiland, Erik	University of the Bundeswehr Munich, Germany
Heshmati-Alamdari, Shahab	Aalborg University, Denmark
Hillmann, Peter	University of the Bundeswehr Munich, Germany
Indiveri, Giovanni	University of Genova, Italy
Jones, Colin N.	EPFL, Switzerland
Karcher, Andreas	University of the Bundeswehr Munich, Germany
Kapoor, Gaurav	University of Luebeck, Germany
Karras, George C.	University of Thessaly, Greece
Kyriakopoulos, Kostas J.	University of Athens, Greece
Maehle, Erik	University of Luebeck, Germany
Munafo, Andrea	University of Pisa, Italy
Negenborn, Rudy	Delft University, The Netherlands
Nienaber, Sören	University of Luebeck, Germany
Pang, Wei	University of Edinburgh, UK
Pang, Yusong	Delft University, The Netherlands
Pascoal, António M.	University of Lisbon, Portugal
Rego, Francisco F.C.	University of Lisbon, Portugal
Simetti, Enrico	University of Genova, Italy
Wanderlingh, Francesco	University of Genova, Italy
Zenz, Lovis J.I.	University of the Bundeswehr Munich, Germany

Chapter 1

Introduction

Frank Ehlers[1]

The maritime domain is of enormous interest to our society, our economy, our health, and our security. Operating in the maritime domain, however, is dangerous, expensive, and most of the time at low execution speed. Unmanned systems "like" such environments, but there has to be an overarching cycle executed prior to the successful realization of Maritime Autonomous Vehicles: starting from the design and modeling of unmanned systems to control and implementation, exhibiting a reliable and explainable "amount" of autonomy and self-organization amid sudden and possibly extreme disturbances occurring in the uncertain and dynamic maritime environment.

Once the unmanned systems are operating at sea, they provide massive data: due to new sensors and platforms with high resolution and accuracy and due to complex oceanographic and biological processes with very high parametric dimensions. This Big Data scenario needs to be handled with specific standards and interfaces, just as the Digitalization Process is running in other domains: the Internet of Things, Digital Twins, and Artificial Intelligence are highly applicable to the maritime domain, especially for measurement and decision-making processes onboard unmanned systems. However, unlike a car that can stop or a drone that can land, a maritime autonomous vehicle needs to "stand a rough night at sea" or a subset of all the dangers such "rough nights" offer: lack of communications, diverging wind and current forces, fish bites, etc. If long endurance missions are in focus, corrosion by salty water and biological activities on the vehicle hulls become challenges, just to name a few.

In this Introduction, we start with a recapitulation of the findings from the Edited IET Book "Autonomous Underwater Vehicles – Design and practice" (in Section 1.1), and then, we really commence our guide tour to embed Maritime Autonomous Vehicles into the Digitalization Process by looking into the Methods during design and application and into the Measurements and their exploitation in an overall optimization cycle.

[1]Bundeswehr Technical Center for Ships and Naval Weapons, Maritime Technology and Research (WTD 71), Kiel, Germany

1.1   Autonomous underwater vehicles as a starting point

As mentioned in the Foreword, this book is a natural follow-on of the edited IET Book on "Autonomous Underwater Vehicles (AUVs) – Design and practice" (published 2020), extending the specific "underwater"-related topic to "maritime," abstracting design processes and best practices, and specifying their interfaces and standards. This said, the natural starting point for this Introduction is the final section of the AUV-book: "Autonomy in and for AUVs – design and practice":

[…], at the first glance, AUVs belong to the class of unmanned or auto-mated systems, which are removing or remotely locating the operator. This reduces cost of the operation, reduces the size of the platform, and it removes risk of life because the maritime operation is a naturally harsh environment. In this sense, AUVs can be viewed as maritime robots. When AUVs are built for carrying sensors only (i.e. without manipulators), they are, in this framework, off-board sensors.

Reading through the chapters of this book [AUVs-Design and prac-tice], however, we have learnt that AUVs can be (much!) more. They can interactively adapt to their environment, which is a sign of autonomy. Onboard processing power will allow them to adapt to machine speed. This enables that we will see in the future that AUVs can even do more: learning to improve with experience.

At present, we are using AUVs to gather more knowledge about the processes in the maritime environment. Characterizing the maritime envi-ronment without a certain level of knowledge regarding these processes turns out to end up in data sets that are too "big" to draw validated con-clusions from. In other words, by just measuring and data storage, we cannot disentangle the various processes in the maritime environment and label them according to their relevance, given a certain operational task. Therefore, a complete picture of the environment and the AUV is needed, including sensors, communications, navigation algorithms, i.e. a holistic view filled with details like those presented in the chapters of this book [AUVs-Design and practice]. Interestingly, the impression from the future work recommendations formulated by the authors of the chapters is that access to information from such a complete picture will help improve the automatic or autonomous capabilities currently implemented in AUVs. This scenario describes a feedback loop: a better picture of the environment leads to better ways to produce To prepare this feedback loop for crossing its real-ity gap, the relevance of details and their connectivity in the design space needs to be explored. To enable this, AUV design and practice have to employ verification and validation mechanisms. The danger, however, is to make institutional decisions (also called standardization) too quickly, for the ease of verification and validation. Instead, first, the community has to generate the data for such institutional decision-making by continuing with the design and practice of AUVs until the ignition of autonomy embedded in (and thereby solving) the design space exploration task generates an eternal flame. Who knows when, or if at all, this will happen?

Meanwhile, just a few years later, following this idea of using autonomy in the aforementioned catalytic manner, the mathematics for solving the Design Space Exploration in Robotics has further evolved (e.g. Frank Ehlers, Design Space Exploration in Robotics, Springer, 2025), and we are theoretically in a position to describe a metric for a reliable and explainable "amount" of autonomy. Practically, we still have to formulate methods that can be applied by the robot in a format in line with the theoretical framework, and we have to synchronize the simulations in such Digital Twins with the real-world measurements.

1.2   A guided tour for embedding MAV into the digitalization process

We start the tour with an overview of the application of autonomy in the maritime domain. In the chapter "Autonomy in the Maritime Domain" by Francesco Wanderlingh, Enrico Simetti and Giovanni Indiveri, the authors explore the technologies that enable advanced maritime autonomous systems, their applications, and finally future developments in maritime autonomy, highlighting their transformative impacts and ongoing research efforts to further improve their functionalities.

Especially in addressing robustness and facing the challenge of cooperative networked marine systems, the chapter "Advances in Remote Estimation and Control with Quantized Communications for Networked Systems subject to Severe Bandwidth Constraints" by Francisco F.C. Rego, Colin N. Jones, and António M. Pascoal addresses the problem of remote state estimation and control for systems subjected to process and measurement noise under stringent emitter–receiver communication bandwidth constraints. At the core of the framework adopted for systems' design and analysis is the adoption of an observer structure for systems that may be operating under state feedback control. Numerical simulation examples illustrate the performance of the proposed algorithm and indicate that the methodologies developed hold potential for further extensions to address challenging problems in the area of cooperative networked marine systems.

Looking into the embodiment of collaborative underwater vehicles with manipulator capability, the chapter "Collaboration of Multiple Underwater Vehicle-Manipulator Systems" by Shahab Heshmati-alamdari, Charalampos P. Bechlioulis, George C. Karras, and Kostas J. Kyriakopoulos defines and addresses the problem of cooperative object transportation by a team of Underwater Vehicle Manipulator Systems operating within a constrained workspace containing static obstacles. Two control strategies are given, where the coordination is achieved solely through implicit communication, which emerges from each robot's onboard sensor measurements and the physical interactions between the robots and the commonly grasped object. Consequently, no explicit online data exchange occurs among the robots.

Finding another solution in the Design Space for underwater manipulation robots, the chapter "Propelling URSULA: Bioinspired Undulating Fins for Robotic

Squid in Underwater Manipulation and Intervention Missions Laboratory Tests and Measurements" by Berke Gur, Ahmad Baaj, and Batuhan Akbulut explains that the project URSULA aims to develop a next-generation robotic system that will enable dexterous underwater manipulation and seabed intervention in ways that current unmanned underwater vehicles cannot. The robot features a biomimetic design inspired by squids, combining a hydrodynamic and agile body with multi-functional robotic manipulators. The robot also integrates numerous unconventional and innovative technologies, such as tendon-driven soft robotic limbs, propeller-less underwater propulsion systems, perception-guided autonomous navigation and posture control, high-bandwidth visible light-based wireless underwater communication, model-mediated teleoperation with haptic feedback, and virtual reality-enhanced operator support systems.

Continuing the collection of new measurements and exploration of new application areas with the Micro-AUV-MONSUN, the chapter "Application of the Swarm-Capable Micro-AUV MONSUN for Monitoring of Vegetation and Water Quality in Lakes" by Erik Maehle, Gaurav Kapoor, and Sören Nienaber introduces the concept and experimental findings related to the MONSUN micro Autonomous Underwater Vehicle (micro-AUV) created at the University of Lübeck for the purpose of monitoring environmental conditions in lakes. In addition to assessing physical–chemical parameters, the evaluation of macrophyte vegetation is carried out to determine water quality. The configuration of MONSUN's sensors and the navigation techniques employed for capturing images of macrophytes are detailed, along with the recognition process for their automatic evaluation using AI methods.

Taking Automatic Target Recognition (ATR) as one example for AI methods, in the Chapter "Reference Modeling for Automatic Target Recognition" by Lovis Justin Immanuel Zenz, Erik Heiland, Peter Hillmann, and Andreas Karcher, the authors state that ATR constitutes a fundamental capability of Maritime Autonomous Vehicles, with corresponding services manifesting in a variety of forms, while relying on common core concepts – e.g., an overarching process and the main tasks forming it. Expanding beyond maritime contexts, such services find further application across different domains. A Reference Model for ATR (ATR-RM) facilitates the communication of these core concepts, enabling diverse ATR applications to build upon it. Subsequently, solution models can be derived from the ATR-RM, and knowledge acquired during that process can be reintegrated into the ATR-RM. Altogether, this fosters the cross-domain sharing of knowledge.

Tuning model frameworks to point at explainability is the focus of the chapter "Surrogate Model Framework for Explainable Autonomous Behaviour in Maritime Robotic Systems" by Konstantinos Gavriilidis, Andrea Munafo, Wei Pang, and Helen Hastie, where the authors state that with the surge in application and deployment of robotic and autonomous systems across various sectors, there is an ever-growing demand for transparency and explainability. The maritime domain, which has embraced these technologies for tasks ranging from disaster site inspections to pipeline maintenance, is no exception. The chapter will discuss a novel surrogate model framework that aims to bring explainability to autonomous behaviors in maritime robotic systems, bridging the existing gap between complex robotic decisions and the human understanding of their actions in relation to the desired mission. The chapter will provide an overview of why transparency and explainability are important for robotic and autonomous systems, especially in the maritime sector, exploring the complexity of conveying complex autonomous decisions in understandable terms for various stakeholders with potentially conflicting needs (e.g. in-field operators, remote controllers, and trained/untrained operators). It will introduce the framework of surrogate models for explainable decision-making, focusing on their utility in simplifying and approximating deterministic agent policies and offering explanations that are independent of the underlying autonomy model. Finally, the chapter will discuss the journey from simulated experiments to actual trials with maritime robots, emphasizing the changes, challenges, and learnings experienced during this transition. The chapter will conclude by highlighting the potential growth areas, applications, and refinements for the surrogate model framework in the future.

In the chapter "Towards a structured framework for control performance and safety assessment for Maritime Autonomous Surface Ship" by V. Garofano, Y. Pang, and R.R. Negenborn, the authors address the necessity to assess the control performance and safety of Maritime Autonomous Surface Ships. The chapter presents a structured framework designed to facilitate testing and data collection using autonomous ship systems, thereby supporting the verification that autonomous operations align with International Maritime Organization standards. The authors discuss the integration of key hardware and software components for robust autonomous operation and evaluate these systems using analytical performance criteria in both simulated and real-world scenarios. The chapter concludes by proposing new key performance indicators necessary for the continued development of autonomous maritime systems. Through extensive datasets and collaborative research via Open Science-focused algorithms and designs, the aim of the authors is to set the groundwork for future advancements in this field.

Chapter 2
Autonomy in the maritime domain

Francesco Wanderlingh[1,2], Enrico Simetti[1] and Giovanni Indiveri[1]

In the last decades, the maritime sector has experienced significant evolution, thanks to the advancements in autonomous systems that enable innovative functionalities in the domains of scientific, commercial, search and rescue, and defense operations. These kind of systems leverage a broad range of technologies, such as sensor integration, advanced perception, navigation, guidance, and machine learning, providing support to autonomous surface and underwater vehicles for handling dynamic and unpredictable marine environments. Autonomous marine platforms are integral to scientific research, allowing for continuous ocean monitoring, underwater inspections, and environmental data collection. In commercial and industrial scenarios, these systems can optimize shipping efficiency by reducing fuel consumption and improving safety. The defense sector also benefits from autonomous maritime systems in surveillance, counter-mine, and security operations, while search and rescue missions take advantage of autonomous vehicles for rapid response and disaster management. This chapter explores the technologies that enable these systems, their applications in the aforementioned fields, and lastly future developments in maritime autonomy, highlighting their transformative impact and ongoing research efforts to improve their functionalities.

2.1 Introduction

Over the past few decades, we have witnessed the fast evolution of autonomous systems, which are reshaping activities in the maritime sector, transforming various application areas including commercial shipping and offshore oil and gas to defense, environmental monitoring, and search and rescue. Several representative projects in the field, demonstrating the progress of autonomy in the marine domain, are presented in [1].

[1]DIBRIS Department, University of Genova (ISME Node), Italy
[2]All the authors are members of ISME, the Interuniversity Research Center of Integrated Systems for the Marine Environment.

Maritime autonomy refers to the capability of vessels and maritime systems to operate with varying degrees of independence from human intervention, enabled by advanced technologies, such as sensor integration, mission planning, and machine learning. It encompasses a spectrum of autonomy levels, ranging from partially automated systems that require human oversight to fully autonomous systems capable of making decisions and executing tasks without human input. An insightful classification of the definition of autonomy for ships is given in [2]. These capabilities involve integrating a series of technological systems designed to overcome the challenges of maritime environments, such as dynamic weather conditions, complex navigation requirements, and communication constraints. As autonomy reshapes the maritime domain, it raises critical questions about safety, regulations, and ethical considerations, making it a multidisciplinary field that intersects technology, law, and policy [3,4].

The diverse domains in which marine systems are used define the relative technological challenges and opportunities. In marine research, for instance, autonomous surface vehicles (ASVs), autonomous underwater vehicles (AUVs), and remotely operated vehicles (ROVs) enable scientists to explore remote and hazardous ocean environments, collect high-resolution data, and monitor ecosystems with minimal human risk and reduce the operational costs of activities typically performed by the aid of large, expensive vessels. These systems are critical for advancing our understanding of climate change, biodiversity, and underwater geology. In the realm of maritime security instead, autonomous systems such as drones and unmanned patrol vessels enhance surveillance, threat detection, and response capabilities, providing cost-effective and scalable solutions for safeguarding territorial waters and combating illegal activities like piracy and smuggling [5]. Furthermore, for the maritime industry, autonomous shipping and underwater vehicle-manipulator systems (UVMS) promise to optimize logistics, reduce operational costs, and improve safety by minimizing human error [6]. Additionally, autonomous systems contribute to sustainability efforts by enabling more efficient fuel use and reducing emissions. These advancements underscore the paradigm-shifting potential of autonomy in addressing global challenges and driving innovation in the maritime sector. An illustration of typical surface and underwater platforms and operations is shown in Figure 2.1.

The distinction between autonomy and automation can give us insight into the evolution of maritime technologies. Automation refers to systems that perform predefined tasks based on human-programmed instructions, operating within a fixed set of parameters without the ability to adapt to unforeseen circumstances. In contrast, autonomy implies a higher degree of decision-making capability, where systems can perceive their environment, interpret data, and make independent decisions to achieve specific goals. In maritime applications, this distinction plays a key role: automated systems might control engine functions or stabilize a vessel in calm conditions, while autonomous systems could collect data, navigate complex routes, avoid collisions, or adapt to dynamic weather conditions without human intervention. Understanding this continuum from automation to full autonomy is essential for designing systems that balance human oversight with technological innovation, ensuring safety and reliability in the unpredictable maritime environment.

Figure 2.1 *Overview of marine robotics platforms. From left to right: a team of torpedo-shaped AUVs performing a coverage path for acoustic surveying, an ASV–ROV cooperative system, an AUV performing an underwater archaeology mission, and a UVMS performing inspection and maintenance on a submerged structure*

The following sections will provide a comprehensive overview of autonomy in the maritime domain. Section 2.2 delves into the key technologies that enable maritime autonomy, such as perception and sensing, navigation, guidance, and control (NGC), communication and networking, and energy and power systems. Subsequently, Section 2.3 explores various applications of autonomous maritime systems, including environmental and scientific research, commercial and industrial uses, search, rescue, and humanitarian efforts, as well as maritime security and defense. Section 2.4 then highlights the role of marine robotics research, focusing on experimental platforms and field deployments for both single robot autonomy and multi-robot systems. Finally, Section 2.5 looks ahead to future directions in maritime autonomy research, suggesting ongoing advancements and emerging trends in the field.

2.2 Key technologies enabling maritime autonomy

2.2.1 Extrinsic perception and sensing

Perception and sensing technologies for the surrounding environment (i.e. exteroception) are key for marine robotics and autonomous vehicles, as they enable situational awareness, environmental mapping, obstacle avoidance, and autonomous decision-making in the highly dynamic maritime environment. These systems rely on a diverse range of sensor modalities, each with distinct advantages and limitations. For example, in above-water environments, the LiDAR (Light Detection and Ranging) sensor provides high-resolution 3D mapping and object detection data, making it valuable for surface vessels navigating near ports or obstacles. However, its performance is severely degraded by fog, rain, and over long distances across water. In contrast, Radar typically offers lower resolution compared to LiDAR but is robust in adverse weather conditions and offers long-range detection of objects and other vessels, making it essential for collision avoidance and maritime surveillance. Moving below the sea surface, sonar (Sound Navigation and Ranging) is the primary sensing modality for underwater perception, with variants such as multi-beam sonar for high-resolution seabed mapping and side-scan sonar for detecting submerged structures or objects. Forward-looking sonar (FLS) is particularly useful for navigation in confined or cluttered underwater environments. Optical cameras (paired with appropriate lighting equipment), including high-resolution RGB and stereoscopic cameras, provide detailed visual information, essential for marine exploration, object recognition, and close-range operations such as underwater inspection. However, their performances are limited by light attenuation, turbidity, and poor visibility at depth. Infrared (IR) cameras enhance perception in low-light conditions, assisting in detecting warm objects, such as vessels or marine life, above and near the surface, but they are ineffective underwater due to infrared absorption by water molecules.

Given the limitations of individual sensors, sensor fusion techniques play a critical role in improving perception accuracy and robustness. Multi-sensor integration combines complementary data from LiDAR, radar, sonar, optical, and infrared sensors to enhance situational awareness and provide redundancy in the case of individual sensor failure [7]. In this context, the general probabilistic approach of recursive Bayesian estimation is one of the most commonly used techniques. The derived algorithm implementations such as Kalman Filter (KF), Extended Kalman Filter (EKF), Unscented Kalman Filter (UKF), and Particle Filter (PF, also known as sequential Monte Carlo) are widely used for sensor fusion in navigation and localization, integrating IMU (Inertial Measurement Unit) data with Global Navigation Satellite System (GNSS) and vision-based localization. Particle filtering is effective for estimating non-Gaussian distributions in dynamic marine environments, particularly when tracking moving objects or navigating through uncertain conditions. More recently, deep learning-based sensor fusion techniques have been explored [8]. A key application of AI in marine robotics is deep learning for object recognition and tracking, which enhances the perception capabilities of autonomous surface and underwater vehicles. Convolutional Neural Networks (CNNs) and Vision Transformers (ViTs) are widely used for object detection, classifying marine targets, such as vessels, buoys, marine life, and underwater structures from optical, infrared, and sonar imagery.

For instance, CNN-based architectures such as YOLO (You Only Look Once) [9] and Faster R-CNN are used for real-time vessel detection and tracking. These systems enable autonomous ships to avoid collisions and navigate safely in busy waterways, as discussed in [10], with the detection output illustrated in Figure 2.2.

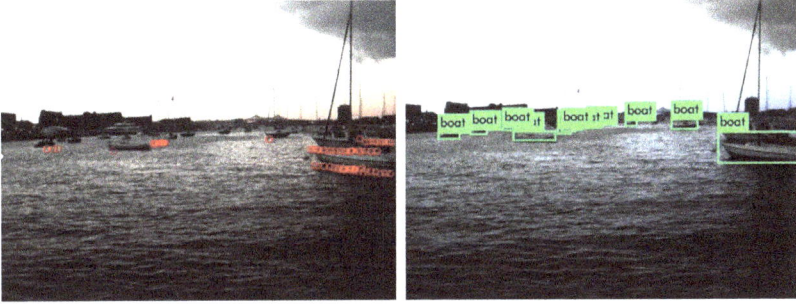

Figure 2.2   Example of object tracking for surface vessels, using the fusion of LiDAR depth data and optical camera images with YOLO-based detection

In underwater environments, where visibility is often poor, deep learning models trained on side-scan sonar, FLS, and multi-beam echo-sounder data enhance object classification and feature extraction [11]. Integrating multi-sensor fusion techniques with deep learning enables robust object recognition, allowing marine robots to make more informed navigation decisions in challenging operational conditions.

2.2.2   Guidance, navigation, and control

The guidance, navigation, and control (GNC) framework is fundamental to the autonomy of marine platforms, ensuring their ability to plan, navigate, and execute maneuvers safely and efficiently in dynamic and uncertain maritime environments [12,13]. These systems integrate multiple path planning algorithms, motion control strategies, and autonomous collision avoidance techniques to enable effective decision-making in real time. A diagram representing the main blocks of a typical GNC framework is shown in Figure 2.3.

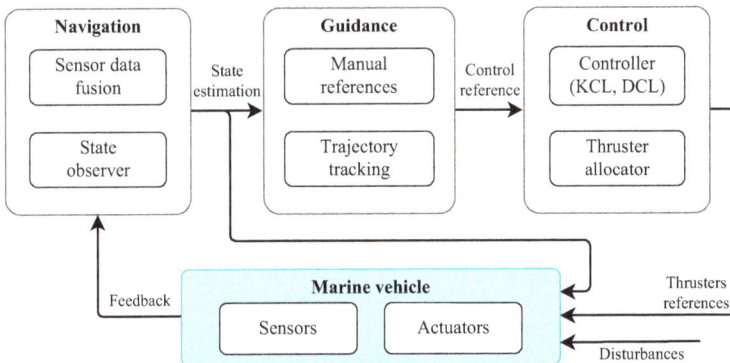

Figure 2.3   Block diagram of a GNC framework. The system consists of three main modules: guidance, which defines the desired trajectory; navigation, responsible for estimating the vehicle's state; and control, which ensures accurate trajectory following

Addressing the challenge of intrinsic sensing (i.e. proprioception), navigation in marine robotics relies on the fusion of data from the GNSS, IMUs, and acoustic positioning systems (such as ultra-short baseline (USBL) and long baseline (LBL) systems), and visual-inertial odometry (VIO). Surface vessels primarily depend on GNSS and IMU-based localization, whereas underwater vehicles must compensate for the absence of GNSS signals by leveraging dead reckoning, Doppler velocity logs (DVL), and SLAM (Simultaneous Localization and Mapping) techniques using sonar or optical sensors. Moreover, given the complexity of marine environments, navigation solutions must account for environmental disturbances, such as ocean currents, waves, and varying seabed topography, while maintaining robust localization accuracy through sensor fusion approaches, including the above-mentioned Bayesian estimation algorithms, and deep learning-based state estimation models [14].

In the context of guidance for autonomous vehicles, motion planning algorithms enable marine robots to compute (optimal) reference paths or trajectories while avoiding obstacles, minimizing energy consumption, and adhering to mission constraints. Graph-based methods like A and Dijkstra's algorithm* provide deterministic, globally optimal paths but struggle with real-time adaptability in highly dynamic environments. More advanced approaches, such as Rapidly exploring Random Trees (RRT) and RRT*, enable efficient path exploration while ensuring obstacle avoidance. Optimization-based planning techniques, including Model Predictive Control (MPC) and Mixed-Integer Linear Programming (MILP), allow for real-time trajectory generation, incorporating vehicle dynamics, environmental constraints, and collision avoidance requirements. Planning strategies are particularly beneficial in congested waterways, where ASVs must comply with the International Regulations for Preventing Collisions at Sea (COLREGs) to ensure safe navigation [15,16].

Once a path is determined, motion control strategies ensure precise execution of planned trajectories. Classical control techniques such as Proportional-Integral-Derivative (PID) control remain widely used due to their simplicity and reliability, especially in maintaining course and speed. However, modern control strategies, such as MPC, adaptive control, and nonlinear control techniques, provide greater robustness to external disturbances and variations in hydrodynamic forces. Another widely adopted strategy is the use of task-priority-based (TP) control frameworks, which enable autonomous marine vehicles to handle multiple competing objectives by dynamically assigning priorities to different tasks based on mission requirements and environmental conditions. These frameworks ensure that higher-priority tasks, such as collision avoidance or maintaining communication links, take precedence over lower-priority objectives, like energy efficiency or secondary data collection. Hierarchical control architectures are commonly used, where primary tasks (e.g. stability and safety constraints) are enforced strictly, while secondary tasks (e.g. optimal path refinement and energy conservation) are executed within the feasible space left by the primary constraints. In dynamic marine environments, adaptive task reallocation allows the system to shift priorities in real time, for instance prioritizing emergency ascent in an AUV when critical faults are detected. This approach is particularly effective for multi-objective missions, such as coordinated operations between surface and underwater vehicles, where communication, localization, and navigation must be continuously balanced to achieve mission success. A more in-depth discussion of this framework within the context of marine robotics research is explored in Section 2.4.

2.2.3   *Communication and networking*

Effective communication and networking are critical for autonomous marine systems, enabling real-time data exchange, coordination, and remote supervision. However, the maritime environment presents unique challenges for connectivity, including vast operational ranges, harsh environmental conditions, and the physical limitations of different communication modalities. Unlike terrestrial wireless networks, where high-speed broadband and cellular infrastructure are widely available, maritime communication must rely on a combination of radio frequency (RF), satellite, acoustic and optical technologies, each with its own trade-offs. For surface communications, radio-based solutions, including very high frequency (VHF) and ultrahigh frequency (UHF) systems, are commonly used for short- to medium-range data transmissions, such as vessel-to-vessel (V2V) and vessel-to-infrastructure (V2I) communication. However, its performance deteriorates over long distances due to signal attenuation and multipath effects caused by sea surface reflections. Satellite communication can be useful for long-endurance ASVs operating in remote oceanic regions. However, it comes with high latency, significant power consumption, and cost constraints, limiting its feasibility for continuous, high-bandwidth operations.

For underwater communication, acoustic waves serve as the primary transmission medium due to the high attenuation of radio and optical signals in water. Underwater acoustic communication enables long-range data exchange between AUVs, ROVs, and seabed sensor networks (such as ocean-bottom seismometers, OBS, for instance). However, acoustic waves propagate relatively slowly with respect to communication protocols, leading to significant latency and Doppler effects, which complicate real-time communication and control. Additionally, acoustic signals suffer from multipath propagation, ambient noise interference, and frequency-dependent absorption, particularly in shallow and turbulent waters. Optical underwater communication is an emerging alternative that offers higher data rates compared to acoustics, using blue–green laser wavelengths to penetrate seawater effectively. However, it has a limited range (tens of meters) and is highly susceptible to turbidity and scattering, making it suitable primarily for short-range AUV-to-AUV or AUV-to-diver communication in clear water conditions. To enhance underwater networking, hybrid communication architectures are being explored, combining acoustic, optical, and electromagnetic (EM) signals, dynamically switching between modalities based on environmental conditions and operational requirements, as in [18], where the authors propose a hybrid acoustic-optical underwater communication networks solution. A graphical representation of a multi-modal based communication is shown in Figure 2.4.

Reliable networking is essential in the context of V2V communication, enabling autonomous vessels to share navigation data, sensor readings, and intent information, facilitating swarm coordination in multi-robot systems. This is particularly

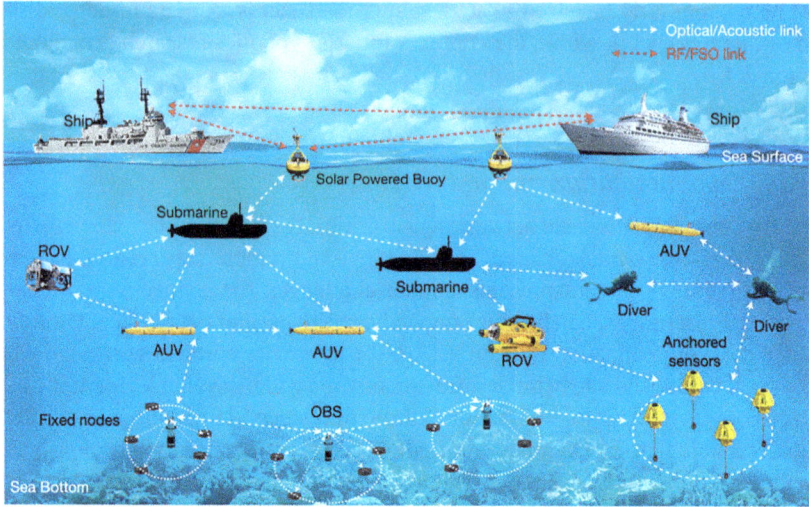

Figure 2.4   Architecture of an underwater wireless communication sensor network (source: [17], used under license CC BY 4.0)

important for formations of ASVs conducting environmental monitoring, search-and-rescue missions, or defense operations. Protocols such as Delay-Tolerant Networking (DTN) are often employed to handle the intermittent connectivity and long propagation delays associated with maritime networks [19]. V2I communication, in contrast, allows autonomous ships to interface with shore-based control centers, ports, and offshore infrastructure, enabling remote mission updates, telemetry uploads, and health monitoring. To this extent, technologies such as Automatic Identification Systems (AIS) and Maritime Wireless Mesh Networks (MWMNs) facilitate connectivity, allowing for enhanced situational awareness and traffic coordination.

2.2.4   Energy and power systems

Energy and power systems play a fundamental role in supporting autonomy in marine robotics, directly influencing the endurance, operational range, and capabilities of autonomous surface and underwater vehicles. The selection of power sources depends on mission requirements, environmental conditions, and the trade-offs between energy density, weight, and reliability. Battery technologies, particularly lithium-ion (Li-ion) and lithium-polymer (LiPo) batteries, dominate the energy storage landscape due to their high energy density, rechargeability, and relatively low maintenance requirements. However, battery limitations, such as energy capacity constraints and degradation over repeated charge cycles, pose challenges for long-duration missions. To extend operational endurance, energy harvesting technologies are increasingly being integrated into marine robotic platforms, as seen in [20], where a vehicle equipped with a wing system exploits wave motion for propulsive thrust. Moreover, solar panels on surface vessels can provide supplemental power, allowing for extended deployments in equatorial and temperate regions, though their efficiency is reduced under cloudy conditions or during night-time operations. Similarly, wave energy converters and hydro-kinetic turbines offer a promising means of harvesting mechanical energy from ocean waves and currents, particularly for long-endurance underwater platforms like autonomous underwater gliders. These vehicles exploit buoyancy-driven propulsion and ocean thermoclines to minimize energy consumption, enabling scientific missions that can last for several months with minimal onboard power storage.

For higher-power applications, fuel cells and hybrid power systems provide alternative solutions. Proton-exchange membrane (PEM) fuel cells, using hydrogen as a fuel source, offer high energy density and relatively clean operation, making them attractive for long-range AUVs and endurance-critical surface vessels. Hybrid power architectures, combining batteries with fuel cells or diesel generators, allow for operational flexibility, balancing power-hungry maneuvers (such as propulsion-intensive tasks) with energy-efficient loitering phases. An example of a combined energy system is shown in the prototype surface vehicle in Figure 2.5. AI-driven energy optimization techniques, including reinforcement learning-based energy management, are being explored to improve efficiency, enabling marine robots to autonomously adjust their power usage based on environmental conditions and operational priorities. As advances in battery technology, energy harvesting, and power management continue, the next generation of marine autonomous systems will achieve greater endurance, enabling persistent ocean monitoring, deep-sea exploration, and long-duration scientific and commercial missions.

Figure 2.5   *The ULISSE ASV prototype, equipped with a hybrid power supply system, combining in-hull LiPo batteries for self-propulsion and external diesel generators for bathymetric acoustic sparker operation, prepares for a long-endurance, energy-demanding bathymetric survey mission*

2.3   Applications of autonomous maritime systems

2.3.1   *Environmental and scientific research*

Autonomous maritime systems have become fundamental tools for environmental and scientific research, enabling persistent, cost-effective, and high-resolution data collection across vast and often inaccessible marine environments. These systems, including ASVs, AUVs, gliders, and robotic drifters, are transforming oceanographic studies by allowing scientists to gather continuous, real-time information on climate change, marine biodiversity, and oceanographic processes without the logistical constraints and high operational costs of crewed research vessels. In long-duration oceanographic monitoring, autonomous platforms equipped with temperature, salinity, dissolved oxygen, pH, and nutrient sensors track key physical and biogeochemical changes over extended periods. AUVs and underwater gliders, such as Slocum Gliders and Spray Gliders, can operate for months at a time, profiling the water column and transmitting data back via satellite when surfacing [21]. Unmanned surface vehicles (USVs) like Saildrones and Wave Gliders, powered by solar or wave energy, offer an additional layer of real-time, near-surface observations, complementing satellite remote sensing efforts and improving climate models used to study ocean–atmosphere interactions, hurricane intensification, ocean acidification, and global climate phenomena dynamics. The deployment of autonomous networks of oceanographic sensors, such as the Argo float program [22], has already revolutionized global ocean monitoring, providing invaluable insights into deep-sea warming, thermohaline circulation, and carbon sequestration processes that were previously difficult to measure at scale.

Beyond large-scale climate studies, autonomous maritime systems are starting to play a relevant role in marine biodiversity assessment and ecosystem monitoring. Traditional methods of biodiversity research, such as ship-based surveys and diver-led studies, are time-consuming, expensive, and often limited to small regions. In contrast, ASVs and AUVs outfitted with high-resolution cameras, multi-beam sonar, passive acoustic recorders, and environmental DNA (eDNA) samplers enable noninvasive, continuous monitoring of marine life across extensive spatial and temporal scales. AI-powered image recognition algorithms integrated into these platforms can autonomously identify and classify marine species, from plankton and fish populations to larger megafauna, such as whales, sharks, and sea turtles, providing critical data for conservation efforts. Passive acoustic monitoring (PAM) using autonomous vehicles has been particularly effective in studying marine mammal migration patterns, fish spawning aggregations, and the impact of underwater noise pollution on aquatic ecosystems. Additionally, eDNA analysis from robotic platforms is emerging as a game-changer for detecting rare, endangered, or cryptic species by analyzing genetic material shed into the water, offering a powerful, noninvasive method for assessing biodiversity in deep-sea and remote environments [23]. These technologies are especially vital for monitoring marine-protected areas (MPAs), coral reef resilience, and the effects of human activities such as overfishing, habitat destruction, and climate-induced changes on marine ecosystems.

Another key application of autonomous maritime systems is in rapid environmental response and disaster mitigation, where real-time, high-resolution oceanographic data is critical for mitigating ecological threats, such as oil spills, harmful algal blooms (HABs), hypoxic dead zones, and microplastic pollution. ASVs and AUVs can be rapidly deployed to track pollutant dispersal, analyze chemical composition, and monitor ecosystem recovery, providing instrumental data to inform remediation efforts. A significant work reporting the results from a subsurface hydrocarbon survey using an autonomous underwater vehicle and a ship-cabled sampler is presented in [24], where the author's findings indicate the presence of a continuous plume of oil, extended over 35 km at a depth of approximately 1100 m that persisted for months without substantial biodegradation. Autonomous sensor networks can also assist in early warning systems for tsunamis, undersea earthquakes, and coastal flooding, using seafloor pressure sensors and water column profiling to detect anomalies in ocean dynamics.

2.3.2   Commercial and industrial applications

Autonomous maritime systems are also revolutionizing a wide range of commercial and industrial applications, enhancing efficiency, reducing costs, and improving safety across multiple sectors, including shipping and logistics, offshore energy, fisheries and aquaculture, and underwater infrastructure inspection. One of the most impactful applications is in autonomous shipping, where fully or semiautonomous vessels are being developed to optimize global supply chains by reducing crew requirements, fuel consumption, and operational risks. Companies such as Kongsberg and Wärtsilä are pioneering autonomous cargo ships equipped with advanced navigation systems, AI-driven collision avoidance, and real-time route optimization algorithms, enabling more efficient and safer maritime transport. Notable projects such as Yara Birkeland, the world's first fully electric and autonomous container ship, are demonstrating the potential of these technologies to reduce emissions, improve fuel efficiency, and enhance logistical operations [25].

In the offshore energy sector, autonomous maritime systems are integral to inspection, maintenance, and monitoring operations of marine structures such as offshore oil and gas platforms, wind farms, and subsea pipelines. Traditionally, these tasks require crewed vessels and divers, posing significant risks and operational costs. Autonomous underwater vehicles and surface vehicles equipped with multi-beam sonar, LiDAR, high-resolution cameras, and AI-driven defect detection algorithms can conduct detailed inspections of subsea pipelines, offshore rigs, and wind turbine foundations without human intervention, reducing downtime, and improving safety [26]. For instance, companies like Saipem and Ocean Infinity are deploying AUV fleets for deep-sea infrastructure surveys, allowing for efficient and cost-effective pipeline integrity assessments, leak detection, and seabed mapping. In this context, Saipem has developed an autonomous underwater architecture for long-term deep-ocean inspection, designed to robustly plan activities and deliberate efficiently without human intervention [27].

Another significant area of commercial application is fisheries and aquaculture, where autonomous systems are transforming traditional practices through real-time monitoring, automation, and precision management [28]. Autonomous surface vehicles equipped with hydro-acoustic fish-finding sensors, AI-driven stock assessment models, and water quality monitoring systems are helping commercial fisheries track fish populations, optimize catch strategies, and prevent overfishing. In aquaculture, underwater drones and robotic feeding systems are automating key processes, such as net cleaning, fish health monitoring, and feed optimization, reducing labor costs and improving sustainability. Autonomous vehicles are also being deployed for environmental monitoring around fish farms, ensuring compliance with regulations and minimizing the impact of nutrient run-off, algal blooms, and disease outbreaks.

Furthermore, autonomous maritime systems are revolutionizing seabed exploration, underwater mining, and deep-sea resource extraction. As demand for critical minerals such as cobalt, nickel, and rare earth elements grows, companies are exploring deep-sea mining as an alternative to traditional land-based extraction. Autonomous underwater vehicles equipped with geophysical mapping sensors and robotic sampling systems are being used to identify and assess deep-sea mineral deposits [29]. These systems allow for more precise exploration, reducing the environmental footprint of underwater mining activities. In addition, autonomous dredging systems and robotic excavation platforms are being developed to extract resources with minimal human intervention, improving operational efficiency and worker safety.

2.3.3   Search and rescue applications

Autonomous maritime systems are increasingly being leveraged for search and rescue (SAR) missions, disaster response, and humanitarian aid, offering rapid deployment, enhanced situational awareness, and improved efficiency in challenging and often dangerous environments [30]. Traditional SAR operations rely on crewed vessels, helicopters, and aircraft, which can be slow to mobilize, resource-intensive, and constrained by adverse weather, poor visibility, and hazardous sea conditions. In contrast, ASVs, AUVs, and aerial drones can significantly improve the speed and effectiveness of rescue efforts. USVs equipped with thermal imaging cameras, LiDAR, radar, and real-time video feeds can autonomously scan large maritime areas, detecting distressed vessels, overboard survivors, or floating debris with a level of endurance and coverage that far surpasses human search teams. An example of USV developed within the ICARUS project [31] is shown in Figure 2.6. Autonomous aerial drones deployed from ASVs or coastal bases further enhance SAR missions by providing aerial reconnaissance, relaying real-time images to command centers, and even dropping emergency survival supplies such as flotation devices, medical kits, or communication beacons.

In addition to SAR operations, autonomous maritime systems are playing a critical role in disaster response and humanitarian aid efforts following natural disasters such as hurricanes, tsunamis, and floods, where traditional infrastructure is often severely damaged or inaccessible. Unmanned vehicles can rapidly assess damage to coastal regions, survey submerged infrastructure, and provide real-time maps of affected areas, allowing response teams to plan evacuations and resource distribution more effectively. In disaster zones where contamination of drinking water is

Figure 2.6   Roaz II, a search and rescue oriented USV (source: [30], used under license CC BY 3.0)

a major concern, ASVs equipped with water quality sensors can rapidly analyze contamination levels and guide relief efforts to the most affected areas.

Another emerging application of autonomous maritime systems in humanitarian operations is migrant search and rescue in the Mediterranean, Gulf of Mexico, and other high-risk migration corridors. Thousands of migrants attempt dangerous sea crossings each year, often in overcrowded and unseaworthy vessels, leading to frequent capsizing incidents and mass drownings. Autonomous maritime systems can provide continuous, real-time monitoring of migration routes, detecting and tracking distressed vessels.

2.3.4   Maritime security and defense

Autonomous maritime systems are playing an increasingly critical role in maritime security and defense, enhancing situational awareness, force projection, threat detection, and autonomous deterrence in naval and law enforcement operations. These systems offer a cost-effective and persistent solution for border patrol, anti-piracy operations, mine countermeasures (MCM), and maritime domain awareness. Unlike traditional manned naval assets, which require significant logistical support and manpower, autonomous platforms can be deployed for extended duration, operating in high-risk environments without exposing human personnel to danger. One of the most prominent applications of autonomous maritime systems in security is in coastal and port surveillance, where ASVs equipped with radar, infrared cameras, AIS transponders, and machine learning-based anomaly detection algorithms continuously monitor vessel traffic, identifying suspicious activities, such as illegal fishing, smuggling, and unauthorized vessel intrusions. These systems integrate with maritime command and control centers, enabling real-time data fusion from multiple sensors, including satellite imagery, sonar arrays, and airborne reconnaissance, to provide a comprehensive picture of the operational maritime theater.

In naval defense, autonomous maritime systems are also increasingly used for MCM operations, reducing the risk to human divers and crewed vessels. Traditionally, mine detection and neutralization require slow, methodical sweeps using manned minesweepers and specialized divers, but autonomous systems can perform these tasks with greater efficiency and reduced exposure to danger. AUVs, such as the General Dynamics Bluefin-21 [32], are capable of deep-sea mine detection, classification, and neutralization, using high-resolution sonar to distinguish mines from natural seabed features. Additionally, ASVs equipped with remotely operated mine-disposal payloads can deploy neutralizing charges once a threat is identified.

Another major defense application of autonomous maritime systems is in antisubmarine warfare (ASW), where unmanned platforms provide persistent, stealthy surveillance of enemy submarines and underwater threats. Traditional ASW relies on manned surface ships, aircraft, and submarines using passive and active sonar systems, but these approaches are resource-intensive and require significant coordination. Autonomous ASVs and AUVs equipped with towed sonar arrays and acoustic modems can track submarines more efficiently by operating in multi-agent swarm configurations, covering large areas of the ocean, as in the DAMPS project [33], where the authors present underwater acoustic source localization using a multi-robot system.

Autonomous maritime systems are also being deployed in anti-piracy and counter-terrorism operations, providing an advanced layer of security for commercial shipping lanes, naval convoys, and offshore energy installations. In piracy-prone regions such as the Horn of Africa, the Strait of Malacca, and the Gulf of Guinea, ASVs equipped with electro-optical sensors, loudspeakers, and non-lethal deterrence weapons can patrol shipping corridors, detect suspicious vessel movements, and warn approaching threats before escalation. These systems can relay live data to naval forces or port authorities, enabling rapid response interventions when necessary.

2.4   Marine robotics research: experimental platforms and field deployments

In advancing autonomous and semiautonomous systems, marine robotics research plays a fundamental role. Experimental platforms serve as test beds for developing and validating novel algorithms, sensing techniques, and control strategies. Field deployments in real-world marine environments are essential for assessing system performance under dynamic and unpredictable conditions, bridging the gap between theoretical research and practical application. This paragraph explores some significant projects and their field trials, highlighting key challenges and technological advancements.

In marine robotics, autonomous systems often need to handle multiple control objectives simultaneously, including maintaining stability, following a predefined path, and avoiding obstacles and optimizing energy consumption. Managing these competing demands requires a structured approach that ensures critical tasks are prioritized while allowing secondary objectives to be addressed without compromising overall system performance. One effective way to achieve this is through hierarchical control strategies, where different tasks are assigned varying levels of importance based on mission requirements and environmental conditions.

The task priority strategy, outlined in Section 2.2.2, is a widely employed control strategy in robotics that facilitates the simultaneous management of multiple objectives while ensuring that critical tasks are executed with higher precedence [34]. This approach is particularly useful in dynamic and uncertain marine environments, where autonomous vehicles must balance competing control objectives, such as stability, obstacle avoidance, path following, and energy efficiency. By assigning hierarchical priorities to different control tasks, task priority frameworks enable robots to fulfill primary mission objectives while accommodating secondary goals without compromising overall performance. Typically implemented using constrained optimization or hierarchical control architectures, this strategy ensures that high-priority tasks, such as safety and collision avoidance, take precedence over lower-priority objectives like energy minimization or data collection.

In the following sections, a series of experimental platforms that take advantage of the TP control strategy will be presented, starting from single robot systems development to more complex multi-agent systems.

2.4.1   Single robot autonomy

Within the context of single robot autonomy in the maritime robotics domain, two notable projects, ROBUST and DexROV, exemplify efforts to enhance autonomous functionality in challenging marine environments. Both projects leverage the task priority framework.

The EU-funded ROBUST project [35] aimed to develop an autonomous robotic system for the exploration of deep-sea mining sites, particularly manganese nodule fields. The project's primary goal was to perform in situ measurements of these nodules to identify the presence of valuable rare earth elements, thus enabling more targeted and efficient mining operations. The main challenges of the project include: developing a system capable of autonomously navigating deep-sea environments and identifying promising areas with manganese nodules after an initial bathymetric survey; conducting detailed low-altitude surveys of selected sub-areas; precisely landing a UVMS on the seafloor in front of an identified nodule to enable subsequent fixed-based manipulation; integrating a Laser Induced Breakdown Spectroscopy (LIBS) system on the manipulator's end-effector to perform in situ material identification; coordinating the vehicle and manipulator to bring the LIBS within a few centimeters of the nodule for the measurement after landing.

To this aim, the project built upon a task priority kinematic inversion control framework, extending the work carried out in the TRIDENT [36] and MARIS [37] projects. Specific operations, like safe navigation or landing, were defined as a list of prioritized control objectives and tasks. A specific landing action was developed with a priority order of objectives, including vehicle horizontal attitude, vehicle longitudinal alignment to the nodule, vehicle distance to the nodule, and vehicle altitude. The control framework could activate and deactivate equality/inequality control objectives of any dimension depending on the system's needs, allowing safety and operational-enabling objectives to have the highest priority. Activation functions were used to ensure smooth transitions between different control actions.

Prior to the on-field experiments, the developed control architecture and its applications, such as landing and in situ inspection, were preliminarily validated through dynamic simulations, taking into account rigid body dynamics, added mass, and damping. A picture of the ROBUST platform in operation is shown in Figure 2.7. A key achievement of this project was the development of a unifying control framework for the kinematic control layer of underwater vehicle manipulator systems, applicable to deep-sea mining exploration, extending the task priority framework beyond grasping scenarios to include operations such as safe navigation and landing.

On another front, the DexROV project [38] had the main objective of enabling the remote operation of a UVMS via satellite communication to execute intervention tasks in an Oil&Gas industry scenario. This involved delocalizing the manned support for ROV operations to onshore sites to reduce costs and risks. The project faced several key challenges due to the nature of remote operation via satellite. The satellite link introduced a substantial delay, making direct teleoperation unfeasible for effective manipulation. This necessitated the development of more autonomous capabilities for the UVMS to also handle potential interruptions. Executing tasks like turning valves required the UVMS to interact with its surroundings, necessitating the ability to exert appropriate forces and torques, and these actions required the system to meet specific conditions, such as maintaining the target object within the camera's field of view. The use of the TP framework allowed the system to handle classical inverse kinematics was extended to manage inequality constraints, together with the objective of providing compliance during interaction and handling unexpected forces, utilizing measurements from a force/torque sensor on the manipulator's wrist. This allowed the system to react safely to contact with the environment. The operator could provide high-level commands, such as end-effector trajectories, while the UVMS control system autonomously managed safety, interaction, and prerequisite tasks. Field trials were conducted in Marseilles, France, with the onshore control center located in Brussels, Belgium, demonstrating the feasibility of remote supervision via satellite communication for underwater intervention. A picture of the DexROV platform in operation is shown in Figure 2.7.

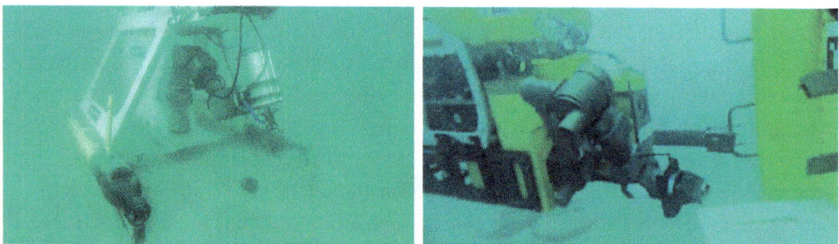

Figure 2.7   Example of two UVMS: during exploration of deep-sea mining sites within the ROBUST project (on the left), and intervention tasks on an underwater structure within the DexROV project (on the right)

2.4.2   *Multi-robot autonomy*

Broadening the scope, multi-robot autonomy enables coordinated operations that improve efficiency, coverage, and robustness in complex marine environments. To this extent, the results achieved within the WIMUST (Widely scalable Mobile Underwater Sonar Technology) project and an ongoing research project on underwater localization and tracking are presented in this section.

The WIMUST project [39] had the main objective of developing a system of cooperative autonomous underwater and surface vehicles for geotechnical surveying. The innovative concept was to replace traditional seismic surveys, which use a large manned vessel towing acoustic sources and long streamers, with a team of autonomous marine robots. Specifically, ASVs would carry acoustic sources, and AUVs would tow short streamers with hydrophones to collect sub-bottom acoustic data. The project faced several key challenges in achieving this objective, the first being the management of heterogeneous platforms. The system involved different types of AUVs and ASVs, requiring the development of coordinated control solutions. Moreover, for effective seismic data acquisition, the AUVs needed to maintain a desired spatial formation while tracking the motion of the leader ASVs. Developing a robust communication and navigation framework was essential for localizing the AUVs, sending control data, and monitoring acoustic data acquisition in real time. This framework needed to be scalable with the number of AUVs used. Precise time synchronization and vehicle positioning were also fundamental for generating high-quality seismic images without artifacts.

To address these challenges, within the WIMUST project, a cooperative marine multi-robot system was developed comprising AUVs and ASVs, successfully conducting a 2 h and 15 min-long survey in the Atlantic Ocean, using a team of seven robots, covering an area of approximately 100 m × 200 m. The seismic data acquired was processed to generate images of the sub-seafloor, revealing geological features without any major hardware or positioning artifacts. During the mission, AUVs successfully tracked spatially shifted versions of the leader ASV's trajectory, maintaining the desired formation. Several integration campaigns, including the final survey in Sines, Portugal, validated the developed technologies and methodologies, paving the way for more efficient and cost-effective surveys. A picture of the WIMUST project concept is shown in Figure 2.8.

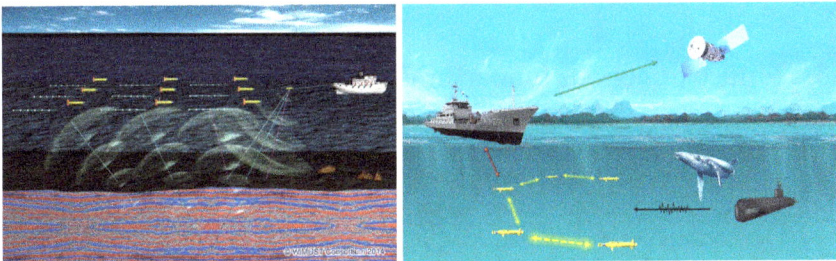

Figure 2.8   Example of two cooperative multi-agent systems, during a autonomous geotechnical survey (on the left), and a passive acoustic monitoring mission (on the right)

Shifting focus, in the work presented in [40] addressed the problem of underwater passive acoustic monitoring using bearing measurements acquired by a team of AUVs. A key aim in the work is to optimize the path of an AUV team's formation with respect to the motion of the target, to enhance the quality of information gathered for tracking. The research sets out to design a distributed control framework for source localization and tracking that facilitates both the coordinated motion control of the AUV agents and their cooperative target tracking efforts. A fundamental challenge in Bearing-Only Tracking (BOT) is that it is strictly related to the relative motion of the observer and the target, and when vehicles are positioned close to each other relative to the target, observability problems arise, appearing aligned and measuring approximately the same bearing angle. This issue is exacerbated by uncertainties in real-world scenarios, especially when dealing with large distances. In light of this, simply following a predetermined path for the AUV formation does not guarantee the observability of the target or good overall tracking performances, and the latencies introduced by the underwater acoustic communication channel pose a considerable challenge for real-time cooperative tracking and control. Moreover, the low bandwidth available for underwater acoustic communication limits the amount of data that can be exchanged between the AUV agents, requiring efficient communication strategies.

To this end, an MPC scheme was implemented to improve target tracking performances and the overall autonomy of the system, explicitly accounting for the limitations associated with underwater acoustic communication, such as latency and low bandwidth, within the proposed framework. The paper proposes a motion optimization strategy based on a receding horizon non-myopic (i.e. not limited to short-term evaluations) policy. This approach aims to increase the system's autonomy and improve tracking performance. A distributed control framework is designed and implemented, which incorporates a version of the Cooperative Path Following (CPF) algorithm [41], tailored to minimize the amount of data exchanged between agents, making it suitable for acoustic communication channels. The framework includes the implementation of cooperative estimation achieved by the AUV agents exchanging bearing observations, the effectiveness of which has been evaluated through simulations, with promising results presented as performance comparisons across several experiments, demonstrating that the proposed motion optimization strategy successfully increases tracking performance. An illustration of the passive acoustic monitoring concept can be seen in Figure 2.7.

2.5   Future directions in maritime autonomy research

Future research in maritime autonomy is expected to enhance the robustness, adaptability, and intelligence of autonomous systems operating in complex marine environments. A significant focus is on the advancement of machine learning techniques to improve decision-making and real-time adaptability of autonomous maritime vehicles. Deep learning methods are being explored to enable these systems to effectively respond to uncertain oceanic conditions and sensor noise [42]. Another critical research direction involves the coordinated control of multiple

ASVs, addressing challenges in communication, navigation, and control to facilitate cooperative missions such as environmental monitoring and search-and-rescue operations [43]. Multi-agent ASV–AUV networks could coordinate large-scale searches, autonomously dividing tasks and optimizing search patterns to cover vast areas efficiently. Furthermore, energy-efficient and solar-powered ASVs will enable continuous monitoring of high-risk areas, ensuring rapid response with the need for minimal human intervention. By reducing response times, increasing operational reach, and minimizing the risk to human rescuers, autonomous maritime systems are set to play an even greater role in saving lives, mitigating disaster impacts, and supporting global humanitarian efforts in the coming years. Advancements in underwater robotics are also notable, with the development of AUVs capable of deep-sea exploration and infrastructure inspection, addressing challenges such as underwater communication and harsh environmental conditions. Furthermore, the rise of autonomous surface vessels is revolutionizing naval capabilities, offering cost-effective and efficient alternatives to traditional manned vessels for various military and commercial applications. Collectively, these research directions determine a transformative period in maritime autonomy, driven by interdisciplinary advancements that bridge robotics, AI, and marine engineering.

References

[1] Zereik E., Bibuli M., Mišković N., *et al.* "Challenges and future trends in marine robotics." *Annual Reviews in Control.* 2018;**46**:350–68.
[2] Rodseth O.J., Wennersberg L.A.L., and Nordahl H. "Levels of autonomy for ships." *Journal of Physics: Conference Series.* 2022;**2311**(1):012018.
[3] Coito J. "Maritime autonomous surface ships: new possibilities—and challenges—in ocean law and policy." *International Law Studies.* 2021; **97**(1):19.
[4] Johansson T.M., Dalaklis D., and Pastra A. "Maritime robotics and autonomous systems operations: exploring pathways for overcoming international techno-regulatory data barriers," *Journal of Marine Science and Engineering.* 2021;**9**(6):594.
[5] Ávila-Zúñiga-Nordfjeld A., Liwång H., and Dalaklis D., "Implications of technological innovation and respective regulations to strengthen port and maritime security: an international agenda to reduce illegal drug traffic and countering terrorism at sea" in Johansson T.M., Dalaklis D., Fernández J.E., *et al.* (ed.) *Smart Ports and Robotic Systems. Studies in National Governance and Emerging Technologies.* Cham: Springer; 2023. pp. 135–47.
[6] Bogue R. "Robots in the offshore oil and gas industries: a review of recent developments." *Industrial Robot: The International Journal of Robotics Research and Application.* 2020;**47**(1):1–6.
[7] Clunie T., DeFilippo M., Sacarny M., *et al.* "Development of a perception system for an autonomous surface vehicle using monocular camera, LIDAR, and Marine RADAR." *2021 IEEE International Conference on Robotics and Automation (ICRA).* IEEE Robotics and Automation Society; 2021. pp. 14112–9.

[8]   Hussain M., O'Nils M., Lundgren J., *et al.* "A comprehensive review on deep learning-based data fusion." *IEEE Access.* 2024;**12**:180093–124.

[9]   Redmon J., Divvala S.K., Girshick R.B., *et al.* "You Only Look Once: Unified, Real-Time object detection." *CoRR.* 2015. https://doi.org/10.48550/arXiv.1506.02640.

[10]   Tarasi L., Wanderlingh F., Noceti N., *et al.* "LiDAR and RGB camera performance for obstacle detection in marine environment." *OCEANS 2024-Halifax.* Piscataway, NJ: IEEE; 2024. pp. 1–10.

[11]   Wang X., Zhang M., Zhou H., *et al.* "Performance analysis and design considerations of the shallow underwater optical wireless communication system with solar noises utilizing a photon tracing-based simulation platform." *Electronics.* 2021;**10**:632.

[12]   Stutters L., Liu H., Tiltman C., *et al.* "Navigation technologies for autonomous underwater vehicles." *IEEE Transactions on Systems, Man, and Cybernetics, Part C (Applications and Reviews).* 2008;**38**(4):581–9.

[13]   Fossen T.I., *Handbook of marine craft hydrodynamics and motion control.* New York: Wiley; 2011.

[14]   Christensen L., de Gea Fernández J., Hildebrandt M., *et al.* "Recent advances in AI for navigation and control of underwater robots." *Current Robotics Reports.* 2022;**3**(4):165–75.

[15]   Du Z., Negenborn R.R., and Reppa V., "COLREGS-compliant collision avoidance for physically coupled multi-vessel systems with distributed MPC." *Ocean Engineering.* 2022;**260**:111917.

[16]   Depalo S., Wanderlingh F., Indiveri G., *et al.* "Protocol-driven A* algorithm for fast ASV motion planning in dynamic scenarios." *OCEANS 2024-Halifax.* Piscataway, NJ: IEEE; 2024. pp. 1–10.

[17]   Wang Y., Liu J., Yu S., *et al.* "Underwater object detection based on YOLO-v3 network." *2021 IEEE International Conference on Unmanned Systems (ICUS).* Piscataway, NJ: IEEE; 2021. pp. 571–5.

[18]   Quintas J., Petroccia R., Pascoal A., *et al.* "Hybrid acoustic-optical underwater communication networks for next-generation cooperative systems: the EUMR experience." *OCEANS 2021: San Diego–Porto.* Piscataway, NJ: IEEE; 2021. pp. 1–7.

[19]   Rahman S., Shaf A., Ali T., *et al.* "An optimal delay tolerant and improved data collection schema using AUVs for underwater wireless sensor networks." *IEEE Access.* 2024;**12**:30146–63.

[20]   Bazzarello L., Collevecchio A., Cosimo D., *et al.* "Energy harvest with WAVE. At sea trials and preliminary results." *2024 IEEE International Workshop on Metrology for the Sea; Learning to Measure Sea Health Parameters (MetroSea).* Piscataway, NJ: IEEE; 2024. pp. 507–11.

[21]   Tian B., Guo J., Song Y., *et al.* "Research progress and prospects of gliding robots applied in ocean observation." *Journal of Ocean Engineering and Marine Energy.* 2023;**9**(1):113–24.

[22]   Jayne S.R., Roemmich D., Zilberman N., *et al.* "The Argo program: present and future." *Oceanography.* 2017;**30**(2):18–28.

[23]   Govindarajan A.F., McCartin L., Adams A., *et al.* "Improved biodiversity detection using a large-volume environmental DNA sampler with in situ filtration and implications for marine eDNA sampling strategies." *Deep Sea Research Part I: Oceanographic Research Papers.* 2022;**189**:103871.

[24]   Camilli R., Reddy C.M., Yoerger D.R., *et al.* "Tracking hydrocarbon plume transport and biodegradation at Deepwater Horizon." *Science.* 2010;**330**(6001):201–4.

[25]   Negenborn R.R., Goerlandt F., Johansen T.A., *et al.* "Autonomous ships are on the horizon: here's what we need to know." *Nature.* 2023;**615**(7950):30–3.

[26]   Petillot Y.R., Antonelli G., Casalino G., *et al.* "Underwater robots: from remotely operated vehicles to intervention-autonomous underwater vehicles." *IEEE Robotics & Automation Magazine.* 2019;**26**(2):94–101.

[27]   Tosello E., Bonel P., Buranello A., *et al.* "Opportunistic (re) planning for long-term deep-ocean inspection: an autonomous underwater architecture." *IEEE Robotics & Automation Magazine.* 2024;**31**(1):72–83.

[28]   Wu Y., Duan Y., Wei Y., *et al.* "Application of intelligent and unmanned equipment in aquaculture: a review." *Computers and Electronics in Agriculture.* 2022;**199**:107201.

[29]   Simetti E., Sartore C., Wanderlingh F., *et al.* "Autonomous Deep Sea Mining Exploration: The EU ROBUST Project Control Framework." *OCEANS 2019 MTS/IEEE Marseille.* Piscataway, NJ: IEEE; 2019. pp. 1–8.

[30]   Matos A., Silva E., Almeida J., *et al.* "Unmanned maritime systems for search and rescue." In: *Search and Rescue Robotics – From Theory to Practice.* London: InTech; 2017. pp. 77–92.

[31]   Serrano D., De Cubber G., Leventakis G., *et al.* "ICARUS and DARIUS approaches towards interoperability." *IARP RISE Workshop, At Lisbon, Portugal. Proceedings of the NATO STO Lecture Series SCI-271.* vol. 1; 2015. pp. 1–10.

[32]   Willcox S., Goldberg D., Vaganay J., *et al.* "Multi-vehicle cooperative navigation and autonomy with the bluefin CADRE system." *Proceedings of IFAC (International Federation of Automatic Control) Conference* Citeseer; 2006. pp. 20–22.

[33]   Allotta B., Antonelli G., Bongiovanni A., *et al.* "Underwater acoustic source localization using a multi-robot system: the DAMPS project." *2021 International Workshop on Metrology for the Sea; Learning to Measure Sea Health Parameters (MetroSea).* Piscataway, NJ: IEEE; 2021. pp. 388–93.

[34]   Simetti E., Casalino G., Wanderlingh F., *et al.* "Task priority control of underwater intervention systems: theory and applications." *Ocean Engineering.* 2018;**164**:40–54.

[35]   Di Vito D., De Palma D., Simetti E., *et al.* "Experimental validation of the modeling and control of a multibody underwater vehicle manipulator system for sea mining exploration." *Journal of Field Robotics.* 2021;**38**(2):171–91.

[36]   Simetti E., Casalino G., Torelli S., *et al.* "Floating underwater manipulation: developed control methodology and experimental validation within the TRIDENT project." *Journal of Field Robotics.* 2014;**31**(3):364–85.

[37]   Simetti E., Wanderlingh F., Torelli S., *et al.* "Autonomous underwater intervention: experimental results of the MARIS project." *IEEE Journal of Oceanic Engineering.* 2018;**43**(3):620–39.

[38]   Di Lillo P., Simetti E., Wanderlingh F., *et al.* "Underwater intervention with remote supervision via satellite communication: developed control architecture and experimental results within the dexrov project." *IEEE Transactions on Control Systems Technology.* 2020;**29**(1):108–23.

[39]   Simetti E., Indiveri G., Pascoal A.M., "WiMUST: a cooperative marine robotic system for autonomous geotechnical surveys." *Journal of Field Robotics.* 2021;**38**(2):268–88.

[40]   Tiranti A., Wanderlingh F., Simetti E., *et al.* "Motion optimization strategy for bearing-only tracking performed with a team of autonomous underwater vehicles navigating in formation." *OCEANS 2023 – Limerick.* IEEE Oceanic Engineering Society; 2023. pp. 1–7.

[41]   Almeida J., Silvestre C., and Pascoal A. "Cooperative control of multiple surface vessels in the presence of ocean currents and parametric model uncertainty." *International Journal of Robust and Nonlinear Control.* 2010;**20**(14):1549–65.

[42]   Qiao Y., Yin J., Wang W., *et al.* "Survey of deep learning for autonomous surface vehicles in marine environments." *IEEE Transactions on Intelligent Transportation Systems.* 2023;**24**(4):3678–701.

[43]   Gantiva Osorio M., Ierardi C., Jurado Flores I., *et al.* "Coordinated control of multiple autonomous surface vehicles: challenges and advances — A systematic review." *Ocean Engineering.* 2024;**312**:119160

Chapter 3

Advances in remote estimation and control with quantised communications for networked systems subject to severe bandwidth constraints

Francisco F.C. Rego[1,2,3], Colin N. Jones[4]
and António M. Pascoal[3]

Motivated by fast-paced progress in the area of networked marine systems communicating over the acoustic channel, we address the problem of remote state estimation and control for systems subjected to process and measurement noise under stringent emitter–receiver communication bandwidth constraints. At the core of the framework adopted for systems design and analysis is the adoption of an observer structure for systems that may be operating under state feedback control. The key strategy proposed consists of having both the emitter, attached to a sensor, and the receiver, at a remote location in the communication channel, run synchronised quantised observers. Quantisation is used with a view to drastically reduce the bandwidth required for communications. We derive a set of conditions on the quantiser design parameters that guarantee ultimate boundedness of the estimation error. These are shown to depend on the least upper bound of the measurement noise and process disturbance signals. To ensure robustness with respect to external disturbances, the latter is assumed to satisfy a persistency of excitation property. This is in contrast with previous results reported in the literature that ensure robustness at the cost of setting limitations on the reset values of the quantisation interval and may yield large ultimate convergence bounds for the estimation error. Numerical simulation examples illustrate the performance of the proposed algorithm and indicate that the methodologies developed hold potential for further extensions to address challenging problems in the area of cooperative networked marine systems.

[1]Escola Superior de Engenharia e Tecnologias, Instituto Politécnico da Lusofonia and COPELABS – Cognitive and People-centric Computing, Lusófona University, Portugal
[2]CTS – Centro de Tecnologia e Sistemas, UNINOVA – Instituto de Desenvolvimento de Novas Tecnologias, Portugal
[3]Institute for Systems and Robotics (ISR), Instituto Superior Técnico (IST), University of Lisbon, Portugal
[4]Automatic Control Laboratory, École Polytechnique Fédérale de Lausanne (EPFL), Switzerland

3.1 Introduction

Whenever a real-valued signal is transmitted through a digital communication channel with finite bandwidth, it undergoes a process called quantisation that codifies the signal into a finite set of values that depend on the number of bits used in the encoding process. Given the proliferation of wireless sensor networks, there is an increasing need to study quantisation effects in control and monitoring systems and to use this knowledge to design systems that can yield acceptable performance while substantially reducing the amount of information sent over the communication channel. It is in this context that this chapter addresses the problems of control and state estimation for a given plant (viewed as a local node), focusing on the acquisition of output sensor data and its processing at a remote node that has access to the transmitted information.

The general setup that we adopt can be simply motivated by considering two representative scenarios that involve a physical plant and the transmission of sensor-related information over a restrictive communication channel. The first scenario encompasses control and estimation and refers to the situation where the plant's output is measured externally by a remote sensor in charge of processing the data acquired and transmitting the results over a limited bandwidth communication channel, back to the plant; based on the information received, the plant then reconstructs its state and uses it in a state feedback arrangement for stabilisation purposes. In scenario 2, which focuses on estimation only, a plant evolves either in open loop or in closed loop, using, in the latter case, feedback from variables obtained using local sensors. In this scenario, the objective is for a remote unit to estimate the state of the plant for tracking purposes. For this purpose, a local sensor unit transmits plant-related data to the remote tracking unit over a limited bandwidth communication channel; based on the information received, the remote unit then reconstructs the state of the plant. In both examples, due to the constraints introduced by the communications channel, the information transmitted must be quantised in an adaptive manner. As we will show, the key strategy to obtain solutions to the above problems consists of having both the emitter and the receiver in the communication channel run synchronised quantised observers.

Practical examples of the above scenarios occur naturally in marine applications where autonomous underwater vehicles (AUVs) work in cooperation with autonomous surface vehicles (ASVs). Both vehicles are equipped with acoustic modems for communications through the unforgiving water medium, which imposes stringent restrictions on the achievable data rates. The first scenario above reflects the situation where the position of the AUV cannot be measured locally but is instead obtained using a remote system consisting of an ultra-short baseline (USBL) device and a GPS unit installed on the ASV capable of measuring the relative position of the AUV with respect to the ASV and the global position of the latter, respectively. For control purposes, this information must be transmitted to the AUV. The second scenario captures the situation where the AUV measures its position and possibly other motion-related variables on-board (i.e. locally) and uses them for motion control purposes. In this case, the objective is to monitor, that is, to estimate the state of the AUV model at the surface, remotely, using information transmitted from the AUV to the ASV through the acoustic channel. In both cases, the fact that information must be exchanged between local and remote systems as parsimoniously as possible poses considerable challenges.

The figure below (Figure 3.1) illustrates two compelling examples of networked marine robots associated with the WiMUST [1] and RAMONES projects [2], in which the first and third authors participated. WiMUST witnessed the development of advanced cooperative and networked control/navigation systems that enabled a group of marine robots (both at the surface and submerged) equipped with acoustic sources and towed acoustic streamers to perform geotechnical seismic surveys in a fully automatic manner. For the first time worldwide, a mission was performed in 2018 at sea in Sines, Portugal with a fleet of seven autonomous marine robots performing high-resolution 3D sub-bottom mapping in cooperation. RAMONES led to the development of a new solution for in situ, continuous, long-term monitoring of radioactivity in harsh subsea environments by resorting to a network of autonomous marine vehicles (surface robot and two underwater gliders) transporting a new generation of submarine radiation-sensing instruments. One of the key components of the project was the study and implementation of the systems in charge of cooperative navigation and control of multiple marine robots equipped with acoustic systems for distributed communication and positioning.

Both projects witnessed the design, development and at sea testing of advanced cooperative control and estimation systems relying on the exchange of information over the acoustic channel. However, no use was made in the course of these projects of quantised communications, a topic that is expected to play an important role in the future for the implementation of advanced cooperative systems involving large numbers of vehicles.

There are two main problems associated with quantised signals used for state estimation, control or both simultaneously. One problem arises if a given signal falls outside the range of the quantiser. In this case, the quantisation errors may be so large as to yield a divergence of the observer's estimate for the estimation problem or an

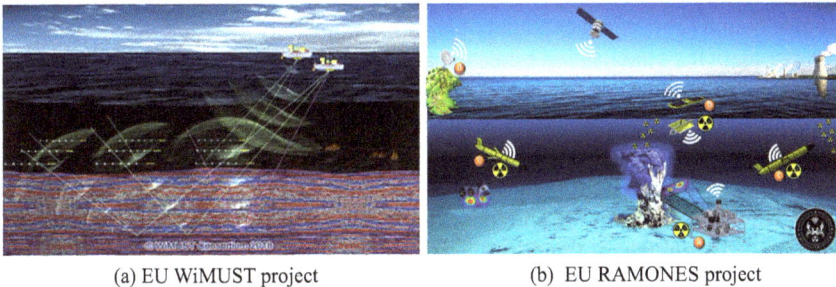

(a) EU WiMUST project   (b) EU RAMONES project

Figure 3.1   *Illustrative examples of acoustically networked marine systems for scientific, commercial and societal applications: (a) automated 3D sub-bottom mapping; (b) radioactivity monitoring in ocean ecosystems*

instability of the closed-loop system in the case of the control problem. Another problem associated with quantisation arises from the requirement to guarantee adequate performance of an estimator when the estimation error nears zero or of a control loop when the state of the system is near an equilibrium point. In fact, in the two situations above, smaller quantisation errors are required to achieve asymptotic convergence of the estimation error and/or closed-loop system state to zero. An earlier study of the non-linear behaviours associated with quantisation can be found in the work of Delchamps [3].

For the case of an infinite quantiser, that is, when the number of transmitted bits is theoretically infinite but the signal is mapped to a countable non-dense set of values, a typical form of addressing the problems associated with quantisation is to use a logarithmic quantiser [4–8], where the precision of the quantiser tends to infinity near its mid-value. In practice, however, with limited bandwidth and also due to computational-related constraints, the range of a quantiser is finite. Therefore, to address this issue, one may dynamically adjust the quantisation levels to achieve better performance; see [9–19], where the strategy adopted is a zoom-out zoom-in approach. The first stage of this approach, the zoom-out, occurs when the state or estimation error is diverging from zero and the transmitted data falls outside the quantiser range, requiring that the quantiser range be increased. The second stage, the zoom-in, occurs during convergence of the estimation error and/or closed-loop system state, and when the transmitted data is already inside the quantisation range. At this stage, the quantisation range is decreased to drive the quantisation error to zero and achieve asymptotic convergence of the estimation error and/or closed-loop system state to zero.

When the objective is to develop dynamic quantisation strategies capable of yielding robustness with respect to external disturbances, asymptotic convergence of the estimation error and/or closed-loop system state is impossible to achieve, and therefore some adaptive strategy must be adopted. One of the most used approaches is to consider the disturbances as random variables and study the behaviour of the state or estimation error in a probabilistic framework [17,20–25]. However, there are situations when deterministic guarantees are preferable. When deterministic bounds on the magnitude of the disturbance are available, a possible strategy is to have a pre-defined quantisation range size, which does not go to zero but rather to a small ultimate bound [26]. When explicit bounds on the noise are unavailable, a better strategy is to adapt the quantisation range dynamically [27], where the quantisation range is transmitted along with the actual data, which is often not practical. A practical approach consists of shrinking the quantisation range whenever the signal is within the quantisation range and expanding it when the signal escapes the range [15,28,29]. As will be seen later, the problem of obtaining input-to-state stability properties, that is, ultimate bounds of the estimation error and/or closed-loop system state proportional to the magnitude of the disturbances acting on the system is very challenging. For this purpose, in a number of publications [15,28,29], a lower bound on the quantisation interval size or a fixed reset interval size is defined to guarantee that the quantisation interval will recapture the signal in finite time. This strategy may not be ideal to achieve good performance near equilibrium since it avoids increasing the quantiser precision to guarantee stability. In this chapter, to overcome the problems inherent to the above-mentioned approaches, we make an assumption on the evolution of the estimation error and the disturbance, which amounts to a persistence of excitation assumption. The motivation for the latter draws from its widespread use in systems theory, namely identification and adaptive control. In this context, the chapter proposes and derives stability conditions for a quantised Luenberger observer operating under limited bandwidth constraints, for which the quantisation interval size does not have a predefined lower bound or returns to a predefined size when zooming out [15,28,29].

The strategy adopted consists of having both the emitter, attached to a sensor, and the receiver in the communication channel run synchronised quantised observers. Given that the emitter has access to the state estimate at the receiver side, it is only necessary to transmit the difference between the expected and the real measurement. We adopt a zoom-in zoom-out approach for an adaptive quantisation interval and consider the case where the state estimates are used for feedback control and the system is subjected to disturbances and measurement noise.

In summary, the main objective of this chapter is to address and derive a theoretical framework for the solution of state estimation and control problems for multiple systems operating under limited communication bandwidth. At the core of the methodologies proposed is the adoption of an observer structure for systems that may be operating under state feedback control. The starting point is the design of a state estimator for the case of unquantised information with adequate input-to-state stability guarantees and an adaptive quantiser structure for coding and decoding information at a finite data rate. For the case where external disturbances are persistently exciting, conditions are derived on the adaptive quantiser that will ensure input-to-state stability of the estimator and the overall controller/estimation system.

The main results are threefold:

- We propose an observer and control structure with adaptive quantisation, where both the emitter, attached to a sensor, and the receiver in the communication channel run synchronised quantised observers, enabling state estimation and control.
- We derive a set of conditions on the quantiser design parameters, without prior knowledge of the magnitude of the disturbances acting on the system, that ensure both the ultimate boundedness of the estimation error and the stability of the closed-loop system, with an ultimate bound proportional to the magnitude of the disturbances.
- Two numerical examples of the proposed algorithm are presented, demonstrating its application to systems with different dynamics, including both estimation and control scenarios.

3.1.1 Chapter structure

This chapter is organised as follows. The formulation of the estimation and control problem that is the central point of this chapter, as well as the assumptions required to solve it, is given in Section 3.2. Section 3.3 describes the proposed quantised consensus-based distributed Luenberger observer. Section 3.4 derives theoretical stability results for the proposed algorithm. Section 3.5 contains numerical results which illustrate the performance of the proposed algorithm. Finally, in Section 3.6, we draw the main conclusions and discuss future work.

3.1.2   Notation

This section summarises the notation used throughout the chapter. The symbols $\mathbb{R}$ and $\mathbb{R}^+$ represent the set of real numbers and the set of strictly positive real numbers, respectively. The notation $\mathbb{N}_0$ represents the set of natural numbers and 0. The symbols $\| \cdot \|$ and $\| \cdot \|_\infty$ represent the 2- and ∞-norm, respectively, of a vector of real numbers. Given a positive definite matrix P, we define the ellipsoidal or P-norm of x as $\|x\|_P := \sqrt{x^T P x}$, where x is a vector of appropriate dimensions. We also define the matrix-induced P-norm of a matrix A as $\|A\|_P := \sup_{\|x\|_P = 1} \|Ax\|_P$, where A is a matrix of appropriate dimensions. The symbol $\lfloor \cdot \rfloor$ represents the floor operator, or the rounding down to the closest lower integer, the function $\mathrm{sgn}(\cdot)$ is the sign function, and $\rho(\cdot)$ is the spectral radius of a square matrix. Given a square matrix P, $\sigma_{\max}(P)$ denotes its maximum singular value, which is equal to the norm of P, $\|P\|$, induced by the vector 2-norm. Further, $\sigma_{\min}(P)$ denotes the minimum singular value of P which, if P is non-singular, is equal to $\|P^{-1}\|^{-1}$. I_M represents an $M \times M$ identity matrix, and 1 represents an $N \times 1$ vector with ones in every entry.

3.1.3   Uniform quantiser

For the sake of clarity, the meaning of quantisation is explained formally next. The concept is motivated by the requirement to transmit a message with a limited number of bits and a known precision, given that both the transmitter and the receiver have the same quantiser parameters [30]: mid-value, quantisation interval, and the number of transmitted bits.

Consider the quantisation interval $\left[\bar{x} - \frac{\Lambda}{2}, \bar{x} + \frac{\Lambda}{2}\right]$ of size $\Lambda \in \mathbb{R}^+$ centred at the mid-value $\bar{x}$. A uniform quantiser with quantisation step-size $\Delta \in \mathbb{R}^+$ is given by

$$Q(x) := \bar{x} + \mathrm{sgn}(x - \bar{x})q(\|x - \bar{x}\|), \tag{3.1}$$

where

$$q(x) := \begin{cases} \frac{\Lambda}{2} & \text{if } x > \frac{\Lambda}{2} \\ \Delta \left\lfloor \frac{x}{\Delta} + \frac{1}{2} \right\rfloor & \text{if } x \leq \frac{\Lambda}{2}, \end{cases} \tag{3.2}$$

The parameter Δ is determined by the number n_b of bits of the quantiser as $\Delta := \frac{\Lambda}{2^{n_b} - 2}$. From (3.1), if $x \in \left[\bar{x} - \frac{\Lambda}{2}, \bar{x} + \frac{\Lambda}{2}\right]$, then the quantisation error is upper-bounded by

$$\|x - Q(x)\| \leq \frac{\Delta}{2} = \frac{\Lambda}{2^{n_b + 1} - 4}. \tag{3.3}$$

For the case where the input of the quantiser and the mid-value are vectors with the same dimension, the quantiser Q is taken element-wise.

3.2   Problem statement

This section formulates the problem that is the main topic of the chapter.

3.2.1   Networked control and estimation system

We start by considering the general setup shown in Figure 3.2, consisting of (i) a discrete-time non-linear dynamical system), (ii) a remote node with emitting capabilities (emitter) consisting of a sensor that takes remote measurements of the dynamical system's output, an emitter-side observer and a device that transmits quantised data through a limited bandwidth communication channel and (iii) a node collocated with the dynamical system (receiver) consisting of a receiver-side observer and a controller, with the latter acting upon the system to stabilise it. Notice how the setup shown in Figure 3.2 highlights the fact that the two observers involved are synchronised. We will discuss the question of observer synchronisation further in Section 3.3.

We assume that the discrete-time dynamical system is described by

$$x_{t+1} = f(x_t, u_t) + w_t, \tag{3.4}$$

where $t \in \mathbb{N}_0$ is the discrete-time index, $x_t \in \mathbb{R}^n$, $u_t \in \mathbb{R}^l$ and $w_t \in \mathbb{R}^n$ denote the complete state vector, the control input and the state noise vector, respectively, at discrete time t, and $f : \mathbb{R}^n \times \mathbb{R}^l \to \mathbb{R}^n$ is the state transition function.

The measurement equation is given by

$$y_t = h(x_t) + v_t, \tag{3.5}$$

where $y_t \in \mathbb{R}^m$ and $v_t \in \mathbb{R}^m$ denote the observation vector and the observation noise vector, respectively, considered at time t, and $h : \mathbb{R}^n \to \mathbb{R}^m$ is the output function. The

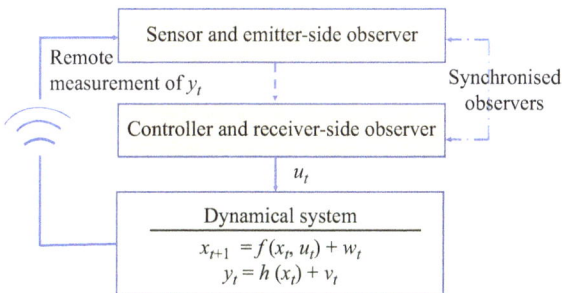

Figure 3.2   *Problem setup consisting of (i) a discrete-time non-linear dynamical system, (ii) a remote node with emitting capabilities (emitter) consisting of a sensor that takes remote measurements of the dynamical system's output, an emitter-side observer and a device that transmits quantised data through a limited bandwidth communication channel and (iii) a node collocated with the dynamical system (receiver) consisting of a receiver-side observer and a controller, with the latter acting upon the system to stabilise it.*

following assumptions are made on the detectability and stabilisability of the system and the intensity of the process and measurement noise.

We assume that there exists a stabilising controller and a stable observer for the non-linear system in the case of perfect, unquantised communications. This may be achieved using classical observation methods such as extended Kalman filtering, high-gain observers or moving horizon estimation, among other methods. In the remainder of the chapter, we will not specify which observer and control methods are used, but assume instead that known Lyapunov decrease properties hold for the unquantised observer and controller. This is expressed by the following assumptions.

Assumption A1. *There exists an observer of the form*

$$x^o_{t+1} = f^o(x^o_t, u_t, y_t), \tag{3.6a}$$
$$\hat{x}_t = h^o(x^o_t), \tag{3.6b}$$

where $x^o_t \in \mathbb{R}^o$ denotes the state of the observer at discrete time t, $\hat{x}_t \in \mathbb{R}^n$ is the state estimate, $f^o : \mathbb{R}^o \times \mathbb{R}^l \times \mathbb{R}^m \to \mathbb{R}^o$ is the observer dynamics function and $h^o : \mathbb{R}^o \times \mathbb{R}^o \to \mathbb{R}^n$ is the observer output function.

Denoting the estimation error as $e_t := \hat{x}_t - x_t$, we assume that observer ((3.6a)– (3.6b)) has the Lyapunov decreasing property

$$V_{t+1}(e_{t+1}) \leq \tilde{\beta} V_t(e_t) + l_1 \|v_t\|_\infty + l_2 \|w_t\|_\infty, \tag{3.7}$$

where V_t satisfies $\alpha_1(\|x\|) \leq V_t(x) \leq \alpha_2(\|x\|)$ for all $x \in \mathbb{R}^n$, $\alpha_1, \alpha_2 \in \mathcal{K}$, l_1 and l_2 are positive scalars and $\tilde{\beta}$ is a known constant with $0 < \tilde{\beta} < 1$.

Assumption A2. *There exists a controller of the form*

$$u_t = k(\hat{x}_t), \tag{3.8}$$

where $k : \mathbb{R}^n \to \mathbb{R}^l$ is a control law which yields the Lyapunov decreasing property

$$V^c(x_{t+1}) \leq \beta^c V^c(x_t) + l_3 \|e_t\|_\infty + l_4 \|w_t\|_\infty, \tag{3.9}$$

where V^c satisfies $\alpha_3(\|x\|) \leq V^c(x) \leq \alpha_4(\|x\|)$ for all $x \in \mathbb{R}^n$, $\alpha_3, \alpha_4 \in \mathcal{K}$, l_3 and l_4 are positive scalars and $0 < \beta^c < 1$.

Assumption A3. *Given any $x_1, x_2 \in \mathbb{R}^n$, $\|h(x_1) - h(x_2)\|_\infty \leq cV_t(x_1 - x_2)$ for all $t \in \mathbb{N}_0$, where $c \in \mathbb{R}^+$.*

Assumption A4. *The $\mathcal{L}_\infty$ norms of the noise signals satisfy*

$$\|w_t\|_\infty \leq \varepsilon_w, \quad \|v_t\|_\infty \leq \varepsilon_v, \tag{3.10}$$

for some unknown positive constants ε_w and ε_v, where $\|\cdot\|_\infty$ represents the ∞-norm of a vector of real numbers.

Remark 1. *For linear systems of the form*

$$x_{t+1} = Ax_t + Bu_t + w_t, \tag{3.11a}$$
$$y_t = Cx_t + v_t, \tag{3.11b}$$

assumption A2 is satisfied with a state feedback control input of the form $u_t := K\hat{x}_t$, *where* $K \in \mathbb{R}^{l \times n}$ *is the state feedback gain matrix. Defining the estimation error as* $e_t := \hat{x}_t - x_t$, *the closed-loop dynamics can be written as*

$$x_{t+1} = (A + BK)x_t + BKe_t + w_t \tag{3.12}$$

and, if the pair (A, B) *is stabilisable, a gain matrix* K *can be computed such that* $\rho(A + BK) < 1$, *thus showing that assumption A2 holds.*

Under the assumption of a linear state feedback law $u_t = K\hat{x}_t$, *assumption A1 is satisfied with a Luenberger observer algorithm described by*

$$\hat{x}_{t+1} = (A + BK - LC)\hat{x}_t + Ly_t, \tag{3.13}$$

where L *is a gain matrix of appropriate dimensions such that* $\rho(A - LC) < 1$, *which can always be found if* (A, C) *is detectable.*

In the above case, the observer Lyapunov function is given by a P-norm of the estimation error, $V_t(e_t) := \|e_t\|_P$ *such that* $\tilde{\beta} := \|A - LC\|_P < 1$. *Since* $\rho(A - LC) < 1$, *it is well known [31] that a P-norm can be found such that* $\tilde{\beta} < 1$ *by choosing an arbitrary positive definite matrix* S *and computing a symmetric positive definite matrix such that* $(A - LC)^T P(A - LC) - P = -S$, *yielding* $\tilde{\beta} = \sqrt{1 - \sigma_{min}\left(Q^{\frac{1}{2}}P^{-\frac{1}{2}}\right)^2} < 1$. *Therefore, noting that*

$$e_{t+1} = (A - LC)e_t - w_t + Lv_t, \tag{3.14}$$

assumption A1 holds with

$$\alpha_1(x) := \sqrt{\sigma_{\min}(P)}x, \tag{3.15a}$$

$$\alpha_2(x) := \sqrt{\sigma_{\max}(P)}x, \tag{3.15b}$$

$$l_1 := \sqrt{n\sigma_{\max}(P)}\|L\|_\infty, \tag{3.15c}$$

$$l_2 := \sqrt{n\sigma_{\max}(P)}, \tag{3.15d}$$

and assumption A3 holds with

$$c := \frac{\|C\|_\infty}{\sqrt{\sigma_{\min}(P)}}. \tag{3.16}$$

Remark 2. *For systems that do not satisfy the Lipschitz conditions of assumption A3 or the linear gains* l_1 *and* l_2 *of assumption A1 but instead satisfy*

$$V_{t+1}(e_{t+1}) \leq \tilde{\beta}V_t(e_t) + \sigma_1(\|v_t\|_\infty) + \sigma_2(\|w_t\|_\infty), \tag{3.17}$$

for some class $\mathcal{K}$ *functions* σ_1 *and* σ_2, *it may not be possible to demonstrate results analogous to those derived here. In general, different adaptive quantisation strategies than those in (3.20), (3.55) or (3.56) may be required to prove stability, which will depend on the particular class of systems under consideration. Notice that [28] considers linear systems, and in [29], similar Lipschitz conditions must hold.*

We anticipate that to extend the results of this chapter to more general classes of systems, we should consider adaptive strategies that are faster than geometric growth when zooming out and slower than geometric decay when zooming in. This, however, would yield worse performance for systems that satisfy the conditions of theorem 1, where strategy (3.20) is considered. Therefore, for more general adaptive strategies, we would be trading off performance for generality.

3.2.2   Monitoring system

In what follows, we consider a different setup where the objective is to estimate the state of a system remotely, relying on quantised information transmitted from a sensor and an emitter-side observer collocated with the system, sent through a limited bandwidth communication channel. A receiver-side observer receives the latter information and estimates the state of the system. This setup is shown in Figure 3.3, where, again, we stress that the two observers are synchronised.

In this case, the dynamics of the system are given by

$$x_{t+1} = f(x_t) + w_t. \tag{3.18}$$

Since the purpose of the observer is simply to monitor the system without acting upon it, we do not require the system to be stable or stabilisable, that is, we do not require assumption A2 to be satisfied.

In the remainder, since both cases can be analysed in the same manner, we will consider the case of Figure 3.2, where the objective is to stabilise the system. However, because assumption A2 is not required to derive the main result in theorem 1, the same framework is also suitable to tackle the case depicted in Figure 3.3. That is, assumption A2 is not required to guarantee stability of the estimation error but is required to show stability of the overall system, given a stable observer.

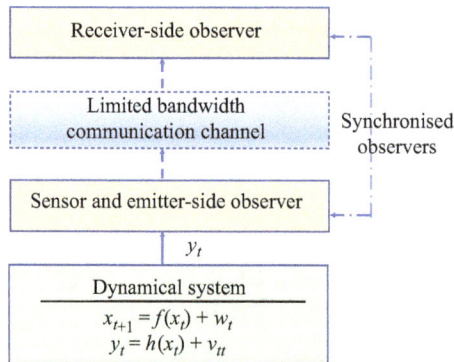

Figure 3.3   Problem setup consisting of (i) a discrete-time non-linear dynamical system, (ii) a (local) node collocated with the system with emitting capabilities (emitter) consisting of a sensor that takes local measurements of the system's output, an emitter-side observer and a device that transmits quantised data through a limited bandwidth communication channel and (iii) a remote node (receiver) with an observer that monitors the system.

3.2.3   Problem statement

Given the system described in Section 3.2.1, the main problem solved in this chapter is described next.

Assume in (3.4)–(3.5) that the process and measurement noise w_t and v_t are uniformly bounded over time. Consider also that synchronously, at each discrete time t, the sensor and emitter-side observers are allowed to transmit a quantised message with a fixed number of bits n_b to the receiver-side observers. In this work, we assume a simple communication model that does not consider packet losses, message overheads and assumes that the computations are instantaneous in comparison with the intervals between communications. For more complex models and a more thorough analysis of digital communications, see [32].

The problem of *quantised state estimation* consists of reconstructing, at the receiver-side observer that receives quantised data from the sensor, the state of the system (3.4), with the estimation error converging to an ultimate bound proportional to the magnitude of the measurement and process noise.

Stated mathematically, the objective is to compute at each discrete time t a state estimate $\hat{x}_t$ such that, for every initial condition, there exists a time T, which depends on the initial conditions, such that for $t \geq T$, $\|\hat{x}_t - x_t\| \leq b$, where b is proportional to ε_v and ε_w defined in assumption A4.

3.3   Proposed quantised Luenberger observer

To solve the problem formulated in Section 3.2, we propose the general architecture shown in Figure 3.4. To have more effective communications, the emitter-side observer computes a state estimate η_t using the quantised transmitted data given by $Q_t(\theta_t)$, where $\theta_t := y_t - h(\eta_t)$. The receiver-side observer, which also estimates the state using the quantised data transmitted from the sensor $Q_t(\theta_t)$, is synchronised with the emitter-side observer, that is, for all discrete times t $\hat{x}_t = \eta_t$. To ensure that the

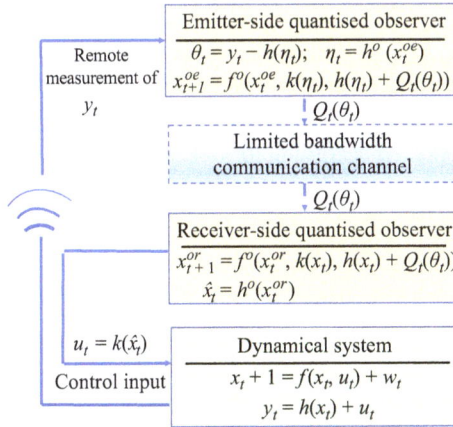

Figure 3.4   Estimation and control architecture diagram; $k(\cdot)$ is a state feedback function

receiver-side observer is an exact copy of the emitter-side observer, both the receiver- and emitter-side observers should be initialised at the same time with the same values, i.e. $x_0^{oe} = x_0^{or}$ and there should not be packet losses, that is, the transmitted data $Q_t(\theta_t)$ should reach the local observer at all times. This procedure can be simplified for the set-up shown in Figure 3.2, as explained in remark 3. This architecture is based on previous work [33] in a distributed setting, where the quantisation interval size is a pre-defined function of time.

Remark 3. *In case there are packet losses, or if the observers are initialised with different initial state estimates, it is not possible in general to guarantee that both observers are synchronised, that is, $\hat{x}_t = \eta_t$, for all times t. However, if the quantisation interval Λ_t is transmitted periodically by the emitter and the unquantised closed-loop system is contractive, that is, if all the trajectories of the system converge to the same trajectory, then it is easy to show that the difference $\hat{x}_t - \eta_t$ converges asymptotically to 0. Contractiveness is an inherent property of stable linear systems that becomes important for the analysis of the problem stated in Section 3.2.3, when the dynamical system in Figure 3.2 is itself stable. Notice, however, that the above difference does not converge to zero if the system in Figure 3.3, which arises in the case of the monitoring problem, is not stable. To deal with packet losses, a possibility is to devise a simple communication protocol that informs the emitter which messages were not received. This would allow for the removal of the effect of the lost messages on the emitter-side observer and quantiser and thus recover synchronisation instantaneously, that is, $\hat{x}_t = \eta_t$, for the cases of Figure 3.2 or Figure 3.3.*

Given assumption A2, if the process noise w_t and the estimation error e_t are ultimately bounded, then the system state x_t is also ultimately bounded. For this reason, a central concept in this section is that of state estimation, where the objective is to estimate the state of the system x_t with a bounded error using an estimator of the form

$$x_{t+1}^{or} = f^o(x_t^{or}, k(\hat{x}_t), h(\hat{x}_t) + Q_t(\theta_t)), \tag{3.19a}$$
$$\hat{x}_t = h^o(x_t^{or}), \tag{3.19b}$$

where $\theta_t := y_t - h(\eta_t)$ is the value before quantisation, and η_t is the synchronised estimated state at the emitter. The mid value of the quantiser is 0, and the quantisation interval length Λ_t is updated according to

$$\Lambda_{t+1} := \begin{cases} \beta\Lambda_t, & \text{if } \|\theta_t\|_\infty \leq \frac{\Lambda_t}{2} \\ \beta_{out}\Lambda_t, & \text{otherwise} \end{cases}, \tag{3.20}$$

where $\beta > \tilde{\beta}$ and $\beta_{out} > \frac{1}{\beta}$ are appropriately chosen constants. As mentioned earlier, the emitter also contains a copy of the observer that is initialised at the same time, with the same value $x_0^{oe} = x_0^{or}$ and has the dynamics

$$x_{t+1}^{oe} = f^o(x_t^{oe}, k(\eta_t), h(\eta_t) + Q_t(\theta_t)), \tag{3.21a}$$
$$\eta_t = h^o(x_t^{oe}). \tag{3.21b}$$

It follows from (3.19a)–(3.19b), assumption A1 and the fact that $h(\hat{x}_t) + Q(\theta_t) = h(x_t) + v_t + \xi_t$, where $\xi_t := Q_t(\theta_t) - \theta_t$, that the estimation error $e_t := \hat{x}_t - x_t$ satisfies the Lyapunov property

$$V_{t+1}(e_{t+1}) \leq V_t(e_t) + l_1 \|v_t\|_\infty + l_1 \|\xi_t\|_\infty + l_2 \|w_t\|_\infty. \tag{3.22}$$

Since, from assumption A4, the measurement and process noises v_t^i and w_t are bounded, it follows that the estimation error e_t is ultimately bounded if ξ_t is also ultimately bounded.

Remark 4. *To understand the challenges that the presence of disturbances poses, we consider the simple system described by*

$$x_{t+1} = x_t + w_t, \tag{3.23a}$$

$$y_t = x_t, \tag{3.23b}$$

with $x_t \in \mathbb{R}$. Consider also the observer

$$\hat{x}_{t+1} = \hat{x}_t + Q_t(y_t - \hat{x}_t), \tag{3.24}$$

with a mid value of the quantiser of 0, and the quantisation interval length Λ_t is updated as in (3.20). Suppose also, for simplicity, that $\frac{\Lambda_0}{2} > \|e_0\|$ and that $\frac{1}{2^{n_b}-2} < \beta$. This will be equivalent to condition (3.51) since $\tilde{\beta} = 0$ and $l_1 c = 1$.

As we will show, with this setup, we can drive the estimation error to an arbitrarily large value with a disturbance satisfying $w_t \leq \varepsilon$ for all $t \geq 0$. That is, for any $\varepsilon_e > 0$, we can design the disturbance such that $w_t \leq \varepsilon$ for all $t \geq 0$, and $\|e_{t_m}\| \geq \varepsilon_e$ for some $t_m > 0$.

To see this, we start by writing the estimation error dynamics as

$$e_{t+1} = e_t - Q_t(e_t) - w_t. \tag{3.25}$$

We then choose the disturbance as

$$w_t := \begin{cases} 0 & \text{if } t < t_s \\ \varepsilon & \text{if } t \geq t_s \end{cases}, \tag{3.26}$$

for some $t_s \geq 0$, which implies that

$$x_t := x_0 + \max(0, t - t_s)\varepsilon. \tag{3.27}$$

From the fact that $\frac{1}{2^{n_b}-2} < \beta$ we have that, for $t < t_s$, if $\frac{\Lambda_t}{2} > \|e_t\|$ then $\frac{\Lambda_{t+1}}{2} > \|e_{t+1}\|$. Therefore, $\frac{\Lambda_{t_s}}{2} = \beta^{t_s}\frac{\Lambda_0}{2} \geq \|e_{t_s}\|$.

If $\|e_t\| > \frac{\Lambda_t}{2}$ for $t_s + 1 \leq t \leq t_s + \tau$, for some integer $\tau \geq 1$, then

$$\hat{x}_{t_s+\tau} < x_0 + \tau \beta_{out}^\tau \beta^{t_s} \frac{\Lambda_0}{2}, \tag{3.28}$$

which implies, from (3.27), that

$$\|e_{t_s+\tau}\| > \tau \left(\varepsilon - \beta_{out}^\tau \beta^{t_s} \frac{\Lambda_0}{2} \right). \tag{3.29}$$

Selecting

$$\tau < -t_s \log_{\beta_{out}} (\beta) - \log_{\beta_{out}} (\Lambda_0), \tag{3.30}$$

we have that

$$\frac{\Lambda_{t_s + \tau}}{2} = \beta_{out}^{\tau} \beta^{t_s} \frac{\Lambda_0}{2} < \frac{\varepsilon}{2}, \tag{3.31}$$

and therefore, if $\tau \geq 1$, $\|e_{t_s + \tau}\| > \tau \frac{\varepsilon}{2} > \frac{\Lambda_{t_s + \tau}}{2}$.
  Selecting

$$t_s = -\frac{2\frac{\varepsilon_e}{\varepsilon} + \log_{\beta_{out}} (\Lambda_0)}{\log_{\beta_{out}} (\beta)} + 1, \tag{3.32}$$

we can choose $\tau = 2\frac{\varepsilon_e}{\varepsilon}$, *yielding* $\|e_{t_m}\| \geq \varepsilon_e$ *for* $t_m = t_s + 2\frac{\varepsilon_e}{\varepsilon}$ *as we wanted to show.*

  To overcome this issue, some authors [15,29] propose schemes where there is a minimum value of the quantisation interval when zooming out, which guarantees that there is a limit on the number of consecutive times the quantisation interval length increases, that is, a limit on τ *in our example, which depends on the magnitude of the disturbance* ε.

  In this chapter, we pursue a different strategy. We assume that there is a persistency of excitation property of the disturbance, which is often found in practice, expressed by assumption A5. In our example, this is equivalent to not being able to select t_s *freely as in (3.32). This approach sacrifices generality but allows for the quantisation interval to decrease freely, which potentially allows for lower ultimate bounds on the estimation error.*

3.4 Theoretical guarantees on the quantised Luenberger observer with adaptive quantisation

We now provide a theoretical result that shows what conditions are required for the use of adaptive quantisation in the context of state estimation.

  The infinity norm of the value of θ_t before quantisation can be bounded, using assumption A3, as

$$\|\theta_t\|_{\infty} = \|h(x_t) - h(\hat{x}_t) + v_t\|_{\infty}$$
$$\leq \bar{\theta}_t := cV_t(e_t) + \frac{\varepsilon}{l_1}, \tag{3.33}$$

where $\varepsilon := l_1 \varepsilon_v + l_2 \varepsilon_w$.

  The following assumption is also required.

Assumption A5. *There exists a constant* $0 < k < 1$ *and two non-negative integers N and T such that for all time t, there exists two non-negative integers* $0 \leq \tau_1 \leq N$ *and* $0 \leq \tau_2 \leq T$ *such that*

$$\|\theta_{t + \tau_1 + \tau_2}\|_{\infty} \geq k\bar{\theta}_{t + \tau_1},$$

To understand the rationale for assumption A5, consider the linear time-invariant system

$$x_{t+1} = Ax_t + Bu_t + w_t, \tag{3.34a}$$
$$y_t = Cx_t + v_t, \tag{3.34b}$$

and the quantised Luenberger observer

$$\hat{x}_{t+1} = A\hat{x}_t + Bu_t + LQ_t(\theta_t), \tag{3.35}$$

with $\rho(A - LC) < 1$. The estimation error dynamics and the output estimation error are given by

$$e_{t+1} = (A - LC)e_t + \tilde{w}_t, \tag{3.36a}$$
$$\theta_t = Ce_t + v_t, \tag{3.36b}$$

where

$$\tilde{w}_t := Lv_t - w_t + L\xi_t. \tag{3.37}$$

Applying (3.36) recursively yields

$$\theta_{t+\tau} = C(A - LC)^\tau e_t + \tilde{v}_{t,\tau}, \tag{3.38a}$$
$$\tilde{v}_{t,\tau} := v_{t+\tau} + C\bar{w}_{t,\tau}, \tag{3.38b}$$
$$\bar{w}_{t,\tau} := \sum_{\mu=0}^{\tau-1}(A - LC)^{\tau-1-\mu}\tilde{w}_{t+\mu}. \tag{3.38c}$$

If the number of transmitted bits is large then the quantisation error ξ_t, as well as the measurement noise v_t and the process disturbance w_t can be considered as random disturbances; see [34]. Therefore, it may be reasonable to make the following persistence of excitation assumption.

Assumption A6. *There exists a positive scalar constant k_p and a positive integer constant N such that for all $\tau, t \geq 0$ and $e \in \mathbb{R}^n$, there exist two integers t_1 and t_2 with $t \leq t_1 < t_2 \leq t + N$ such that*

$$\|\tilde{v}_{t_1,\tau}\| \geq k_p\varepsilon, \tag{3.39}$$
$$\|\tilde{v}_{t_2,\tau}\| \geq k_p\varepsilon, \tag{3.40}$$
$$\tilde{v}_{t_2,\tau}^T C(A - LC)^\tau \bar{w}_{t_1,\Delta t} \geq 0, \tag{3.41}$$
$$\left(\tilde{v}_{t_1,\tau}^T C(A - LC)^\tau e\right)\left(\tilde{v}_{t_2,\tau}^T C(A - LC)^{\tau+\Delta t}e\right) \leq 0, \tag{3.42}$$

with $\Delta t := t_2 - t_1$.

Then, the following lemma establishes the reason for assumption A5 for linear systems.

Lemma 1. *Given the LTI system (3.34) and the quantised Luenberger observer (3.35), if assumption A6 holds, it follows that assumption A5 also holds.*

Proof.   From the observability of the pair $(C, A - LC)$ and using the Lyapunov function obtained in remark 1, there exists an integer $T \geq 0$ and a scalar k_1 satisfying $0 < k_1 < 1$ such that for all $x \in \mathbb{R}^n$, there exists an integer τ_2 satisfying $0 \leq \tau_2 \leq T$ and

$$\|C(A - LC)^{\tau_2} x\| \geq k_1 c \|x\|_P. \tag{3.43}$$

If assumption A6 is satisfied, there exists an integer $N \geq 0$ and a scalar $0 < k_2 < 1$ such that for all $t \geq 0$, there exists an integer τ_1 satisfying $0 \leq \tau_1 \leq N$ such that for $t' = t + \tau_1$

$$\|\tilde{v}_{t', \tau_2}\| \geq k_2 \frac{\varepsilon}{l_1}, \tag{3.44}$$

and

$$\tilde{v}_{t', \tau_2}^T C(A - LC)^{\tau_2} e_{t'} \geq 0. \tag{3.45}$$

Then,

$$\|\theta_{t+\tau_1+\tau_2}\|_\infty \geq \frac{\|C(A - LC)^{\tau_2} e_{t+\tau_1} + \tilde{v}_{t+\tau_1, \tau_2}\|}{\sqrt{m}} \tag{3.46}$$

$$\geq \frac{\|C(A - LC)^{\tau_2} e_{t+\tau_1}\| + \|\tilde{v}_{t+\tau_1, \tau_2}\|}{\sqrt{m}} \tag{3.47}$$

$$\geq \frac{1}{\sqrt{m}} \left(k_1 c \|e_{t+\tau_1}\|_P + k_2 \frac{\varepsilon}{l_1} \right) \tag{3.48}$$

$$\geq k \bar{\theta}_{t+\tau_1}, \tag{3.49}$$

for $k := \frac{1}{\sqrt{m}} \min(k_1, k_2)$ as we wanted to show. $\qquad\square$

We now establish the following stability result for the Luenberger observer with adaptive quantisation.

Theorem 1. *Consider the Luenberger observer* (3.19a)–(3.19b) *and let Assumptions A1, A3, A4 and A5 hold. Suppose that at each discrete time instant $t \in \mathcal{N}$, the sensor transmits a constant number of bits n_b, where n_b satisfies*

$$n_b > \log_2 \left(\frac{l_1 c}{1 - \tilde{\beta}} + 2 \right). \tag{3.50}$$

Suppose also that the zoom-in coefficient β satisfies

$$1 > \beta > \bar{\beta} := \tilde{\beta} + \frac{l_1 c}{2^{n_b} - 2}, \tag{3.51}$$

and the zoom-out coefficient β_{out} satisfies

$$\beta_{out} > \frac{\psi}{\beta^{T+N+1}}, \tag{3.52}$$

where

$$a := 1 + cl_1,$$

(3.53a)

$$\psi' := \frac{a^{T+N}}{k} + \frac{\beta^{T+N}\left(1 - \left(\frac{a}{\beta}\right)^{T+N}\right)l_1 c}{(\beta - a)\,(2^{n_b} - 2)},$$

(3.53b)

$$\psi := (1 + 2cl_1)\psi'.$$

(3.53c)

Then, the estimation error converges to within the following neighbourhood of the origin:

$$\limsup_{t \to \infty} \|e_t\| < \alpha_1^{-1}\left(\left(\beta_{out}\frac{1 - \tilde{\beta} + l_1 c}{\beta - \tilde{\beta}} - 1\right)\frac{\varepsilon}{l_1 c}\right).$$

(3.54)

The proof is given in Appendix A.1.

Remark 5. *In theorem 1, the lower bound on the number of bits transmitted n_b in (3.50) depends only on the design of the observer. To minimise the lower bound in (3.50), one must address the trade-off between minimising $\tilde{\beta}$ and l_1.*

From (3.51), one can see that the lower bound on β, $\bar{\beta}$, tends to $\tilde{\beta}$ as n_b increases.

From (3.54), we also see that if conditions (3.50)–(3.52) hold, the estimation error converges to an ultimate bound which is proportional to the magnitude of disturbances acting on the system and the measurement noise, given by ε. Thus, the adopted observer solves the problem stated in Section 3.2.3.

Remark 6. *Assumption A5 avoids the pathological case, exemplified in remark 4, where the magnitude of the noise is small for a long time but increases suddenly. In this case, the quantisation interval may be so small that when the magnitude of the noise increases and consequently also θ_t, the quantisation interval takes an arbitrarily long time to increase to an acceptable size. To prevent this situation, possible solutions would be to have a predefined quantisation interval size when zooming out [29], that is,*

$$\Lambda_{t+1} := \begin{cases} \beta\Lambda_t, & \text{if } \|\theta_t\|_\infty \le \frac{\Lambda_t}{2} \\ \max\left(\beta_{out}\Lambda_t, \Lambda_0\right), & \text{otherwise} \end{cases},$$

(3.55)

or to have a lower bound of the quantisation interval size [28], that is,

$$\Lambda_{t+1} := \begin{cases} \max\left(\beta\Lambda_t, \Lambda_0\right), & \text{if } \|\theta_t\|_\infty \le \frac{\Lambda_t}{2} \\ \beta_{out}\Lambda_t, & \text{otherwise} \end{cases}.$$

(3.56)

Theorem 1 shows that if assumption A5 holds, then these solutions are not necessary to ensure stability, and one can decrease the quantisation interval freely as in (3.20). This approach has the potential to yield better performance since it allows for the quantisation interval, and consequently the quantisation errors, to shrink without any constraint.

Remark 7. *Another method to avoid the pathological case mentioned in remark 6 would be to introduce an artificial measurement noise with a small covariance. That is, the measurement equation in (3.5) becomes*

$$y_t = h(x_t) + v_t + v_t^a, \tag{3.57}$$

where $v_t^a \sim \mathcal{N}(0, V_a)$ is a randomly generated artificial measurement noise and V_a is an appropriately chosen covariance matrix. This guarantees a persistence of excitation property of the measurement noise, therefore avoiding the case mentioned in remark 6.

Remark 8. *An alternative strategy to (3.20) is one where the emitter selects the scale of the quantiser and sends this information to the receiver in real time. However, this would increase the amount of data to be transmitted, and the number of allowable quantisation intervals would be finite. This would result in an upper bound on the disturbance magnitudes that the observer would be capable of handling and a fixed lower scale of the quantiser, which means that it would not be possible for the precision of the observer estimates to increase above a certain point.*

Suppose the communication limitations are not very stringent or the quantisation interval is not required to change very often and an upper bound on the magnitude of the disturbance is known. In that case, this alternative strategy is expected to yield a better performance. However, in this chapter, we aim for global results, where there is no prior information on the magnitude of the disturbance.

3.5   Numerical results

3.5.1   Control and estimation system

To illustrate the performance of the proposed control scheme, we performed numerical simulations on an example of the controlled system of Figure 3.2 with dynamics given by

$$x_{t+1} = Ax_t + Bg(Cx_t) + Bu_t + w_t, \tag{3.58a}$$
$$y_t = Cx_t + v_t, \tag{3.58b}$$

where g is a globally Lipschitz continuous function which satisfies $\|g(x) - g(y)\|_\infty \leq d\|x - y\|_\infty$ for all admissible x and y for some positive constant d. This example represents a linear system with well-behaved non-linearity.

We consider a state feedback law of the form $u_t = K\hat{x}_t - g(C\hat{x}_t)$, which satisfies assumption A2, and an observer of the form

$$\hat{x}_{t+1} = (A - LC)\hat{x}_t + L\bar{y}_t + Bu_t + Bg(\bar{y}_t), \tag{3.59a}$$
$$\bar{y}_t = C\hat{x}_t + Q_t(\theta_t), \tag{3.59b}$$

which, for an appropriate choice of L, satisfies assumption A1, where V_t and l_2 are given in remark 1 and $l_1 := \sqrt{n\sigma_{max}(P)}\|L\|_\infty + l_2\|B\|_\infty d$.

In particular, the matrices A, B and C are

$$A = \begin{bmatrix} 1 & 0 \\ 1 & 1.1 \end{bmatrix}, \quad B = \begin{bmatrix} 1 \\ 0 \end{bmatrix}, \quad C = \begin{bmatrix} 0 & 1 \end{bmatrix}, \tag{3.60}$$

and the function g is $g(x) := \sin(x)$, with a global Lipschitz constant $d := 1$.

The process noise w_t follows a Gaussian distribution with covariance equal to identity, the measurement noise v_t also follows a Gaussian distribution with covariance $R = 1$ and the initial state is drawn from a normal distribution with a standard deviation of 5 for each element, that is, $x_0 \sim \mathcal{N}(0, 5^2 I_2)$. The Luenberger gain matrix L and the state feedback control gain matrix K are computed following classical linear-quadratic-Gaussian control design methods.

The quantiser was set with an initial quantisation interval of $\Lambda_0 = 10^4$ and zoom-in–zoom-out parameters set as $\beta = \tilde{\beta} + 0.5(1 - \tilde{\beta})$ and $\beta_{\text{out}} = 1.1a$, where a is defined in (3.53a). To illustrate the dependence of the estimation error and the system's state on the number of bits used to transmit information, we computed the average estimation error norm and state norm when the number of transmitted bits varies from 3 to 20. The results are shown in Figures 3.5 and 3.6.

According to (3.50), the minimum number of transmitted bits for which one can guarantee stability is $n_b = 6$. However, Figures 3.5 and 3.6 show that this is a conservative lower bound since we obtain stable systems for $n_b = 3$. From Figures 3.5 and 3.6, we can observe that above $n_b = 10$, performance of the system does not improve and is thus similar to a situation when the data is unquantised, where the estimation error and state norm do not converge to zero due to the measurement and process noise.

The evolution of the state norm $\|x_t\|$, for different numbers of bits is plotted in Figure 3.7. One can observe that for higher numbers of transmitted bits, the state converges faster to its ultimate bound. This is due to the fact that during the transient phase, the quantisation error converges slower than the convergence rate of the unquantised Luenberger observer.

With $n_b = 3$, Figure 3.8 shows, for a single run, the evolution of the system's state norm using different methodologies to select the length of the quantisation interval,

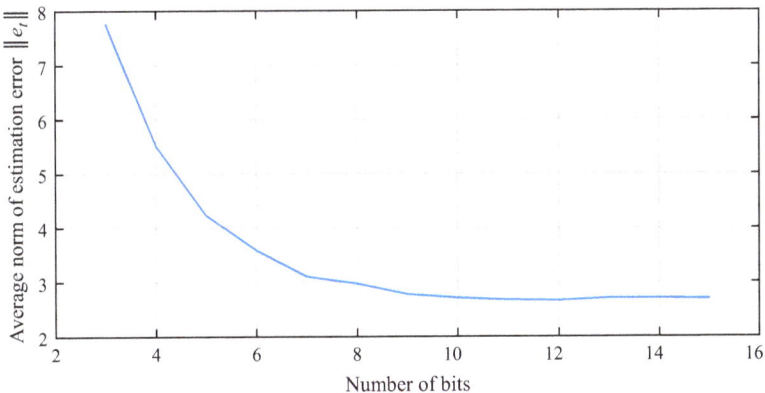

Figure 3.5   *Root mean square of the estimation error as a function of the number of transmitted bits, starting at 3, with an average of 40 samples, with an initial state drawn from the distribution $x_0 \sim \mathcal{N}(0, 10^4 I_2)$, in a logarithmic scale*

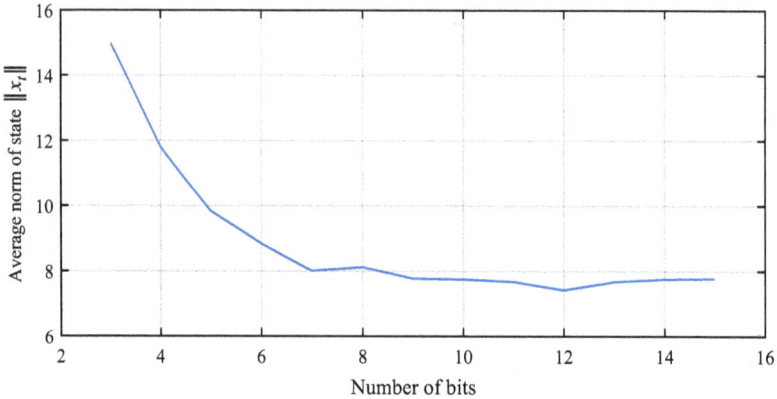

Figure 3.6   *Root mean square of the state norm as a function of the number of transmitted bits, starting at 3, with an average of 40 samples, with an initial state drawn from the distribution $x_0 \sim \mathcal{N}((0, 10^4 I_2))$, in a logarithmic scale*

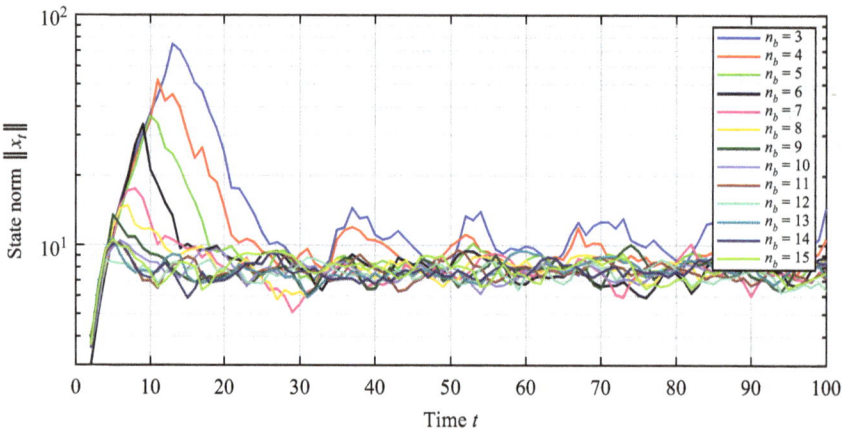

Figure 3.7   *Average of 40 samples of the system state norm $\|x_t\|$ for different values of n_b with an initial state drawn from the distribution $x_0 \sim \mathcal{N}(0, 10^4 I_2)$, in a logarithmic scale*

with $x_0 \sim \mathcal{N}(0, 5^2 I_2)$. Specifically, the yellow line corresponds to the newly proposed scheme described by (3.20), while the blue line and the red line are obtained using (3.55) and (3.56), based on earlier works [28,29], respectively. The green line corresponds to the solution with the introduction of an additional measurement noise as in (3.57) with $V_a = 0.1$. For comparison, the purple line corresponds to the results obtained with a fixed quantisation interval length of $\Lambda_t = 37$ for all t, which corresponds to the best possible quantiser with prior knowledge of the disturbance

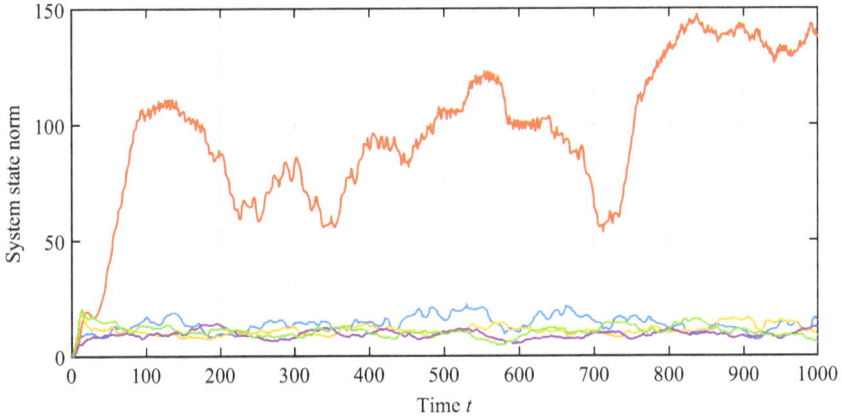

Figure 3.8   *Moving average of the system state norm x_t during a numerical simulation, on a logarithmic scale. The yellow line corresponds to the proposed scheme using (3.20). The green line corresponds to an observer also with (3.20) but with the introduction of an artificial measurement noise as in (3.57). The blue line corresponds to a simulation with (3.55), the purple line uses (3.56) and for the red line $\Lambda_t = 37$ for all t*

characteristics. At each time, we present the average of the previous 50 times of the system state.

One can observe from Figure 3.8 that for Gaussian noise, the proposed zoom-in zoom-out scheme is more advantageous since the system state norm remains inside a lower bound than the other solutions. This is because there is a defined lower bound on the estimation error that depends on the magnitudes of the process disturbance and measurement noise and therefore assumption A5 holds for some k almost all the time.

It should be noted that the ultimate bounds of the other schemes depend on the choice of Λ_0, and therefore if Λ_0 is chosen to be sufficiently small, one would obtain similar performance of the proposed zoom-in zoom-out scheme. However, this would assume that we know a priori the magnitudes of the process disturbance and measurement noise.

To understand better when the other solutions are necessary, we show in Figure 3.9 a case where there are no disturbances for a period of time before the noise is reintroduced to the system. In this case, assumption A5 does not hold.

As expected, after the disturbances are reintroduced in the system the norm of the state using the proposed scheme is greater than with solutions (3.55), (3.56) and (3.57) for a brief period of time. This is because when there is no noise, the quantisation interval shrinks to a value close to zero and when the noise is reintroduced, the quantisation interval takes a large amount of time until it reaches an acceptable size. The magnitude of this effect depends on the duration of the time interval during which the noise is absent from the system.

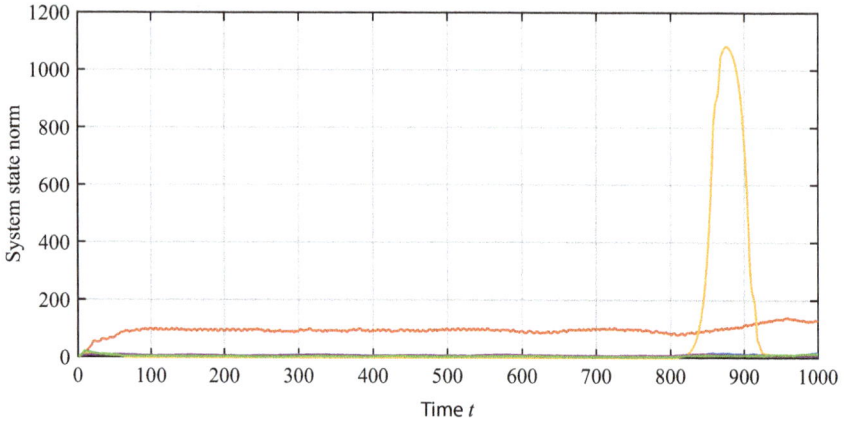

*Figure 3.9   System state norm x_t during a numerical simulation on a linear scale,
            when the process disturbance and measurement noise are zero, i.e.
            $w_t = 0$ and $v_t = 0$, for $t <= 800$. The yellow line corresponds to the
            proposed scheme using (3.20). The green line corresponds to an
            observer also with (3.20) but with the introduction of an artificial
            measurement noise as in (3.57). The blue line corresponds to a
            simulation with (3.55), the red line uses (3.56), and for the purple line,
            $\Lambda_t = 37$ for all t*

3.5.2   Monitoring system

As an example of a non-linear observer with the proposed architecture for monitoring
illustrated in Figure 3.3, we consider the Van der Pol oscillator given in [35].

   In this example, we consider a discretised version of the Van der Pol oscillator
subjected to process and observation noise, described by

$$x_{t+1} = x_t + \tau f(x_t) + \tau w_t, \tag{3.61a}$$

$$y_t = [1 \ 0]x_t + v_t, \tag{3.61b}$$

where τ is the discretisation step, $w_t := [w_{1,t} \ w_{2,t}]^T$ is the process noise, v_t is the
measurement noise and

$$f\left(\begin{bmatrix} x_1 \\ x_2 \end{bmatrix}\right) := \begin{bmatrix} x_2 \\ -\omega^2 x_1 + \mu\left(1 - x_1^2\right)x_2 \end{bmatrix}, \tag{3.62}$$

where μ is the non-linear damping and ω is the intrinsic frequency, given by $\mu = 10\,\mathrm{s}^{-1}$ and $\omega = 9\,\mathrm{rad\,s}^{-1}$, respectively.

   In this example, we take $\tau = 10^{-3}$ s and $w_{1,t}$, $w_{2,t}$ and v_t are drawn from uniform distributions with $w_{1,t}, w_{2,t} \in [-10, 10]$ and $v_t \in [-10^{-2}, 10^{-2}]$. Accordingly,
the discretised, quantised observer we consider is given by

$$\hat{x}_{t+1} = \hat{x}_t + \tau f\left(\begin{bmatrix} \hat{x}_{1,t} + Q_t(\theta_t) \\ \hat{x}_{2,t} \end{bmatrix}\right) + \tau L Q_t(\theta_t), \tag{3.63a}$$

$$\theta_t = y_t - \hat{x}_{1,t}, \tag{3.63b}$$

where $L := [g_1 \; g_2]^T$ and $g_1 = \mu + a$ and $g_2 = \mu g_1 + b$ with $a, b > 0$. In this example, we take $a = b = 1$. The quantiser parameters used in this example are $\beta = 0.95$ and $\beta_{\text{out}} = 2$.

Figure 3.10 shows the trajectories of the system state and its error, Figure 3.11 shows the evolution of the estimation error and Figure 3.12 shows the evolution the transmitted variable θ_t, its quantised value $Q_t(\theta_t)$ and the quantisation bounds $\pm \frac{\Delta_t}{2}$.

Figures 3.10–3.12 show that the quantised observer behaves as expected, with an ultimately bounded estimation error.

Figure 3.13 illustrates the impact of the parameter β on the system by displaying the average of the quantisation error across 2000 simulations for different values of β. The plot reveals that the system performs optimally when $\beta = 0.9$.

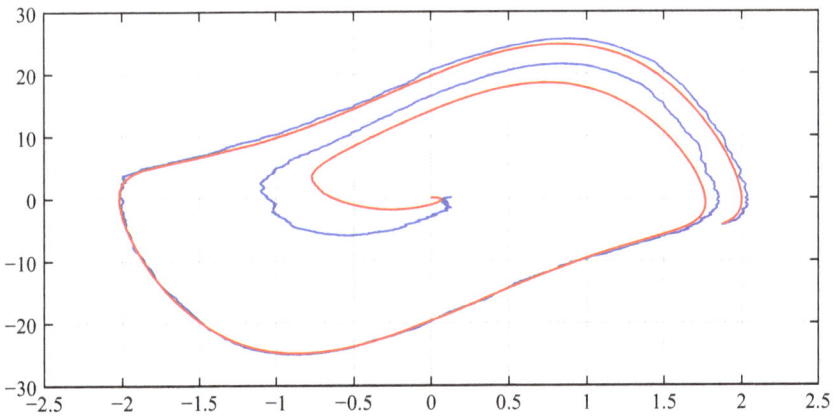

Figure 3.10   *System state (blue) and state estimate (red) of the Van der Pol oscillator system*

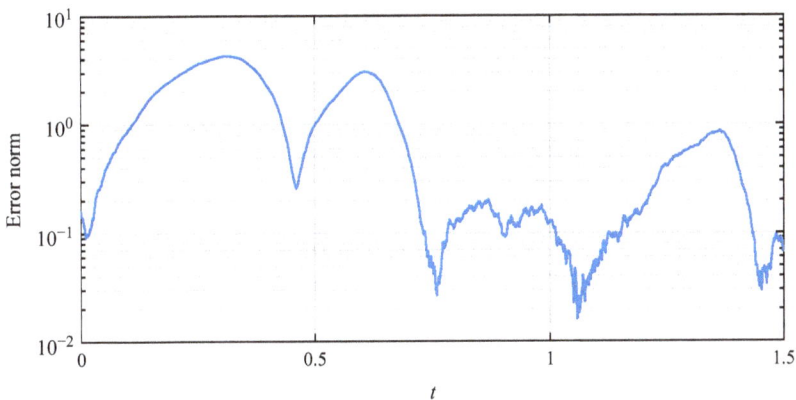

Figure 3.11   *Evolution of the estimation error norm $\|x_t - \hat{x}_t\|$ for the Van der Pol example*

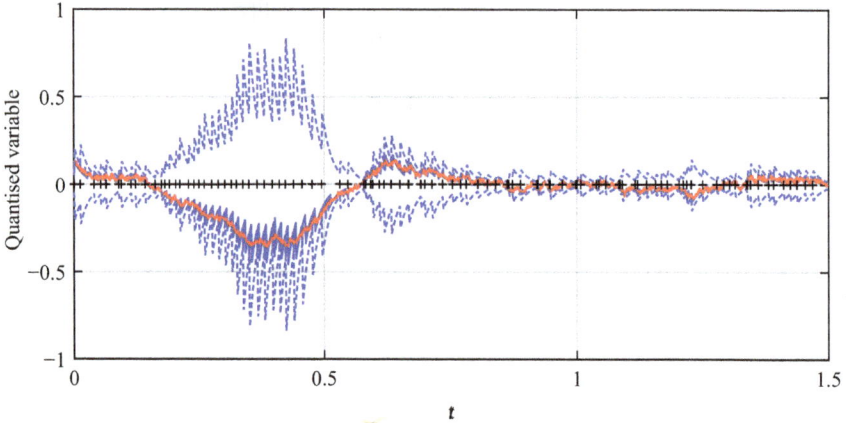

Figure 3.12   Evolution of θ_t (in red), $Q_t(\theta_t)$ (in blue) and $\pm\frac{\Lambda_t}{2}$ (dashed blue lines). The black '+' markers correspond to zoom-out events

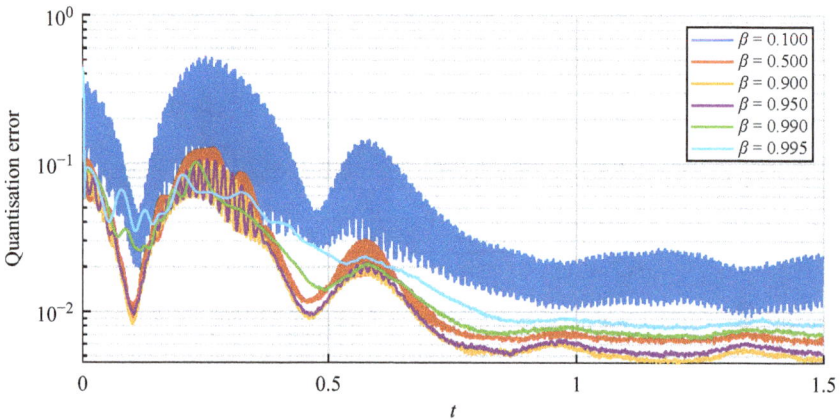

Figure 3.13   Average of the quantisation error over 2000 sample simulations for various values of β

3.6   Conclusions and future work

This chapter proposed a quantised observer and controller with adaptive quantisation for limited bandwidth environments, under strong emitter–receiver communication bandwidth constraints, that is applicable to systems operating under state feedback control. The proposed method is shown to yield a bounded estimation error, whose bound depends on the $\mathscr{L}_\infty$ norm of the measurement noise and process disturbance signals. We derived a set of conditions on the quantiser design parameters that guarantee ultimate boundedness of the estimation error. Numerical simulations showed that when the disturbance is persistently exciting, the proposed method is more advantageous than the other solutions found in the literature.

Future work will focus on refining the quantiser parameters, including the 'zoom-in' and 'zoom-out' coefficients, to further enhance system performance under various noise and disturbance conditions. Another key direction involves applying the proposed methods to real-world networked marine systems, such as autonomous underwater and surface vehicles. This will allow for the evaluation of their effectiveness in environments characterised by severe communication bandwidth constraints. Additionally, the focus will be on leveraging distributed navigation and control strategies to enable coordinated operations in challenging settings. Furthermore, efforts will be directed toward improving robustness in the presence of packet loss, a critical challenge in acoustic communication networks.

Acknowledgements

This work was supported in part by the Foundation for Science and Technology (FCT) of Portugal under the projects UIDB/04111/2020 and CTS/00066. The work of the third author was supported by the Laboratory of Robotics and Engineering Systems (LARSyS) of the Foundation for Science and Technology (FCT) of Portugal under Grant DOI: 10.54499/LA/P/0083/2020 and FCT project RAIECO.

A.1 Proof of theorem 1

To prove theorem 1, we require three previous results:

- Lemma 4 states that if the quantisation interval Λ_t is initially too small, that is if $\frac{\Lambda_0}{2} < k\bar{\theta}_0$, it increases until $\frac{\Lambda_t}{2} > k\bar{\theta}_t$ in finite time.
- Lemma 5 guarantees that if at time t $\frac{\Lambda_t}{2} > k\bar{\theta}_t$, then the next time there is a zoom-out, e.g. at t', the quantisation interval is sufficiently large such that $\frac{\Lambda_{t'}}{2} > \bar{\theta}_{t'} > \|\theta_{t'}\|_\infty$
- Lemma 6 shows that when $\frac{\Lambda_t}{2} > \theta_t$, the quantisation interval converges to an ultimate bound which is proportional to ε.

These results are in turn based on two other results:

- Lemma 2 guarantees that there always exists a zoom-out if for $T + N$ consecutive steps, $\frac{\Lambda_t}{2} < k\bar{\theta}_t$.
- Lemma 3 provides useful bounds on $\bar{\theta}_t$ after a zoom-out event.

Before proceeding further, we first note that from (3.22) and (3.3), one has

$$V_{t+1}(e_{t+1}) \le \tilde{\beta}V_t(e_t) + \varepsilon + l_1 \begin{cases} \frac{\Lambda_t}{2(2^{n_b}-2)}, & \text{if } \|\theta_t\|_\infty \le \frac{\Lambda_t}{2} \\ \bar{\theta}_t, & \text{otherwise} \end{cases} . \tag{A.1}$$

Lemma 2. *Suppose that assumption A5 holds. If $\frac{\Lambda_t}{2} < k\bar{\theta}_t$, then either $\frac{\Lambda_{t+\tau}}{2} \ge k\bar{\theta}_{t+\tau}$ or $\theta_{t+\tau} \ge \frac{\Lambda_{t+\tau}}{2}$ for some τ such that $0 \le \tau \le T+N$.*

Proof. If for all $0 \leq \tau_1 \leq N \frac{\Lambda_{t+\tau}}{2} < k\bar{\theta}_{t+\tau}$, then from assumption A5, there exists $0 \leq \tau_1 \leq N$ and $0 \leq \tau_2 \leq T$ such that

$$\frac{\Lambda_{t+\tau_1}}{2} < k\bar{\theta}_{t+\tau_1} \leq \|\theta_{t+\tau_1+\tau_2}\|_\infty. \tag{A.2}$$

Then, if $\|\theta_{t+\tau}\|_\infty \leq \frac{\Lambda_{t+\tau}}{2}$ for $\tau_1 \leq \tau \leq \tau_1 + \tau_2$,

$$\frac{\Lambda_{t+\tau_1+\tau_2}}{2} = \beta^{\tau_2} \frac{\Lambda_{t+\tau_1}}{2} < \frac{\Lambda_{t+\tau_1}}{2} < \|\theta_{t+\tau_1+\tau_2}\|_\infty. \tag{A.3}$$

Thus, by contradiction, the lemma is proven. □

Lemma 3. *If for some τ satisfying $0 \leq \tau \leq T + N$ for all t' satisfying $t \leq t' \leq t + \tau - 1$, $\|\theta_t\|_\infty \leq \frac{\Lambda_t}{2}$ and $\|\theta_{t+\tau}\|_\infty > \frac{\Lambda_{t+\tau+1}}{2}$, then if $k\bar{\theta}_t \geq \frac{\Lambda_t}{2}$*

$$\bar{\theta}_{t+\tau+1} \leq \psi \bar{\theta}_t, \tag{A.4}$$

or if $k\bar{\theta}_t \leq \frac{\Lambda_t}{2}$

$$\bar{\theta}_{t+\tau+1} \leq \psi \frac{\Lambda_t}{2}. \tag{A.5}$$

Proof. From (A.1) and the definition of $\bar{\theta}_t$, one has that

$$\bar{\theta}_{t+1} \leq \begin{cases} (1 + cl_1)\bar{\theta}_t + \frac{l_1 c}{2^{n_b} - 2} \frac{\Lambda_t}{2}, & \text{if } \|\theta_t\|_\infty \leq \frac{\Lambda_t}{2} \\ (1 + 2cl_1)\bar{\theta}_t, & \text{otherwise} \end{cases}. \tag{A.6}$$

Also, from (A.6) and the properties of geometric series, one has that if $\|\theta_{t'}\|_\infty \leq \frac{\Lambda_{t'}}{2}$ holds for all t' such that $t \leq t' \leq t + \tau$ for some $\tau \geq 0$, then

$$\bar{\theta}_{t+\tau} \leq (1 + cl_1)^\tau \bar{\theta}_t + \beta^{\tau-1} \frac{1 - \left(\frac{1+cl_1}{\beta}\right)^\tau}{1 - \frac{1+cl_1}{\beta}} \frac{l_1 c}{2^{n_b} - 2} \frac{\Lambda_t}{2}, \tag{A.7}$$

and the results ensue from (A.6) and the fact that $k < 1$. □

Lemma 4. *Suppose that assumption A5 holds and the zoom-out coefficient β_{out} satisfies (3.52). If $\frac{\Lambda_0}{2} < k\bar{\theta}_0$, then $\frac{\Lambda_t}{2} > k\bar{\theta}_t$ for some time t satisfying*

$$t \leq t_0 := (T + N + 1) \left\lceil \log_{\frac{\beta_{out}\beta^{T+N}}{\psi}} \left(\frac{2k\bar{\theta}_0}{\Lambda_0} \right) \right\rceil. \tag{A.8}$$

Proof. From lemma 3, one has that if $\frac{\Lambda_t}{2} < k\bar{\theta}_t$ for $0 \leq t \leq t_0$, then there exists at least

$$o = \frac{t_0}{T + N + 1} = \left\lceil \log_{\frac{\beta_{out}\beta^{T+N}}{\psi}} \left(\frac{2k\bar{\theta}_0}{\Lambda_0} \right) \right\rceil. \tag{A.9}$$

zoom-out events. Denoting t'_p as the time after the pth zoom-out and $t'_0 = 0$, then $t'_o \leq t_0$. From lemma 3, one has also that if $\frac{\Lambda_t}{2} < k\bar{\theta}_t$ for $0 \leq t \leq t_0$, then

$$\bar{\theta}_{t'_o} \leq \psi^o \bar{\theta}_0. \tag{A.10}$$

and

$$\Lambda_{t_o'} \geq \left(\beta^{T+N}\beta_{out}\right)^o \Lambda_0. \tag{A.11}$$

Therefore

$$\frac{\Lambda_{t_o'}}{2k\bar{\theta}_{t_o'}} \geq \left(\frac{\beta_{out}\beta^{T+N}}{\psi}\right)^o \frac{\Lambda_0}{2k\bar{\theta}_0} \geq 1. \tag{A.12}$$

Thus, by contradiction, it follows that $\frac{\Lambda_t}{2} < k\bar{\theta}_t$ for some time t satisfying (A.8), which proves the lemma. ☐

Lemma 5. *Suppose that assumption A5 holds and the zoom-out coefficient β_{out} satisfies (3.52). If $k\bar{\theta}_t < \frac{\Lambda_t}{2}$, then, if the next zoom-out event occurs at time t_o, $\bar{\theta}_{t_o+1} < \frac{\Lambda_{t_o+1}}{2}$.*

Proof. From lemma 2, there exists τ such that $0 \leq \tau \leq T+N+1$ and $k\bar{\theta}_{t_o-\tau} \leq \frac{\Lambda_{t_o-\tau}}{2}$. Therefore from lemma 3, one obtains

$$\bar{\theta}_{t_o+1} < \psi \frac{\Lambda_{t_o-\tau}}{2} \leq \frac{\psi}{\beta_{out}\beta^{T+N+1}} \frac{\Lambda_{t_o+1}}{2}, \tag{A.13}$$

and the result follows from (3.52). ☐

Lemma 6. *Suppose that the zoom-in coefficient β satisfies (3.51). If $\frac{\Lambda_t}{2} > \bar{\theta}_t$ and*

$$\frac{\Lambda_t}{2} > \frac{1 - \tilde{\beta} + l_1 c}{\beta - \bar{\beta}} \frac{\varepsilon}{l_1}, \tag{A.14}$$

then $\frac{\Lambda_{t+1}}{2} > \bar{\theta}_{t+1}$.

Proof. From the definition of $\bar{\theta}_t$, (3.33), since $\frac{\Lambda_t}{2} > \bar{\theta}_t$, one has

$$cV_t(e_t) < \frac{\Lambda_t}{2} - \frac{\varepsilon}{l_1}. \tag{A.15}$$

Applying (A.15) to (A.1), we obtain

$$cV_{t+1}(e_{t+1}) < \bar{\beta}\frac{\Lambda_t}{2} + \left(l_1 c - \tilde{\beta}\right)\frac{\varepsilon}{l_1}.$$

Noting that $\frac{\Lambda_{t+1}}{2} = \frac{\beta\Lambda_t}{2}$ yields

$$\frac{\Lambda_{t+1}}{2} > \frac{\beta}{\bar{\beta}}\left(cV_{t+1}(e_{t+1}) - \left(l_1 c - \tilde{\beta}\right)\frac{\varepsilon}{l_1}\right),$$

and therefore, from the definition of $\bar{\theta}_t$, (3.33), and noting that $\frac{\bar{\beta}}{\beta}\Lambda_{t+1} = \Lambda_{t+1} - (\beta - \bar{\beta})\Lambda_t$,

$$\frac{\Lambda_{t+1}}{2} - (\beta - \bar{\beta})\frac{\Lambda_t}{2} > \bar{\theta}_{t+1} - \frac{\left(1 - \tilde{\beta} + l_1 c\right)\varepsilon}{l_1},$$

and the lemma is proven. ☐

Proof of theorem 1.   From Lemmas 4 and 5, we have that for $t > t_0$, there are no consecutive zoom-outs.

Lemmas 5 and 6 then guarantee that for $t > t_0$, there can only be a zoom-out if (A.14) does not hold or if $\bar{\theta}_t > \frac{\Lambda_t}{2}$. In the latter case, either the quantisation interval keeps decreasing until (A.14) does not hold or there is a single zoom-out which causes $\frac{\Lambda_t}{2} > \bar{\theta}_t$.

After an eventual first zoom-out after t_0, we have that if (A.14) holds, then $\frac{\Lambda_t}{2} > \bar{\theta}_t > \|\theta_t\|_\infty$ and thus $\Lambda_{t+1} = \beta\Lambda_t < \Lambda_t$. Therefore,

$$\limsup_{t \to \infty} \frac{\Lambda_t}{2} < \beta_{\text{out}} \frac{1 - \tilde{\beta} + l_1 c}{\beta - \bar{\beta}} \frac{\varepsilon}{l_1},$$

and accordingly, from (A.15), the conclusion of the theorem follows.   □

References

[1]   Simetti E., Indiveri G., and Pascoal A.M.: 'WiMUST: A cooperative marine robotic system for autonomous geotechnical surveys'. *Journal of Field Robotics*. 2021;**38**(2):268–288.

[2]   RAMONES Consortium: *Radioactive Monitoring in Ocean Ecosystems*; 2024. Available from: https://ramones-project.eu/. [Accessed: 2025-01-19].

[3]   Delchamps D.F.: 'Stabilizing a linear system with quantized state feedback'. *IEEE Transactions on Automatic Control*. 1990;**35**(8):916–924.

[4]   Elia N. and Mitter S.K.: 'Stabilization of linear systems with limited information'. *IEEE Transactions on Automatic Control*. 2001;**46**(9):1384–1400.

[5]   Fu M. and Xie L.: 'The sector bound approach to quantized feedback control'. *IEEE Transactions on Automatic Control*. 2005;**50**(11):1698–1711.

[6]   You K., Su W., Fu M., *et al.*: 'Attainability of the minimum data rate for stabilization of linear systems via logarithmic quantization'. *Automatica*. 2011;**47**(1):170–176.

[7]   Hu J., Wang Z., Shen B., *et al.*: 'Quantised recursive filtering for a class of nonlinear systems with multiplicative noises and missing measurements'. *International Journal of Control*. 2013;**86**(4):650–663.

[8]   Dong H., Wang Z., Ding S.X., *et al.*: 'Finite-horizon reliable control with randomly occurring uncertainties and nonlinearities subject to output quantization'. *Automatica*. 2015;**52**:355–362.

[9]   Brockett R.W. and Liberzon D.: 'Quantized feedback stabilization of linear systems'. *IEEE Transactions on Automatic Control*. 2000;**45**(7):1279–1289.

[10]   Petersen I.R. and Savkin A.: 'Multi-rate stabilization of multivariable discrete-time linear systems via a limited capacity communication channel'. *Proceedings of the 40th IEEE Conference on Decision and Control* (Cat. No. 01CH37228). Vol. 1. Piscataway, NJ: IEEE; 2001. pp. 304–309.

[11]   Liberzon D.: 'On stabilization of linear systems with limited information'. *IEEE Transactions on Automatic Control*. 2003;**48**(2):304–307.

[12]   Tatikonda S. and Mitter S.: 'Control under communication constraints'. *IEEE Transactions on Automatic Control*. 2004;**49**(7):1056–1068.

[13]   Nair G.N., Evans R.J., Mareels I.M., *et al.* 'Topological feedback entropy and nonlinear stabilization'. *IEEE Transactions on Automatic Control.* 2004;**49**(9):1585–1597.

[14]   Ling Q. and Lemmon M.D.: 'Stability of quantized control systems under dynamic bit assignment'. *IEEE Transactions on Automatic Control.* 2005;**50**(5):734–740.

[15]   Liberzon D. and Nesic D., 'Input-to-state stabilization of linear systems with quantized state measurements'. *IEEE Transactions on Automatic Control.* 2007;**52**(5):767–781.

[16]   Liberzon D.: 'Observer-based quantized output feedback control of nonlinear systems'. *IFAC Proceedings Volumes.* 2008;**41**(2):8039–8043.

[17]   You K. and Xie L.: 'Minimum data rate for mean square stabilizability of linear systems with Markovian packet losses'. *IEEE Transactions on Automatic Control.* 2010;**56**(4):772–785.

[18]   Liu K., Fridman E. and Johansson K.H.: 'Dynamic quantization of uncertain linear networked control systems'. *Automatica.* 2015;**59**:248–255.

[19]   Wakaiki M., Zanma T. and Liu K.Z.: 'Quantized output feedback stabilization by Luenberger observers'. *IFAC-PapersOnLine.* 2017;**50**(1):2577–2582.

[20]   Nair G.N. and Evans R.J.: 'Stabilizability of stochastic linear systems with finite feedback data rates'. *SIAM Journal on Control and Optimization.* 2004;**43**(2):413–436.

[21]   Gupta V., Dana A.F., Murray R.M., *et al.*: 'On the effect of quantization on performance at high rates'. *2006 American Control Conference.* Piscataway, NJ: IEEE; 2006. p. 6.

[22]   Sukhavasi R.T. and Hassibi B.: 'The Kalman like particle filter: Optimal estimation with quantized innovations/measurements'. *Proceedings of the 48h IEEE Conference on Decision and Control (CDC) held jointly with 2009 28th Chinese Control Conference.* Piscataway, NJ: IEEE; 2009. pp. 4446–4451.

[23]   Tanaka T., Johansson K.H., Oechtering T., *et al.*: 'Rate of prefix-free codes in LQG control systems'. *2016 IEEE International Symposium on Information Theory (ISIT).* Piscataway, NJ: IEEE; 2016. pp. 2399–2403.

[24]   Stavrou P.A., Østergaard J. and Charalambous C.D.: 'Zero-delay rate distortion via filtering for vector-valued Gaussian sources'. *IEEE Journal of Selected Topics in Signal Processing.* 2018;**12**(5):841–856.

[25]   Huang C.C., Amini B. and Bitmead R.R.: 'Predictive coding and control'. *IEEE Transactions on Control of Network Systems.* 2018;**6**(2):906–918.

[26]   Hespanha J., Ortega A. and Vasudevan L.: 'Towards the control of linear systems with minimum bit-rate'. *Proceedings of the 15th International Symposium on Mathematical Theory of Networks and Systems (MTNS).* Citeseer; 2002.

[27]   Niu Y. and Ho D.W.: 'Control strategy with adaptive quantizer's parameters under digital communication channels'. *Automatica.* 2014;**50**(10):2665–2671.

[28]   Fu M. and Xie L.: 'Finite-level quantized feedback control for linear systems'. *IEEE Transactions on Automatic Control.* 2009;**54**(5):1165–1170.

[29]   Sharon Y. and Liberzon D.: 'Input to state stabilizing controller for systems with coarse quantization'. *IEEE Transactions on Automatic Control.* 2011;**57**(4):830–844.

[30]   Gray R.M. and Neuhoff D.L.: 'Quantization'. *IEEE transactions on information theory.* 1998;**44**(6):2325–2383.

[31]   Anderson B.D.O. and Moore J.B.: *Optimal Filtering.* Englewood Cliffs: Prentice-Hall; 1979;**21**:22–95.

[32]   Cristi R.: *Modern Digital Signal Processing.* Pacific Grove, CA: Thomson/Brooks/Cole Pacific Grove; 2004.

[33]   Castro Rego F., Pu Y., Alessandretti A., *et al.*: 'A distributed Luenberger observer for linear state feedback systems with quantized and rate-limited communications'. *IEEE Transactions on Automatic Control.* 2021;**66**: 3922–3937.

[34]   Xiao J.J. and Luo Z.Q.: 'Decentralized estimation in an inhomogeneous sensing environment'. *IEEE Transactions on Information Theory.* 2005;**51**(10):3564–3575.

[35]   Gholizade-Narm H.: 'A new state observer for two coupled Van Der Pol Oscillators'. *International Journal of Control, Automation and Systems.* 2011;**9**(2):410.

Chapter 4
Collaboration of multiple underwater vehicle–manipulator systems

Shahab Heshmati-Alamdari[1], Charalampos P. Bechlioulis[2], George C. Karras[3] and Kostas J. Kyriakopoulos[4]

4.1 Introduction

Unmanned underwater vehicles (UUVs) have become prevalent in various fields such as marine science and offshore maintenance over the past decades [1]. These applications often require intervention abilities [2], driving increased attention toward underwater vehicle manipulator systems (UVMS) [3], a subset of Floating Base Mobile Manipulator System (FBMMS) [4]. Underwater interventions are typically conducted using remotely operated vehicles (ROVs), equipped with manipulators for object manipulation and controlled by human operators through a master–slave teleoperation setup [5]. Recognizing human–robot teleoperation limitations, the scientific community is prioritizing the development of reliable autonomous control systems for UVMS in challenging tasks [6].

Underwater manipulation tasks can be performed more effectively when multiple UVMSs collaborate [7]. For example, using two or more UVMSs to transport large objects enhances safety and efficiency compared to a single UVMS, which is often limited by constraints related to shape, actuation, and payload capacity [8,9]. In [8], researchers explored the modeling of two UVMSs jointly carrying a rigid object, assuming a rigid robot–object contact. This assumption resulted in a singular system of differential equations governing the system dynamics [10]. Further analysis of the system's kinematic redundancy and manipulability was conducted in [11,12]. In addition, a centralized cooperative control strategy for multiple UVMSs jointly handling an object was introduced in [9]. Nevertheless, these studies did not account for the significant challenges and constraints inherent to underwater environments.

[1] Section of Automation and Control, Department of Electronic Systems, Aalborg University, Denmark
[2] Division of Systems and Control, Department of Electrical and Computer Engineering, University of Patras, Greece
[3] Department of Informatics and Telecommunications, University of Thessaly, Greece
[4] Control Systems Lab, School of Mechanical Engineering, National Technical University of Athens, Greece

Underwater operations present significant challenges, with one of the most critical being the strict limitations on communication [13,14]. In general, communication in multi-robot systems falls into two primary categories: explicit and implicit. The explicit mode is specifically designed to transmit information, such as control signals or sensor data, directly between robots [15]. In contrast, implicit communication emerges naturally from interactions, either physical (e.g., forces exerted during object manipulation) or non-physical (e.g., visual observation [16]). In such cases, information is gathered through onboard sensors, such as vision systems or force/torque sensors. Explicit communication is the most widely studied and commonly used method in multi-robot systems, as it facilitates theoretical analysis and improves coordination efficiency. However, while inter-robot communication plays a crucial role in collaborative manipulation tasks, the use of explicit communication in underwater environments often leads to severe performance bottlenecks due to the restricted bandwidth and slow update rates of underwater acoustic systems. Additionally, as the number of collaborating robots increases, the communication protocols become more complex, struggling to manage bandwidth congestion [17]. As a result, the number of operating underwater robots, involved in cooperative schemes that exploit explicit communication protocols, is strictly limited owing to the narrow bandwidth of acoustic communication devices. To mitigate these issues, recent research on underwater cooperative manipulation has shifted toward developing control approaches that reduce communication demands by utilizing implicit interactions between agents. While this introduces additional complexity in theoretical modeling, it simplifies communication protocols and optimizes bandwidth and energy efficiency by minimizing the exchange of explicit data.

Cooperative manipulation has been well-studied in the literature, especially the centralized schemes [18–21]. While centralized approaches can be efficient, they tend to be less robust, as all robots depend on a central unit, and their complexity escalates significantly as the number of robots increases. Conversely, decentralized cooperative manipulation strategies offer greater robustness and reduced complexity but often rely on explicit communication between robots, such as real-time transmission of the desired trajectory [22,23] or pre-defined knowledge of the object's trajectory [24–26]. For example, to ensure collision avoidance, the desired trajectory of the object must either be continuously shared among the underwater robots or be mutually agreed upon beforehand. This necessitates a precise global localization system for all robots [27], which is particularly challenging in underwater environments and, in the best-case scenario, significantly increases mission costs. As a result, the need for decentralized cooperative manipulation strategies that minimize explicit communication while integrating implicit coordination has become evident. Recent research [27,28] has explored potential field methods within a multi-layered control framework to regulate robot swarm coordination, guide and navigate UVMSs, and manage manipulation tasks. To address localization and consensus issues, some studies have designated the object as the reference frame for the swarm. However, this method requires each robot to communicate with the entire team, imposing limitations on the number of robots that can participate due to bandwidth constraints.

Further progress in this domain has been demonstrated in [29–31], where a priority-based control strategy [32] was introduced. Specifically, a three-step decentralized cooperative control scheme was proposed: first, each robot independently computes an optimal task-space control velocity; second, this velocity is exchanged among the team members to establish a consensus through a fusion mechanism; and finally, the agreed-upon velocity is mapped to each UVMS's joint space using a task-priority approach [32], this time with increased priority. Additional safety constraints, such as joint limits and manipulability, can also be incorporated when only two UVMSs are involved. However, scaling this approach to larger teams introduces further challenges, particularly in managing bandwidth congestion. Moreover, when operating in a constrained workspace with obstacles, achieving consensus on a universally safe trajectory becomes even more complex.

In this chapter, we define and address the problem of cooperative object transportation by a team of UVMSs operating within a constrained workspace containing static obstacles. Two control strategies are given where the coordination is achieved solely through implicit communication, which emerges from each robot's onboard sensor measurements and the physical interactions between the robots and the commonly grasped object. Consequently, no explicit online data exchange occurs among the robots. This chapter is organized as follows: The first section of this chapter presents preliminaries on the coupled kinematic and dynamic modeling of the system, as well as geometrical modeling for a team of UVMSs rigidly grasping a common object in constrained environment. The second section based on the results given in [33] presents a decentralized impedance control scheme for the scenario where UVMSs are equipped with force/torque sensors at their end-effectors. In this case, coordination is achieved exclusively through implicit communication, utilizing the physical interaction between the robots and the commonly grasped object. Finally, the third section based on the results given in [34,35] explores a distributed predictive control approach for the case where UVMSs lack force/torque sensors at their end-effectors. In addition, this approach explicitly accounts for system constraints, including control input saturation, arm joint limits, as well as kinematic and representation singularities.

It is important to note that while underwater vehicles are equipped with acoustic modems to facilitate communication with the surface control station*, the use of cooperative control protocols based on implicit communication is strongly motivated by the inherent bandwidth limitations of underwater acoustic systems. Furthermore, to ensure collision avoidance, either the leading robot must continuously transmit the desired object trajectory to the followers, or all UVMSs must establish a mutually agreed trajectory beforehand. Achieving this consensus requires an accurate and shared localization system [27], which is particularly difficult to implement in underwater environments and remains susceptible to errors. In this way, although the control strategies discussed in this chapter do not entirely eliminate the necessity for

---

*For instance, all cooperating UVMSs must be aware of the initial position of the object to be transported to successfully reach and grasp it. They also need to coordinate various task phases under higher-level specifications [36–38] using simple high-level messages (e.g., "Ok, I've grasped it," "Let us proceed").

communication in underwater intervention tasks; especially in terms of safety, adaptability, and efficiency; they significantly reduce the need for continuous inter-robot communication during task execution. This reduction enhances the robustness of the cooperative framework while also mitigating the constraints imposed by the limited bandwidth of acoustic communication channels, such as restrictions on the number of UVMSs that can effectively participate.

4.2   Preliminaries on kinematic and dynamic modeling

This section provides a comprehensive modeling framework for UVMSs and the object's kinematics and dynamics, as well as a detailed examination of their coupled dynamics. Additionally, it includes the necessary geometric representations that will serve as a foundation for the analyses and methodologies discussed in the following sections.

4.2.1   UVMSs and object kinematics

Consider N UVMSs operating in a bounded workspace $\mathscr{W} \subseteq \mathbb{R}^3$. We denote the coordinates of the commonly agreed body-fixed frame on the object as well as the leader's and followers' task space (i.e., end-effector) coordinates by $x_O = [\eta_{1,O}^\top, \eta_{2,O}^\top]^\top$, $x_L = [\eta_{1,L}^\top, \eta_{2,L}^\top]^\top$, and $x_{F_i} = [\eta_{1,F_i}^\top, \eta_{2,F_i}^\top]^\top$, $i \in \mathscr{F} = \{1, \ldots, N_F\}$, respectively. We have $N = N_F + 1$, where N_F is the number of following UVMSs and we denote by $\mathscr{N} = \{1, \ldots, N\}$. More specifically, $\eta_{1,i} = [x_i, y_i, z_i]^\top$ and $\eta_{2,i} = [\phi_i, \theta_i, \psi_i]^\top$, $i \in \{O, L, F_1, \ldots, F_{N_F}\}$ denote the position and the orientation expressed in Euler angles representation with respect to the inertial frame. Alternatively, the orientation coordinates $\eta_{2,i}$ $i \in \{O, L, F_1, \ldots, F_{N_F}\}$ expressed in Euler angles may be described by a rotation matrix $R_i = [n_i, o_i, \alpha_i] \in \mathbb{R}^3$ that is mainly employed owing to its physical meaning (i.e., the inertial coordinate frame after three successive rotations of ψ_i, θ_i, ϕ_i angles about its z, y and x axes, respectively, ends up parallel to the object-fixed coordinate) [39]. Thus, the rotation matrix R_i $i \in \{O, L, F_1, \ldots, F_{N_F}\}$ may be expressed via $\eta_{2,i}$ as follows:

$$R_i = \begin{bmatrix} c_{\psi_i} c_{\theta_i} & c_{\psi_i} s_{\theta_i} s_{\phi_i} - s_{\psi_i} c_{\phi_i} & c_{\psi_i} s_{\theta_i} c_{\phi_i} + s_{\psi_i} s_{\phi_i} \\ s_{\psi_i} c_{\theta_i} & s_{\psi_i} s_{\theta_i} s_{\phi_i} + c_{\psi_i} c_{\phi_i} & s_{\psi_i} s_{\theta_i} c_{\phi_i} - c_{\psi_i} s_{\phi_i} \\ -s_{\theta_i} & c_{\theta_i} s_{\phi_i} & c_{\theta_i} c_{\phi_i} \end{bmatrix} \tag{4.1}$$

where $s_\star = \sin(\star)$ and $c_\star = \cos(\star)$. Let $q_i = [q_{v,i}^\top, \; q_{m,i}^\top]^\top \in \mathbb{R}^{6+n}$, $i \in \mathscr{N} = \{L, F_1, \ldots, F_{N_F}\}$ be the joint state variables of each UVMS, where $q_{v,i} \in \mathbb{R}^6$ is the vector that involves the position and the orientation of the vehicle and $q_{m,i} \in \mathbb{R}^n$ is the vector of the angular positions of the manipulator's joints. In addition, the position and orientation of the UVMS end-effector with respect to inertial frame are given by the forward kinematics of the complete system (arm and vehicle base) as follows:

$$x_i = \mathscr{F}(q_i), \quad i \in \mathscr{N} \tag{4.2}$$

Let also define the object as well as the leader's and followers' end effector generalized velocities by $v_O = [\dot{\eta}_{1,O}^\top, \omega_O^\top]^\top$, $v_L = [\dot{\eta}_{1,L}^\top, \omega_L^\top]^\top$ and $v_i = [\dot{\eta}_{1,i}^\top, \omega_i^\top]^\top$, $i \in$

$\{F_1, \ldots, F_{N_F}\}$ respectively, where $\dot{\eta}_{1,i}$ and ω_i denote the linear and angular velocities, respectively. Without any loss of generality, for the augmented UVMS system, we get [40]:

$$v_i = J_i(q_i)\zeta_i, \quad i \in \mathcal{N} \tag{4.3}$$

where $\zeta_i = [v_i^\top, \dot{q}_{m,i}^\top]^\top \in \mathbb{R}^{6+n}$ is the velocity vector involving the body velocities of the vehicle as well as the joint velocities of the manipulator with v_i to be the velocity of the vehicle expressed in the body-fixed frame and $J_i(q_i)$ is the geometric Jacobian matrix [40]. Moreover, for the vehicle we have:

$$\dot{q}_{v,i} = J_{v,i}(q_{v,i})v_i, \quad i \in \mathcal{N} \tag{4.4}$$

Where $J_{v,i}(q_{v,i})$ is the Jacobian matrix transforming the velocities from the body-fixed to the inertial frame. Comparing (4.4) and (4.3), it can be state that:

$$\dot{q}_i = \begin{bmatrix} J_{v,i}(q_{v,i}) & 0_{3\times3} \\ 0_{3\times3} & I_{3\times3} \end{bmatrix} \zeta_i, \quad i \in \mathcal{N} \tag{4.5}$$

Furthermore, owing to the rigid grasp of the object, the following equations hold:

$$x_i = x_O + \begin{bmatrix} {}^I R_O l_i \\ \alpha_i \end{bmatrix}, \quad i \in \mathcal{N} \tag{4.6}$$

where the vectors $l_i = [l_{ix}, l_{iy}, l_{iz}]^\top$ and $\alpha_i = [\alpha_{ix}, \alpha_{iy}, \alpha_{iz}]^\top$, $i \in \mathcal{N}$ represent the *constant* relative position and orientation of the end-effector w.r.t. the object, expressed in the object's frame and ${}^I R_O$ denotes the rotation matrix, which describes the orientation of the object expressed in the inertial frame $\{I\}$. Thus, using (4.6), each UVMS can compute the object's position w.r.t. inertial frame $\{I\}$, since the object geometric parameters are considered known. Furthermore, along with the fact that, due to the grasping rigidly, it holds that $\omega_i = \omega_O$, $i \in \mathcal{N}$, one obtains:

$$v_i = J_{iO}v_O, \quad i \in \mathcal{N} \tag{4.7}$$

where J_{iO}, $i \in \mathcal{N}$ denotes the Jacobian from the end-effector of each UVMS to the object's center of mass, that is defined as follows:

$$J_{iO} = \begin{bmatrix} I_{3\times3} & -S(l_i) \\ 0_{3\times3} & I_{3\times3} \end{bmatrix} \in \mathbb{R}^{6\times6}, \quad i \in \mathcal{N}$$

where $S(l_i)$ is the skew-symmetric matrix of vector $l_i = [l_{ix}, l_{iy}, l_{iz}]^\top$ defined as follows:

$$S(l_i) = \begin{bmatrix} 0 & -l_{iz} & l_{iy} \\ l_{iz} & 0 & -l_{ix} \\ -l_{iy} & l_{ix} & 0 \end{bmatrix} \in \mathbb{R}^{3\times3}, \quad i \in \mathcal{N}$$

Notice that J_{iO}, $i \in \mathcal{N}$ are always full-rank owing to the grasp rigidity and hence obtain a well defined inverse. Thus, the object's velocity can be easily computed via the inverse of (4.7). Moreover, from (4.7), one obtains the acceleration relation:

$$\dot{v}_i = J_{iO}\dot{v}_O + \dot{J}_{iO}v_O, \quad i \in \mathcal{N} \tag{4.8}$$

which will be used in the subsequent analysis.

4.2.2   Dynamics

4.2.2.1   UVMS dynamics

The dynamics of a UVMS after straightforward algebraic manipulations can be written as follows [40]:

$$M_{q_i}(q_i)\dot{\zeta}_i + C_{q_i}(\zeta_i, q_i)\zeta_i + D_{q_i}(\zeta_i, q_i)\zeta_i + g_{q_i}(q_i) + d_{q_i}(\zeta_i, q_i, t) = \tau_i + J_i^\top \lambda_i \quad (4.9)$$

for $i \in \mathcal{N}$, where λ_i is the vector of *measured* interaction forces and torques exerted at the end-effector by the object, τ_i denotes the vector of control inputs (forces and torques), $M_{q_i}(q_i)$ is the inertial matrix, $C_{q_i}(\zeta_i, q_i)$ represents coriolis and centrifugal terms, $D_{q_i}(\zeta_i, q_i)$ models dissipative effects, $g_i(q_i)$ encapsulates the gravity and buoyancy effects, and $d_{q_i}(\zeta_i, q_i, t)$ is a bounded vector representing unmodeled friction, uncertainties and external disturbances. In view of (4.3), we have:

$$\dot{v}_i = J_i(q_i)\dot{\zeta}_i + J_i^d(\zeta_i, q_i)\zeta_i, \quad i \in \mathcal{N} \quad (4.10)$$

where $J_i^d(\zeta_i, q_i) \in \mathbb{R}^{6 \times (6+n)}$ represents the Jacobian derivative function, with $J_i^d(\zeta_i, q_i) \triangleq \dot{J}_i(q_i)$. Then, by employing the differential kinematics as well as (4.10), we obtain from (4.9) the transformed task space dynamics [41]:

$$M_i(q_i)\dot{v}_i + C_i(\zeta_i, q_i)v_i + D_i(\zeta_i, q_i)v_i + g_i(q_i) + d_i(\zeta_i, q_i, t) = u_i + \lambda_i \quad (4.11)$$

for all $i \in \mathcal{N}$, with the corresponding task space terms $M_i \in \mathbb{R}^{6 \times 6}$, $C_i \in \mathbb{R}^{6 \times 6}$, $D_i \in \mathbb{R}^{6 \times 6}$, $g_i \in \mathbb{R}^6$, $d_i \in \mathbb{R}^6$, and u_i to be the vector of task space generalized forces/torques. It is worth noting that the vector of control inputs τ_i, $i \in \mathcal{N}$ can be related to the task space wrench $u_i \in \mathbb{R}^6$, $i \in \mathcal{N}$ via

$$\tau_i = J_i^\top(q_i)u_i + (I_{6+n} - J_i^\top(q_i)\tilde{J}_i^\top(q_i))\tau_{i0} \quad (4.12)$$

where $\tilde{J}_i^\top(q_i)$ is the generalized pseudo-inverse of J_i [41] and the vector τ_{i0} does not contribute to the end effector's wrench u_i (i.e., they belong to the null space of the Jacobian J_i) and can be regulated independently to achieve secondary tasks (e.g., maintaining manipulator's joint limits, increasing the manipulability)[†]. Moreover, the UVMS task space dynamics (4.11) can be written in vector form as follows:

$$M(q)\dot{v} + C(\dot{q}, q)v + D(\dot{q}, q)v + g(q) = u - \lambda \quad (4.13)$$

where $v = [v_1^\top, \dots, v_N^\top]^\top \in \mathbb{R}^{6N}$, $M = \text{diag}\{[M_i]\} \in \mathbb{R}^{6N \times 6N}$, $C = \text{diag}\{[C_i]\} \in \mathbb{R}^{6N \times 6N}$, $D = \text{diag}\{[D_i]\} \in \mathbb{R}^{6N \times 6N}$, $\lambda = [\lambda_1^\top, \dots, \lambda_N^\top]^\top$, $u = [u_1^\top, \dots, u_N^\top]^\top$, $g = [g_1^\top, \dots, g_N^\top]^\top \in \mathbb{R}^{6N}$.

4.2.2.2   Object dynamic

Without any loss of generality, we consider the following second-order dynamic for the object, which can be derived based on the Newton–Euler formulations:

$$\dot{x}_O = J_O(\eta_{2,O})v_O \quad (4.14)$$
$$M_O(x_O)\dot{v}_O + C_O(\dot{x}_O, x_O)v_O + D_O(\dot{x}_O, x_O)v_O + g_O = \lambda_O + \lambda_e \quad (4.15)$$

[†]For more details on task priority based control and redundancy resolution for UVMSs the reader is referred to [32] and [42].

where $M_O(x_O)$ is the positive definite inertia matrix, $C_O(\dot{x}_O, x_O)$ is the Coriolis matrix, g_O is the vector of gravity and buoyancy effects, $D_O(\dot{x}_O, x_O)$ models dissipative effects, λ_O is the vector of generalized forces acting on the object's center of mass, and λ_e is a vector representing uncertainties and external disturbances. Moreover, $J'_O(\eta_{2,O})$ is the object representation Jacobian $J_O(\eta_{2,O}) = \text{diag}\{I_3, J'_O(\eta_{2,O})\}$:

$$J'_O(\eta_{2,O}) = \begin{bmatrix} 1 & \sin(\phi_O)\tan(\theta_O) & \cos(\phi_O)\tan(\theta_O) \\ 0 & \cos(\phi_O) & -\sin(\theta_O) \\ 0 & \frac{\sin(\phi_O)}{\cos(\theta_O)} & \frac{\cos(\phi_O)}{\cos(\theta_O)} \end{bmatrix}, \qquad (4.16)$$

Moreover, the kineto-statics duality along with the grasp rigidity suggest that the force λ_O acting on the object's center of mass and the generalized forces λ_i, $i \in \mathcal{N}$, exerted by UVMSs at the grasping points, are related through

$$\lambda_O = G^\top \lambda \qquad (4.17)$$

where

$$G = \left[[J_{LO}]^\top, [J_{F_1 O}]^\top, \dots, [J_{F_N O}]^\top \right]^\top \in \mathbb{R}^{6(N+1) \times 6} \qquad (4.18)$$

is the full column-rank grasp matrix and $\lambda = [\lambda_L^\top, \lambda_{F_1}^\top, \dots, \lambda_{F_{N_F}}^\top]^\top$ is the vector of overall interaction forces and torques.

Remark 1. *Wrenches that lie on the null space of the grasp matrix $G^\top$ do not contribute to the object dynamics. Therefore, we may incorporate in the control scheme an extra component $\lambda_{int,i} = (I - (G^\top)^\# G^\top)\lambda_{int}^d$, $i \in \mathcal{N}$, that belongs to the null space of $G^\top$, to regulate the steady-state internal forces, where $(G^\top)^\#$ denotes the generalized inverse of $G^\top$. Notice that owing to the rigid grasp, l_i, $i \in \mathcal{N}$ remain constant. Thus, since l_i, $i \in \mathcal{N}$ are considered known to the team of UVMSs[‡], if λ_{int}^d is chosen constant, no communication is needed during task execution to compute $G^\top$, $(G^\top)^\#$, and $\lambda_{int,i}$.*

4.2.3 Description of the workspace

Consider the team of N UVMSs operating in a bounded workspace $\mathcal{W} \subseteq \mathbb{R}^3$ with boundary $\partial \mathcal{W}$. The object of interest is a rigid body, which is required to be transported cooperatively by the robot team from an initial to a goal position. Without any loss of the generality, the obstacles, the robots, as well as the workspace are all modeled by spheres (i.e., we adopt the spherical world representation [43]). However, the proposed control strategy could be extended to more general and complex geometries following the analysis in [43]. In this spirit, let $\mathcal{B}(x_O, r_O)$ be a closed ball that covers the volume of the object and has radius r_O. We also define the closed balls

---

[‡]This can be achieved by using the acoustic modems before beginning the task execution.

$\mathscr{B}(\boldsymbol{x}_i, \bar{r}), i \in \mathscr{N}$, centered at the end-effector of each UVMS that cover the robot volume for all possible configurations. Notice that the value of $\bar{r}$ can be calculated easily for each UVMS based solely on its own design parameters. We also assume that the distance among the grasping points on the given object is at least $2\bar{r}$. In particular, the distance $2\bar{r}$ denotes the minimum allowed distance at which two bounding spheres $\mathscr{B}(\boldsymbol{x}_i, \bar{r})$ and $\mathscr{B}(\boldsymbol{x}_j, \bar{r})$, $i, j \in \mathscr{N}, i \neq j$ do not collide (see Figure 4.1). Furthermore, we define a ball area $\mathscr{B}(\boldsymbol{x}_O, R)$ located at $\boldsymbol{x}_O$ with radius $R = \bar{r} + r_o$ that includes the whole volume of the robotic team and the object (see Figure 4.2). Finally, the $\mathscr{M}$ static obstacles within the workspace are defined as closed spheres described by $\pi_m = \mathscr{B}(\boldsymbol{p}_{\pi_m}, r_{\pi_m})$, $m \in \{1, \ldots, \mathscr{M}\}$, where $\boldsymbol{p}_{\pi_m} \in \mathbb{R}^3$ is the center and $r_{\pi_m} > 0$ is the radius of the obstacle π_m. Obviously, the ultimate goal of the proposed cooperative control strategy is to transport the object from the initial configuration to the

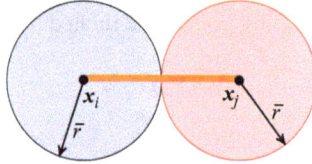

Figure 4.1   Graphical representation of the minimum allowed distance $\bar{r}$

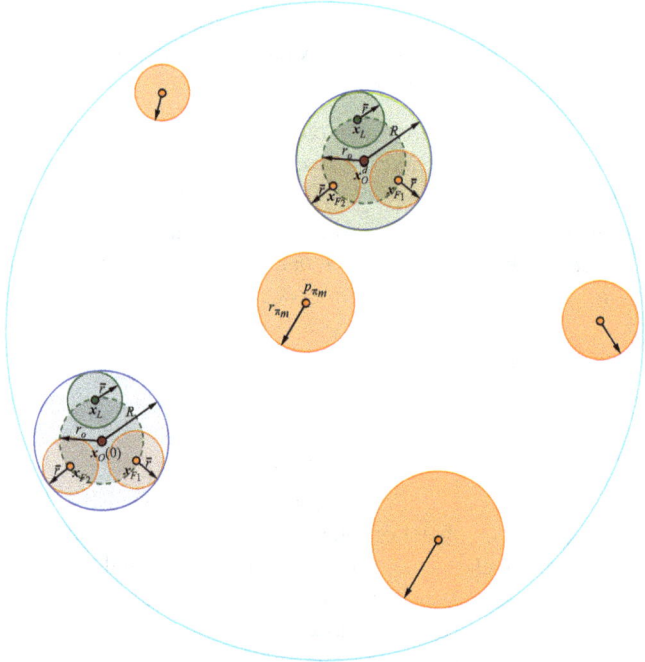

Figure 4.2   Graphical representation of a feasible trajectory of the team of UVMS carrying object from the initial position $x_O(t_0)$ to the desired position x_O^d. The boundary of workspace $\partial \mathscr{W}$ is illustrated in cyan. Red circles indicate the obstacles within the workspace $\mathscr{W}$. A feasible trajectory of the whole team is depicted in green.

desired one, without colliding with the obstacles and the boundary of workspace. Additionally, based on the property of spherical world [43], for each pair of obstacles $m, m' \in \{1, \ldots, \mathcal{M}\}$, the following inequality holds:

$$\|\boldsymbol{p}_{\pi_m} - \boldsymbol{p}_{\pi'_m}\| > 2R + r_{\pi_m} + r_{\pi'_m}$$

which intuitively means that the obstacles m and m' are disjoint in a such a way that the whole team of UVMSs including the object can pass through the free space between them. Therefore, there exists a feasible trajectory $\boldsymbol{x}_O(t)$ for the whole team that connects the initial configuration $\boldsymbol{x}_O(t_0)$ with $\boldsymbol{x}_O^d$ such as:

$$\mathcal{B}(\boldsymbol{x}_O(t), R) \cap \{\mathcal{B}(\boldsymbol{p}_{\pi_m}, r_{\pi_m}) \cup \partial \mathcal{W}\} = \emptyset, \quad \forall m \in \{1, \ldots, \mathcal{M}\} \qquad (4.19)$$

4.2.4 Dynamical systems

Consider the initial value problem:

$$\dot{\xi} = H(t, \xi), \xi(0) = \xi^0 \in \Omega_\xi, \qquad (4.20)$$

with $H : \mathbb{R}_{\geq 0} \times \Omega_\xi \to \mathbb{R}^n$, where $\Omega_\xi \subseteq \mathbb{R}^n$ is a non-empty open set.

Definition 1. *[44] A solution $\xi(t)$ of the initial value problem (4.20) is maximal if it has no proper right extension that is also a solution of (4.20).*

Theorem 1. *[44] Consider the initial value problem (4.20). Assume that $H(t, \xi)$ is: (a) locally Lipschitz in ξ for almost all $t \in \mathbb{R}_{\geq 0}$, (b) piecewise continuous in t for each fixed $\xi \in \Omega_\xi$ and (c) locally integrable in t for each fixed $\xi \in \Omega_\xi$. Then, there exists a maximal solution $\xi(t)$ of (4.20) on the time interval $[0, \tau_{\max})$, with $\tau_{\max} \in \mathbb{R}_{>0}$ such that $\xi(t) \in \Omega_\xi, \forall t \in [0, \tau_{\max})$.*

Proposition 1. *[44] Assume that the hypotheses of Theorem 1 hold. For a maximal solution $\xi(t)$ on the time interval $[0, \tau_{\max})$ with $\tau_{\max} < \infty$ and for any compact set $\Omega'_\xi \subseteq \Omega_\xi$, there exists a time instant $t' \in [0, \tau_{\max})$ such that $\xi(t') \notin \Omega'_\xi$.*

4.3 Decentralized impedance control for cooperative object transportation with multiple UVMSs under implicit communication

This section presents a decentralized leader–follower control approach for cooperative object transportation using multiple UVMSs in a constrained workspace with static obstacles. The challenge lays in replacing explicit communication with implicit, by incorporating sensor data that result from the physical interaction of the robots with the commonly grasped object (i.e. we assume that each UVMS is equipped with a force/torque sensor attached on its end-effector). The leader UVMS, which has knowledge of the object's desired trajectory, tries to achieve the desired tracking behavior via an impedance control law, navigating in this way, the overall formation toward the goal configuration while avoiding collisions with the obstacles. All UVMSs implement similar impedance-based control, ensuring accurate trajectory tracking despite uncertainties in dynamics and external disturbances in the object and the UVMS dynamics, respectively. The control scheme also distributes load among UVMSs according to their payload capacities. Additionally, adaptive control laws compensate for parametric uncertainties and disturbances, improving robustness. The follower UVMSs locally estimate the object's desired trajectory using an estimation law designed based on the prescribed performance technique [45]. This estimation relies on local force/torque measurements and onboard sensors, without explicit data exchange. It is worth noting that, unlike other methods that require a precise common localization reference for all agents, the proposed scheme enables each follower to independently and distributively estimate the desired object trajectory relative to its own inertial frame. This estimation relies solely on its own measurements, including position, velocity, and force/torque, without the need for external references. This approach eliminates the need for intra-team explicit communication, instead relying on force/torque sensing at the end-effector and fused position/velocity data from onboard sensors (e.g., IMU, USBL, and DVL). Additionally, we extend the current state of art in implicit communication-based cooperative manipulation [46,47], via a more robust estimation algorithm that converges even though the desired object's acceleration profile is nonzero (i.e. for an arbitrary object's desired trajectory profile as long as it is bounded and smooth). Finally, the customizable ultimate bounds allow us to achieve practical stabilization of the estimation error, with accuracy limited only by the sensors' resolution. Finally, simulation studies confirm the effectiveness of the method.

4.3.1   Problem formulation

Consider N UVMSs under a single leader–multiple followers architecture, rigidly grasping an object[§] within a constrained workspace with static obstacles. We also assume that each vehicle is actuated in all six degrees of freedoms (DoFs) and is equipped with an n DoF manipulator. Thus, each UVMS is fully actuated at its end-effector frame. This assumption implies that all UVMSs are able to exert arbitrary forces and torques on the object along and around any direction. It should also be noted that in the proposed scheme, only the leading robot is aware of the obstacles' position in the workspace and the object's desired configuration x_o^d. However, the followers estimate locally in a distributed way the object's desired trajectory profile and manipulate the object in coordination with the leader based solely on their own sensory information. Moreover, we assume that UVMSs are equipped with appropriate sensors, that allow them to measure their position and velocity (e.g., employing a fusion technique based on measurements by various onboard sensors such as USBL, IMU, DVL, and depth-sensor), as well as the interaction forces/torques with the object via a force/toque sensor. Additionally, the geometric parameters of the both UVMSs and the commonly grasped object are considered known, whereas their dynamic parameters are completely unknown. Moreover, the control of each UVMS will be designed based on a commonly agreed frame on a specific feature of the object, which could be identified employing a visual detection system [48], owing to the fact that the limited underwater visibility is not an issue when all robots are

---

[§]The end-effector frame of each UVMS is always constant relative to the object's body fixed frame.

close to the object of interest. Finally, we highlight the main challenges owing to (i) the strict communication constraints (i.e., online inter-robot communication is not permitted), (ii) the model uncertainties of UVMSs (common problem in underwater robotics), and (iii) the constrained workspace. Hence, the problem that we aim to solve in this section is stated as follows.

Problem: Given N UVMSs operating in a constrained workspace $\mathcal{W}$, design distributed control protocols $\boldsymbol{u}_i$, $\in \mathcal{N}$ that navigate safely the whole robotic team to the desired configuration without colliding with the obstacles and the boundary of the workspace, while satisfying the following specifications: (i) impose no strict requirements regarding the underwater communication bandwidth; (ii) enforce robustness against the parametric uncertainty of the UVMS dynamic model.

4.3.2 Control methodology

We assume that the leading UVMS is aware of both the desired configuration of the object and the position of the obstacles in the workspace. Thus, its control objective is to navigate the overall formation toward the goal configuration while avoiding collisions with the static obstacles that lie within the workspace. Toward this direction and in view of (4.19), we assume that there is a feasible trajectory within the workspace which is known only for the leader. In contrast, the followers are not aware of the object's desired configuration. However, even though explicit inter-robot communication is not permitted, the followers will estimate the object's desired trajectory profile via their own state measurements (sensor fusion of locally onboard navigation system sensors, e.g., DVL, IMU, and USBL). Toward this direction, acceleration residuals owing to the lack of acceleration measurements for the object will be compensated by adopting a robust prescribed performance estimator that guarantees ultimate boundedness of the estimation error with predefined transient and steady-state specifications. Finally, an adaptive control scheme will be designed to achieve the asymptotic tracking of the estimated trajectory profile, thus increasing greatly the robustness of the overall control scheme and avoiding high interaction forces among the object and the robots.

Remark 2. *The desired/feasible object trajectory within the workspace $\mathcal{W}$ can be generated based on the Navigation Functions concept originally proposed by Rimon and Koditschek in [43] as follows:*

$$\phi_O(\boldsymbol{x}_O; \boldsymbol{x}_O^d) = \frac{\gamma(\boldsymbol{x}_O - \boldsymbol{x}_O^d)}{[\gamma^k(\boldsymbol{x}_O - \boldsymbol{x}_O^d) + \beta(\boldsymbol{x}_O)]^{\frac{1}{k}}} \tag{4.21}$$

where $\phi_O : \dfrac{\mathcal{W} - \overset{\mathcal{M}}{\underset{m=1}{\cap}} \mathcal{B}(\boldsymbol{p}_{\pi_m}, r_{\pi_m})}{} \longrightarrow [0, 1)$ denotes the potential that derives a safe motion vector field within the free space $\mathcal{W} - \overset{\mathcal{M}}{\underset{m=1}{\cap}} \mathcal{B}(\boldsymbol{p}_{\pi_m}, r_{\pi_m})$. Moreover, $k > 1$ is a design *constant,* $\gamma(x_O - x_O^d) > 0$ *with* $\gamma(0) = 0$ *represents the attractive potential field to the goal configuration* x_O^d *and* $\beta(x_O) > 0$ *with*

$$\lim_{x_O \to \begin{cases} Boundary \\ Obstacles \end{cases}} \beta(x_O) = 0$$

represents the repulsive potential field by the workspace boundary and the obstacle regions. In that respect, it was proven in [43] that $\phi_O(x_O, x_O^d)$ *has a global minimum at* x_O^d *and no other local minima for sufficiently large k. Thus, a feasible path that leads from any initial obstacle-free configuration*[l] *to the desired configuration might be generated by following the negated gradient of* $\phi_O(x_O, x_O^d)$*. Consequently, the desired velocity profile at leader's side is designed as follows:*

$$v_{O_L}^d(t) = -K_{NF} J_O^{-1}(\eta_{2,O}) \nabla_{x_O} \phi_O(x_O(t), x_O^d) \tag{4.22}$$

where $K_{NF} > 0$ *is a positive gain. Moreover, given the initial configuration, the leading UVMS may easily calculate the desired trajectory and velocity profile denoted by* $x_{O_L}^d(t)$ *and* $v_{O_L}^d(t)$*, respectively, by propagating the model* $\dot{x}_O^d(t) = J_O(\eta_{2,O}(t)) v_{O_L}^d(t)$*.*

4.3.2.1   Control design

In the sequel, we propose an decentralized control scheme that guarantees the asymptotic stabilization of the object to the goal configuration x_O^d. First we start by invoking the kinematic relations (4.6)–(4.8), we may express the aforementioned dynamics (4.11) with respect to the object's coordinates as follows:

$$M_i(q_i)\dot{v}_O + C_i(\zeta_i, q_i)v_O + D_i(\zeta_i, q_i)v_O + g_i(q_i) + d_i(\zeta_i, q_i, t) = J_{iO}^\top u_i + J_{iO}^\top \lambda_i \tag{4.23}$$

where:

$$M_i(q_i) = J_{iO}^\top M_i(q_i) J_{iO}$$
$$C_i(\zeta_i, q_i) = J_{iO}^\top \left[C_i(\zeta_i, q_i) J_{iO} + M_i(q_i) \dot{J}_{iO} \right]$$
$$D_i(\zeta_i, q_i) = J_{iO}^\top D_i(\zeta_i, q_i) J_{iO}$$
$$g_i(q_i) = J_{iO}^\top g_i(q_i)$$
$$d_i(\zeta_i, q_i, t) = J_{iO}^\top d_i(\zeta_i, q_i, t)$$

Now, the following common properties will be employed in the analysis.

Property 1. *The matrix* $M_i(q_i)$*,* $i \in \mathcal{N}$ *is positive definite, and the matrix* $\dot{M}_i(q_i) - 2C_i(\zeta_i, q_i)$*,* $i \in \mathcal{N}$ *is skew-symmetric. We have that a quadratic form of a skew-symmetric matrix is always equal to 0. Hence, for the matrices* $\dot{M}_i(q_i) - 2C_i(\zeta_i, q_i)$*, the following holds [41,49,50]:*

$$s^\top \left[\dot{M}_i(q_i) - 2C_i(\zeta_i, q_i) \right] s = 0, \quad \forall s \in \mathbb{R}^6$$

[l]Except from a set of measure zero [43].

Property 2. *The uncertainty of the UVMS model appears linearly in the dynamics (4.23), in terms of an unknown but constant parameter vector $\theta_i \in \mathbb{R}^{q_i}$, $i \in \mathcal{N}$ in the following way [50,51]:*

$$M_i(a_i)d_i + C_i(a_i,b_i)c_i + D_i(a_i,b_i)c_i + g_i(a_i) = \Omega_i(a_i,b_i,c_i,d_i)\theta_i$$

for $i \in \mathcal{N}$, where $\Omega_i(a_i,b_i,c_i,d_i) \in \mathbb{R}^{6\times q_i}$, $i \in \mathcal{N}$ is a regressor matrix of known functions of $a_i, b_i, c_i, d_i \in \mathbb{R}^6$ independent of θ_i.

Now, we introduce the following assumption regarding the unmodeled dynamics/external disturbances.

Assumption 1. *There exists positive, finite unknown constant $\theta_{d,i} \in \mathbb{R}^{q_i}$, $i \in \mathcal{N}$ and known bounded function $\Delta_i \in \mathbb{R}^{6\times q_i}$, $i \in \mathcal{N}$, such that*

$$d_i(\zeta_i, q_i, t) = \Delta_i(\zeta_i, q_i, t)\theta_{d,i}, \quad i \in \mathcal{N}$$

Before proceeding the analysis, we introduce the load sharing coefficients c_i, $i \in \mathcal{N}$ that are subject to the following design constraints:

$$c_i \in (0,1), \forall i \in \mathcal{N}, \quad \text{and} \quad \sum_{i\in\mathcal{N}} c_i = 1 \tag{4.24}$$

Thus, before losing any generality, for simplify the analysis we set:

$$c_i = \frac{1}{N+1}, \quad i \in \mathcal{N} \tag{4.25}$$

which satisfy the constraints of (4.24). In view of the object dynamics (4.15), it can be concluded that the vector of external disturbances λ_e is unknown. Thus, in order to design the impedance control scheme, each UVMS must estimate the aforementioned vector in a distributed way (since there is not explicit communication between UVMSs). Moreover, based on the object dynamics (4.15), the vector of external disturbances is impossible to be estimated in a decentralized way by relying only on local measurements (i.e., local force torque measurements at UVMS's end effector), since it depends on the applying force from all member of the UVMS teams on the object. Therefore, an online estimation method based on the object momentum concept [52] is designed in the sequel. First, in view of the load coefficients (4.25) and (4.17), the object dynamic of (4.15) can be rewritten as follows:

$$\sum_{i\in\mathcal{N}} \{M_{O_i}(x_O)\dot{v}_O + C_{O_i}(\dot{x}_O,x_O)v_O + D_{O_i}(\dot{x}_O,x_O)v_O + g_{O_i}\} = \sum_{i\in\mathcal{N}} J_{iO}^\top \lambda_i + \sum_{i\in\mathcal{N}} \lambda_{e_i} \tag{4.26}$$

where $M_{O_i} = c_i M_O$, $C_{O_i} = c_i C_O$, $D_{O_i} = c_i D_O$, $g_{O_i} = c_i g_O$ and $\lambda_{e_i} = c_i \lambda_e$. To estimate locally the λ_{e_i} for the UVMS i, $i \in \mathcal{N}$, we define the object momentum equivalent momentum [52]: $\mu_i = M_{O_i} v_O$ and the vector $\zeta_i(t) \in \mathbb{R}^6$ as:

$$\zeta_i(t) = K_\mu \left(\mu_i(t) + \int_{t_0}^t \left(C_{O_i} v_O + D_{O_i} v_O + g_{O_i} - \zeta_i(d\tau) \right) d\tau \right) \tag{4.27}$$

whose time derivative is given by

$$\dot{\zeta}_i(t) = -K_\mu \zeta_i(t) + c_i K_\mu \left(\sum_{i \in \mathcal{N}} J_{iO}^\top \lambda_i + \lambda_e \right) \tag{4.28}$$

where K_μ is a positive definite matrix gain. Notice that the for a properly large matrix K_μ, we obtain:

$$\zeta_i(t) \approx c_i \left(\sum_{i \in \mathcal{N}} J_{iO}^\top \lambda_i + \lambda_e \right) \tag{4.29}$$

which intuitively means that the $\zeta_i(t)$ represents the effect of overall external forces exerted on the object (i.e., external disturbances and the forces exerted by all the UVMS team on the object). Consequently, an estimation of $\lambda_{e_i} = c_i \lambda_e$ can be given as follows:

$$\lambda_{e_i} \approx \zeta_i(t) - J_{iO}^\top \lambda_i, \quad i \in \mathcal{N} \tag{4.30}$$

Now let us assume that each UVMS is expected to exert the following desired force/torque on the object:

$$\lambda_i^d = \lambda_{\text{int},i}^d - J_{iO}^{-\top} (M_{O_i} y_i^{\text{cmd}} + C_{O_i} v_O + D_{O_i} v_O + g_{O_i} - \lambda_{e_i}) \tag{4.31}$$

where $\lambda_{\text{int},i}^d$ is the desired inertial forces (see Remark 1) and the y_i^{cmd} is a predesigned input given by

$$y_i^{\text{cmd}} = \dot{v}_{O_i}^d + M_{do}^{-1} \left[-D_{do} \tilde{v}_{O_i} - K_{do} \tilde{e}_{O_i} \right] \tag{4.32}$$

where M_{do}, D_{do} and K_{do} are the desired inertia and damping and stiffness matrices for the object dynamics, respectively, $\tilde{v}_{O_i}(t) = v_O - v_{O_i}^d$ denotes the velocity error and $\tilde{e}_{O_i}$ is the object pose error, defined as follows:

$$\tilde{e}_{O_i} = \begin{bmatrix} \eta_{1,O} - \eta_{1,O_i}^d \\ \tilde{\varepsilon}_{O_i} \end{bmatrix} \tag{4.33}$$

where

$$\tilde{\varepsilon}_{O_i} = \frac{1}{2} \left(n_O \times n_{O_i}^d + o_O \times o_{O_i}^d + \alpha_O \times \alpha_{O_i}^d \right) \in \mathbb{R}^3 \tag{4.34}$$

is the orientation error expressed in the outer product formulation [53]. In view of (4.26), it can be concluded that if all robots cooperatively apply the desired wrench vector (4.31) to the object, then

$$M_{do} \dot{\tilde{v}}_{O_i} + D_{do} \tilde{v}_{O_i} + K_{do} \tilde{e}_{O_i} = 0 \tag{4.35}$$

which intuitively means that the aforementioned selection of λ_i^d cancels the object's nonlinearities, ensures adequate internal forces via $\lambda_{\text{int},i}^d$, and achieves the desired dynamics of the object. Thus, the control objective for each UVMS $i \in \mathcal{N}$ is to enforce $\lim_{t \to \infty} \boldsymbol{w}_i(t) = 0$, where the error signal $\boldsymbol{w}(t)$ is constructed as follows:

$$\boldsymbol{w}_i(t) = \boldsymbol{M}_\mathrm{d} \dot{\tilde{\boldsymbol{v}}}_{O_i} + \boldsymbol{D}_\mathrm{d} \tilde{\boldsymbol{v}}_{O_i} + \boldsymbol{K}_\mathrm{d} \tilde{\boldsymbol{e}}_{O_i} - \boldsymbol{J}_{iO}^\top \boldsymbol{\lambda}_i^\mathrm{d}, \quad i \in \mathcal{N} \tag{4.36}$$

where $\boldsymbol{M}_\mathrm{d}, \boldsymbol{D}_\mathrm{d}$, and $\boldsymbol{K}_\mathrm{d}$ are the desired inertia, damping, and stiffness matrices for the robot dynamics, respectively. Thus, we get an augmented impedance error:

$$\tilde{\boldsymbol{w}}_i = \boldsymbol{K}_f \boldsymbol{w}_i = \dot{\tilde{\boldsymbol{v}}}_{O_i} + \boldsymbol{K}_g \tilde{\boldsymbol{v}}_{O_i} + \boldsymbol{K}_p \tilde{\boldsymbol{e}}_{O_i} - \boldsymbol{K}_f \boldsymbol{J}_{iO}^\top \boldsymbol{\lambda}_i^\mathrm{d} \tag{4.37}$$

where $\boldsymbol{K}_f = \boldsymbol{M}_\mathrm{d}^{-1}, \boldsymbol{K}_g = \boldsymbol{K}_f \boldsymbol{D}_\mathrm{d}$, and $\boldsymbol{K}_\mathrm{d} = \boldsymbol{K}_f \boldsymbol{K}_\mathrm{d}$. We also choose two positive-definite matrices $\boldsymbol{F}$ and $\boldsymbol{Y}$ such that:

$$\boldsymbol{F} + \boldsymbol{Y} = \boldsymbol{K}_g$$
$$\dot{\boldsymbol{F}} + \boldsymbol{Y}\boldsymbol{F} = \boldsymbol{K}_p$$

and define the filtered force/torque measurement:

$$\dot{\boldsymbol{f}}_i + \boldsymbol{Y}\boldsymbol{f}_i = \boldsymbol{K}_f \boldsymbol{J}_{iO}^\top \boldsymbol{\lambda}_i^\mathrm{d}, \ i \in \mathcal{N}. \tag{4.38}$$

Thus, we may rewrite (4.37) as follows:

$$\tilde{\boldsymbol{w}}_i = \dot{\tilde{\boldsymbol{v}}}_{O_i} + (\boldsymbol{F} + \boldsymbol{Y})\tilde{\boldsymbol{v}}_{O_i} + (\dot{\boldsymbol{F}} + \boldsymbol{Y}\boldsymbol{F})\tilde{\boldsymbol{e}}_{O_i} - \dot{\boldsymbol{f}}_i - \boldsymbol{Y}\boldsymbol{f}_i. \tag{4.39}$$

Now, we define the auxiliary variables z_i, $i \in \mathcal{N}$ as follows:

$$\boldsymbol{z}_i = \tilde{\boldsymbol{v}}_{O_i} + \boldsymbol{F}\tilde{\boldsymbol{e}}_{O_i} - \boldsymbol{f}_i, \quad i \in \mathcal{N}. \tag{4.40}$$

Hence, the augmented impedance error becomes:

$$\tilde{\boldsymbol{w}}_i = \dot{\boldsymbol{z}}_i + \boldsymbol{Y}\boldsymbol{z}_i, \quad i \in \mathcal{N} \tag{4.41}$$

which represents a stable low pass filter. Therefore, if we achieve $\lim_{t \to \infty} z_i(t) = 0$, then the initial control objective is readily met, i.e., $\lim_{t \to \infty} \boldsymbol{w}_i(t) = 0$. In this respect, let us define the augmented state variable:

$$\boldsymbol{v}_{O_i}^r = \boldsymbol{v}_{O_i}^\mathrm{d} - \boldsymbol{F}\tilde{\boldsymbol{e}}_{O_i} + \boldsymbol{f}_i, \ i \in \mathcal{N} \tag{4.42}$$

Thus, (4.40) and (4.42) immediately result in the following:

$$\boldsymbol{z}_i = \boldsymbol{v}_O - \boldsymbol{v}_{O_i}^r, \quad i \in \mathcal{N} \tag{4.43}$$

from which the dynamics (4.23) becomes

$$\boldsymbol{M}_i \dot{\boldsymbol{z}}_i + \boldsymbol{C}_i \boldsymbol{z}_i + \boldsymbol{D}_i \boldsymbol{z}_i = \boldsymbol{J}_{iO}^\top \boldsymbol{u}_i + \boldsymbol{J}_{iO}^\top \boldsymbol{\lambda}_i - \left[\boldsymbol{M}_i \dot{\boldsymbol{v}}_{O_i}^r + \boldsymbol{C}_i \boldsymbol{v}_{O_i}^r + \boldsymbol{D}_i \boldsymbol{v}_{O_i}^r + \boldsymbol{g}_i + \boldsymbol{d}_i \right].$$

Invoking Property 2 and Assumption 1, we arrive at the open loop dynamics:

$$\boldsymbol{M}_i \dot{\boldsymbol{z}}_i + \boldsymbol{C}_i \boldsymbol{z}_i + \boldsymbol{D}_i \boldsymbol{z}_i = \boldsymbol{J}_{iO}^\top \boldsymbol{u}_i + \boldsymbol{J}_{iO}^\top \boldsymbol{\lambda}_i - \boldsymbol{\Delta}_i(\boldsymbol{\zeta}_i, \boldsymbol{q}_i, t)\boldsymbol{\theta}_{d,i} - \boldsymbol{\Omega}_i(\boldsymbol{q}_i, \boldsymbol{\zeta}_i, \boldsymbol{v}_{O_i}^r, \dot{\boldsymbol{v}}_{O_i}^r)\boldsymbol{\theta}_i, \quad i \in \mathcal{N}. \tag{4.44}$$

Therefore, we design the following impedance control scheme:

$$u_i = -\lambda_i + J_{iO}^{-\top}\left[\Omega_i(q_i, \zeta_i, v_{O_i}^r, \dot{v}_{O_i}^r)\hat{\theta}_i + \Delta_i(\zeta_i, q_i, t)\hat{\theta}_{d,i} - Kz_i\right], \quad i \in \mathcal{N} \quad (4.45)$$

where $K > 0$ is a positive definite gain matrix and $\hat{\theta}_i$ and $\hat{\theta}_{d,i}$ denote the estimates of the unknown parameters θ_i and $\theta_{d,i}$ respectively, provided by the update laws:

$$\dot{\hat{\theta}}_i = -\Gamma_i\Omega_i(q_i, \zeta_i, v_{O_i}^r, \dot{v}_{O_i}^r)z_i, \quad \Gamma_i > 0 \quad (4.46)$$

$$\dot{\hat{\theta}}_{d,i} = -\Gamma_{d_i}\Delta_i(\zeta_i, q_i, t)z_i, \quad \Gamma_{d_i} > 0 \quad (4.47)$$

with Γ_i and Γ_{d_i} being positive diagonal gain matrices.

Theorem 2. *Consider N UVMSs operating in a constrained workspace $\mathcal{W}$ with dynamics given by (4.23) that obey Properties 1 and 2, grasping rigidly a common object. The control scheme for each UVMS i, $i \in \mathcal{N}$ given in (4.45) with adaptive laws (4.46) and (4.47) guarantees $\lim_{t\to\infty} w_i(t) = 0$ and the boundedness of all signals in the closed-loop system.*

Proof. Consider the following Lyapunov function candidate:

$$V = \sum_{i\in\mathcal{N}}\frac{1}{2}z_i^\top M_i z_i + \sum_{i\in\mathcal{N}}\frac{1}{2}\tilde{\theta}_i^\top\Gamma_i^{-1}\tilde{\theta}_i + \sum_{i\in\mathcal{N}}\frac{1}{2}\tilde{\theta}_{d_i}^\top\Gamma_{d_i}^{-1}\tilde{\theta}_{d_i}$$

where $\tilde{\theta}_i = \hat{\theta}_i - \theta_i$ and $\tilde{\theta}_{d_i} = \hat{\theta}_{d_i} - \theta_{d_i}$ denote the parametric errors. Differentiating with respect to time yields:

$$\dot{V} = \sum_{i\in\mathcal{N}}\frac{1}{2}z_i^\top\dot{M}_i z_i + \sum_{i\in\mathcal{N}}z_i^\top M_i\dot{z}_i + \sum_{i\in\mathcal{N}}\tilde{\theta}_i^\top\Gamma_i^{-1}\dot{\hat{\theta}}_i + \sum_{i\in\mathcal{N}}\tilde{\theta}_{d_i}^\top\Gamma_{d_i}^{-1}\dot{\hat{\theta}}_{d_i}$$

Invoking Property 1 and substituting the adaptive laws (4.46) and (4.47), we get the following:

$$\dot{V} = \sum_{i\in\mathcal{N}} -z_i^\top Kz_i - z_i^\top D_i z_i \leq 0 \quad (4.48)$$

Hence, we deduce z_i, $\tilde{\theta}_i$, and $\tilde{\theta}_{d_i} \in L_\infty$. Moreover, from the definition of z_i in (4.43), we conclude that $x_O, v_O \in L_\infty$, and consequently $v_{O_i}^r, \dot{v}_{O_i}^r \in L_\infty$. Furthermore, employing (4.44), we arrive at $\dot{z} \in L_\infty$. Therefore, integrating both sides of (4.48) leads to the following:

$$V(t) - V(0) \leq \sum_{i\in\mathcal{N}}\int_0^t -z_i^\top(\tau)Kz_i - z_i^\top D_i z_i(\tau)d\tau \quad (4.49)$$

Thus, $\sum_{i\in\mathcal{N}}\int_0^t -z_i^\top(\tau)Kz_i - z_i^\top D_i z_i(\tau)d$ is bounded, which results in $z_i \in L_2$. Finally, Barbalat's Lemma leads to $z_i \to 0$, $\forall i \in \mathcal{N}$ as $t \to 0$, since $z_i \in L_2$ and $\dot{z}_i \in \infty$, which completes the proof. $\square$

4.3.2.2 Follower's estimation scheme

It should be noticed that the followers are not aware of either the object's desired configuration x_O^d or the obstacles' position in the workspace. However, even thought explicit communication among the leader and the followers is not permitted, the followers will estimate the object's desired trajectory profile by $\hat{x}_O^{d_i}(t)$ $i \in \mathscr{F}$, via their own state measurements by adopting a novel prescribed performance estimator. Hence, let us define the error:

$$e_i(t) = x_O(t) - \hat{x}_O^{d_i}(t) \in \mathbb{R}^6, \quad i \in \mathscr{F}. \tag{4.50}$$

The expression of prescribed performance for each element of $e_i(t) = [e_{i1}(t), \dots, e_{i6}(t)]^\top$, $i \in \mathscr{F}$ is given by the following inequalities:

$$-\rho_{ij}(t) < e_{ij}(t) < \rho_{ij}(t), \quad j = 1, \dots, 6 \text{ and } i \in \mathscr{F} \tag{4.51}$$

for all $t \geq 0$, where $\rho_{ij}(t)$, $j = 1, \dots, 6$ and $i \in \mathscr{F}$ denote the corresponding performance functions. A candidate exponential performance function could be the following:

$$\rho_{ij}(t) = (\rho_{ij,0} - \rho_{ij,\infty})e^{-\lambda t} + \rho_{ij,\infty}, \quad i \in \mathscr{F} \tag{4.52}$$

where the constant λ dictates the exponential convergence rate, $\rho_{ij,\infty}$, $i \in \mathscr{F}$ denotes the ultimate bound, and $\rho_{ij,0}$ is chosen to satisfy $\rho_{ij,0} > |e_{ij}(0)|$, $i \in \mathscr{F}$. Hence, following the prescribed performance control technique [54], the estimation law is designed as follows:

$$\dot{\hat{x}}_{O_j}^{d_i} = k_{ij} \ln \left(\frac{1 + \frac{e_{ij}(t)}{\rho_{ij}(t)}}{1 - \frac{e_{ij}(t)}{\rho_{ij}(t)}} \right), k_{ij} > 0, \quad j = 1, \dots, 6 \tag{4.53}$$

for $i \in \mathscr{F}$, from which the followers' estimate $\hat{x}_O^{d_i}(t) = [\hat{x}_{O_1}^{d_i}(t), \dots, \hat{x}_{O_6}^{d_i}(t)]^\top$, $i \in \mathscr{F}$ is calculated via a simple integration. Moreover, differentiating (4.53) with respect to time, we acquire the desired acceleration signal:

$$\ddot{\hat{x}}_{O_j}^{d_i} = \frac{2k_{ij}}{1 - \left(\frac{e_{ij}(t)}{\rho_{ij}(t)}\right)^2} \frac{\dot{e}_{ij}(t)\rho_{ij}(t) - e_{ij}(t)\dot{\rho}_{ij}(t)}{\left(\rho_{ij}(t)\right)^2} \tag{4.54}$$

employing only the velocity $\dot{x}_O(t)$ of the object, which can be easily calculated via (4.7), and not its acceleration which is unmeasurable.

Lemma 1. *Consider the error:*

$$e_i(t) = x_O(t) - \hat{x}_O^{d_i}(t) = [e_{i1}(t), \dots, e_{i6}(t)]^\top, \quad i \in \mathscr{F} \tag{4.55}$$

where $x_O(t)$ and $\hat{x}_O^{d_i}(t)$, $i \in \mathscr{F}$ denote the object's actual configuration and the estimation of the object's desired trajectory profile at the followers' side, respectively. Given the appropriately selected performance functions $\rho_{ij}(t)$, $j = 1, \dots, 6$ and $i \in \mathscr{F}$ that satisfy $|e_{ij}(0)| < \rho_{ij}(0)$, $j = 1, \dots, 6$ and $i \in \mathscr{F}$ and incorporate the desired transient and steady-state performance specifications, the estimation law (4.53) guarantees that $|e_{ij}(t)| < \rho_{ij}(t)$, $j = 1 \dots, 6$ and $i \in \mathscr{F}$ for all $t \geq 0$ as well as that $\hat{x}_O^{d_i}$ and $\dot{\hat{x}}_O^{d_i}$ remain bounded.

Proof: The proof follows identical arguments for each element of $e_i(t)$, $i \in \mathscr{F}$. Hence, let us define the normalized errors:

$$\xi_{ij} = \frac{e_{ij}(t)}{\rho_{ij}(t)}, \quad j = 1, \ldots, 6 \text{ and } i \in \mathscr{F}. \tag{4.56}$$

The estimation law (4.53) may be rewritten as a function of the normalized error ξ_{ij} as follows:

$$\dot{x}_{O_j}^{d_i} = k_{ij} \ln \left(\frac{1 + \xi_{ij}}{1 - \xi_{ij}} \right), \quad j = 1, \ldots, 6 \text{ and } i \in \mathscr{F}. \tag{4.57}$$

Hence, differentiating ξ_{ij} with respect to time and substituting (4.57), we obtain the following:

$$\dot{\xi}_{ij} = h_{ij}(t, \xi_{ij}) \equiv \frac{\dot{x}_{O_j}(t) - k_{ij} \ln \left(\frac{1 + \xi_{ij}}{1 - \xi_{ij}} \right)}{\rho_{ij}(t)} - \xi_{ij} \frac{\dot{\rho}_{ij}(t)}{\rho_{ij}(t)} \tag{4.58}$$

We also define the non-empty and open set $\Delta_{\xi_{ij}} = (-1, 1)$. In the sequel, we shall prove that $\xi_{ij}(t)$ never escapes a compact subset of $\Delta_{\xi_{ij}}$ and thus the performance bounds (4.22) are met. The following analysis is divided in two phases. First, we show that a maximal solution exists, such that $\xi_{ij}(t) \in \Delta_{\xi_{ij}} \forall t \in [0, \tau_{\max})$, and subsequently we prove by contradiction that $\tau_{\max}$ is extended to ∞.

Phase I: Since $|e_{ij}(0)| < \rho_{ij}(0)$, we conclude that $\xi_{ij}(0) \in \Delta_{\xi_{ij}}$. Moreover, owing to the smoothness of the object trajectory and the proposed estimation scheme (4.53) over $\Delta_{\xi_{ij}}$, the function $h_{ij}(t, \xi_{ij})$ is continuous for all $t \geq 0$ and $\xi_{ij} \in \Delta_{\xi_{ij}}$. Therefore, the hypotheses of Theorem 1 hold and the existence of a maximal solution $\xi_{ij}(t)$ of (4.58) on a time interval $[0, \tau_{\max})$ such that $\xi_{ij}(t) \in \Delta_{\xi_{ij}}, \forall t \in [0, \tau_{\max})$ is ensured.

Phase II: Notice that the transformed error signal

$$\varepsilon_{ij}(t) = \ln \left(\frac{1 + \xi_{ij}(t)}{1 - \xi_{ij}(t)} \right), \quad j = 1, \ldots, 6 \tag{4.59}$$

for $i \in \mathscr{F}$ is well defined for all $t \in [0, \tau_{\max})$. Hence, consider the positive definite and radially unbounded function $V_{ij} = \frac{1}{2}(\varepsilon_{ij})^2$. Differentiating with respect to time and substituting (4.58), we obtain the following:

$$\dot{V}_{ij} = \frac{2\varepsilon_{ij}}{(1 - \xi_{ij}^2)\rho_{ij}(t)} \left(\dot{x}_{O_j}(t) - k_{ij}\varepsilon_{ij} - \xi_{ij}\dot{\rho}_{ij}(t) \right)$$

Since $\dot{x}_{O_j}(t)$, $j = 1, \ldots, 6$ was proven bounded in Theorem 2 for all $t \geq 0$, and $\dot{\rho}_{ij}(t)$ is bounded by construction, we conclude that

$$|\dot{x}_{O_j}(t) - \xi_{ij}\dot{\rho}_{ij}(t)| \geq \bar{d}_{ij}, \quad j = 1, \ldots, 6 \text{ and } i \in \mathscr{F}.$$

for an unknown positive constant $\bar{d}_{ij}$. Moreover, $\frac{1}{1-(\xi_{ij})^2} > 1, \forall \xi_{ij} \in \Delta_{\xi_{ij}}$ and $\rho_{ij}(t) > 0$ for all $t \geq 0$. Hence, we conclude that $\dot{V}_{ij} < 0$ when $|\varepsilon_{ij}(t)| > \frac{\bar{d}_{ij}}{k_{ij}}$ and consequently that

$$|\varepsilon_{ij}(t)| \geq \bar{\varepsilon}_{ij} = \max \left\{ |\varepsilon_{ij}(0)|, \frac{\bar{d}_{ij}}{k_{ij}} \right\}, \quad \forall t \in [0, \tau_{\max}) \tag{4.60}$$

Thus, invoking the inverse of (4.59), we get the following:

$$-1 < \frac{e^{-\bar{\varepsilon}_{ij}} - 1}{e^{-\bar{\varepsilon}_{ij}} + 1} = \underline{\xi}_{ij} \leq \xi_{ij}(t) \leq \bar{\xi}_{ij} = \frac{e^{\bar{\varepsilon}_{ij}} - 1}{e^{\bar{\varepsilon}_{ij}} + 1} < 1 \qquad (4.61)$$

for $j = 1, \ldots, 6$ and $i \in \mathscr{F}$. Therefore, $\xi_{ij}(t) \in \Delta'_{\xi_{ij}} = [\underline{\xi}_{ij}, \bar{\xi}_{ij}]$, $\forall t \in [0, \tau_{\max})$, which is a non-empty and compact subset of $\Delta_{\xi_{ij}}$. Consequently, assuming $\tau_{\max} < \infty$, Proposition 1 dictates the existence of a time instant $t' \in [0, \tau_{\max})$ such that $\xi_{ij}(t') \notin \Delta'_{\xi_{ij}}$, which is a clear contradiction. Therefore, $\tau_{\max}$ is extended to ∞. As a result $\xi_{ij}(t) \in \Delta'_{\xi_{ij}} \in \Delta_{\xi_{ij}}$, $\forall t \geq 0$. Thus, from (4.56) and (4.59), we conclude that

$$-\rho_{ij}(t) < \underline{\xi}_{ij}\rho_{ij}(t) \leq e_{ij}(t) \leq \bar{\xi}_{ij}\rho_{ij}(t) < \rho_{ij}(t), \forall t \geq 0$$

Finally, invoking (4.51) as well as the boundedness of $\boldsymbol{x}_O(t)$ and $\dot{\boldsymbol{x}}_O(t)$ from Theorem 2, we also deduce the boundedness of $\hat{\boldsymbol{x}}_O^{d_i}(t)$, and $\dot{\hat{\boldsymbol{x}}}_O^{d_i}(t)$ for all $t \geq 0$, which completes the proof. Based on the aforementioned estimation of the object's desired trajectory profile $\hat{\boldsymbol{x}}_O^{d_i}(t), \dot{\hat{\boldsymbol{x}}}_O^{d_i}(t)$ and $\ddot{\hat{\boldsymbol{x}}}_O^{d_i}(t)$, $i \in \mathscr{F}$, we can easily derive the corresponding desired trajectory profile for the follower's End-Effector via (4.14) as follows:

$$\begin{aligned} \boldsymbol{v}_{O_{F_i}}^{d_i}(t) &= \boldsymbol{J}_O^{-1}(\boldsymbol{\eta}_{2,o})\dot{\hat{\boldsymbol{x}}}_O^{d_i}(t) \\ \dot{\boldsymbol{v}}_{O_{F_i}}^{d_i}(t) &= \boldsymbol{J}_O^{-1}(\boldsymbol{\eta}_{2,o})\ddot{\hat{\boldsymbol{x}}}_O^{d_i} + \dot{\boldsymbol{J}}_O^{-1}(\boldsymbol{\eta}_{2,o})\dot{\hat{\boldsymbol{x}}}_O^{d_i} \end{aligned} \qquad (4.62)$$

Remark 3. *The proposed estimation scheme is more robust against trajectory profiles with nonzero acceleration than previous results presented in [46,47]. In particular, our method guarantees bounded closed-loop signals and practical asymptotic stabilization of the estimation errors. Moreover, the aforementioned ultimate bounds depend directly on the design parameters $\rho_{ij,\infty}, j = 1, \ldots, 6$ and $i \in \mathscr{F}$ of the performance functions $\rho_{ij}(t), j = 1, \ldots, 6$ and $i \in \mathscr{F}$, which can be set arbitrarily small to a value reflecting the resolution of the measurement devices, thus achieving practical convergence of the estimation errors to zero. Additionally, the transient response depends on the convergence rate of the performance functions $\rho_{ij}(t), j = 1, \ldots, 6$ and $i \in \mathscr{F}$ that is directly affected by the parameter λ.*

Remark 4. *The appropriate selection of the performance function $\rho_{ij}(t)$ imposes transient and steady-state performance characteristics on the estimation errors $e_{ij}(t)$ irrespective of the design parameters $k_{ij}, i \in \mathscr{F}$, $j = 1 \ldots, 6$. In particular, for an initial estimation of the object's desired trajectory profile $\hat{\boldsymbol{x}}_O^{d_i}(0)$ and given $\boldsymbol{x}_O(0)$, the performance functions $\rho_{ij}(t)$, $i \in \mathscr{F}$ and, $j = 1, \ldots, 6$ are designed such that (i) $-\rho_{ij}(0) < e_{ij}(0) < \rho_{ij}(0)$ and (ii) the desired transient and steady-state performance specifications are met. However, extensive simulation studies have revealed that the selection of the control gains $\boldsymbol{K}$, $\boldsymbol{\Gamma}_i$, $\boldsymbol{\Gamma}_{d_i}$, $i \in \mathscr{N}$ can have positive influence on the closed-loop system response in both the control input characteristics (magnitude and slew rate) and the evolution of the tracking errors. In particular, decreasing the gain values leads to slow convergence, which is improved when adopting higher values, enlarging however the control effort in both magnitude and rate. Thus, an additional fine tuning is needed in real scenarios to retain the required control input signals within the feasible range that can be implemented by the actuators.*

4.3.3   Simulation study

In this section, the theoretical findings of this work are verified in a dynamic simulation environment built in MATLAB®. We consider a scenario involving a 3D motion in a constrained workspace with static obstacles(see Figure 4.3). The object of interest was a pipe grid whose design parameters are given in Table 4.1. The initial and desired configurations of the object are $x_O^{init} = [-4, 0.5, 0.6, 0, 0, 0]^\top$ and $x_O^d = [10, 0, 0, 0, 0, 0]^\top$, respectively. Moreover, the calculation of the interaction force/torque vector λ_i, $i \in \mathcal{N}$ was performed following [10]. The cooperative transportation is performed by 4 UVMSs grasping the object at its corners. The blue UVMS acts as the leader. Thus, we assume that the desired object's configuration and the obstacles' position in the workspace are transferred to the leading UVMS beforehand. The obstacles are modeled as spheres (1 m radius) and are located in the workspace to complicate the transportation task of the object. In this respect, a Navigation Function is constructed following (4.21) to handle the aforementioned constrained workspace. Since, only the leading UVMS (blue) is aware of the obstacles' position in the workspace and the object's desired configuration, the followers will estimate it via the proposed algorithm (4.53), by simply observing the motion of

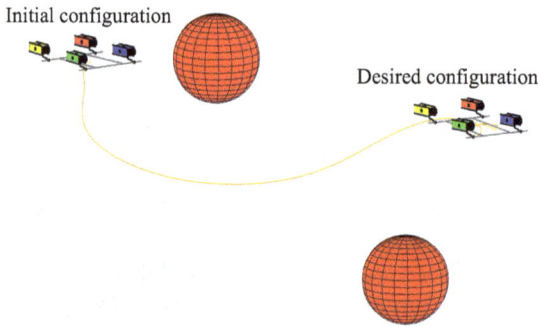

Figure 4.3   Four UVMSs transport a rigidly grasped object in a constrained workspace with static obstacles. Only the leading UVMS (indicated with blue color) is aware of the object's desired configuration and the obstacles' position in the workspace. A safe object trajectory in 3D space is indicated by orange color

Table 4.1   Object characteristics

Parameter	Value	Unit
Length	1.8	m
Pipe diameter	5	cm
Mass in air	1.5	kg

Table 4.2 Control gains of the proposed control scheme

Parameter	Value
M_{d_o}, $i \in \mathcal{N}$	$1 \cdot I_{6 \times 6}$
D_{d_o}, $i \in \mathcal{N}$	$1 \cdot I_{6 \times 6}$
K_{d_o}, $i \in \mathcal{N}$	$1 \cdot I_{6 \times 6}$
M_d, $i \in \mathcal{N}$	$1 \cdot I_{6 \times 6}$
D_d, $i \in \mathcal{N}$	$1 \cdot I_{6 \times 6}$
K_d, $i \in \mathcal{N}$	$1 \cdot I_{6 \times 6}$
k (see (4.21))	12
k_{NF} (see (4.22))	0.5
Γ_i, $i \in \mathcal{N}$	$10 \cdot I_{10 \times 10}$
Γ_{d_i}, $i \in \mathcal{N}$	$10 \cdot I_{6 \times 6}$

Table 4.3 Parameters of the proposed Estimator

Parameter	Value
k_j^i, $i \in \mathcal{F}$, $j = 1, \ldots, 6$	1.2
$\rho_{1,0}^i$, $i \in \mathcal{F}$	5
$\rho_{2,0}^i$, $i \in \mathcal{F}$	4
$\rho_{3,0}^i$, $i \in \mathcal{F}$	4
$\rho_{4,0}^i$, $i \in \mathcal{F}$	1
$\rho_{5,0}^i$, $i \in \mathcal{F}$	1
$\rho_{6,0}^i$, $i \in \mathcal{F}$	1
$\rho_{j,\infty}^i$, $i \in \mathcal{F}$, $j = 1, \ldots, 6$	0.03
λ	1

the object and without communicating explicitly with the leader. To test the robustness of the proposed scheme, in all subsequent simulations studies, the dynamics of the UVMS were affected by external disturbances in the form of slowly time varying sea currents modeled by the corresponding velocities $v_x^c = 0.3 \sin(\frac{\pi}{15}t) \frac{m}{s}$ and $v_y^c = 0.3 \cos(\frac{\pi}{15}t) \frac{m}{s}$. Finally, in all simulations, the control gains and the parameters of the proposed estimator were chosen as shown in Tables 4.2 and 4.3.

4.3.3.1 Simulation study A

The results are illustrated in Figures 4.4–4.8. The evolution of the system under the proposed methodology is given in Figure 4.4. It should be noticed that UVMSs have transported cooperatively the grasped object from the initial configuration to the desired one without colliding with obstacles. By observing the object's tracking error (Figure 4.5), it can be concluded that even under the influence of external disturbances, the errors in all directions converge very close to

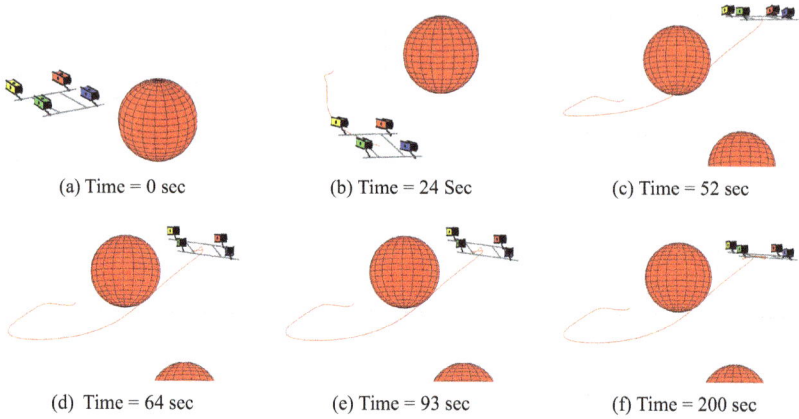

(a) Time = 0 sec   (b) Time = 24 Sec   (c) Time = 52 sec

(d) Time = 64 sec   (e) Time = 93 sec   (f) Time = 200 sec

Figure 4.4   Simulation study A: The evolution of the proposed methodology in six consecutive time instants

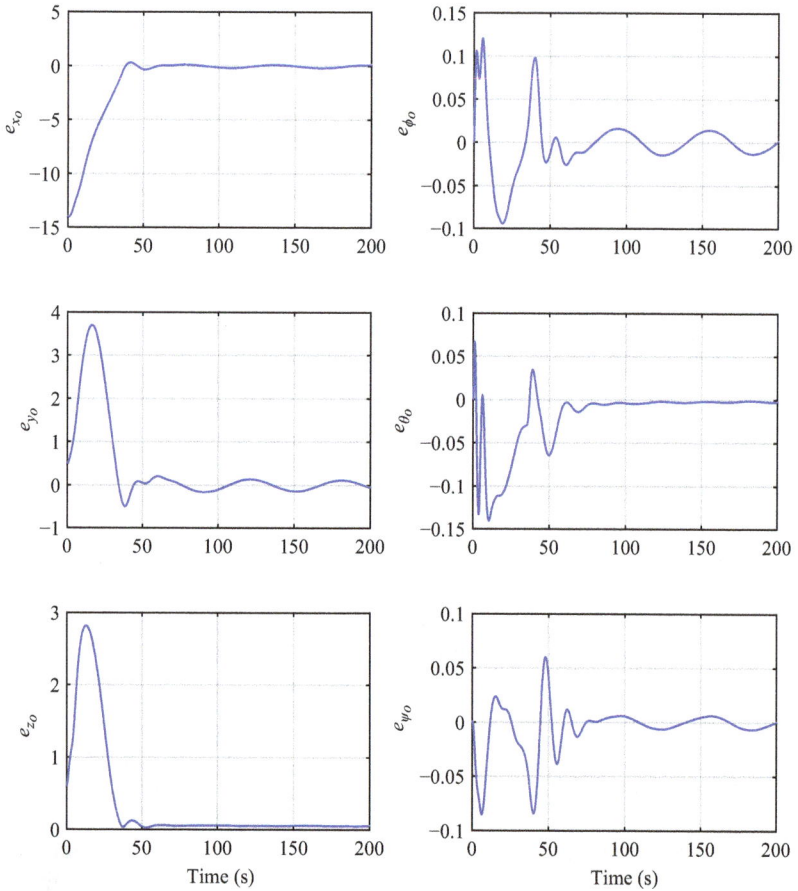

Figure 4.5   Simulation study A: The object tracking errors in all directions

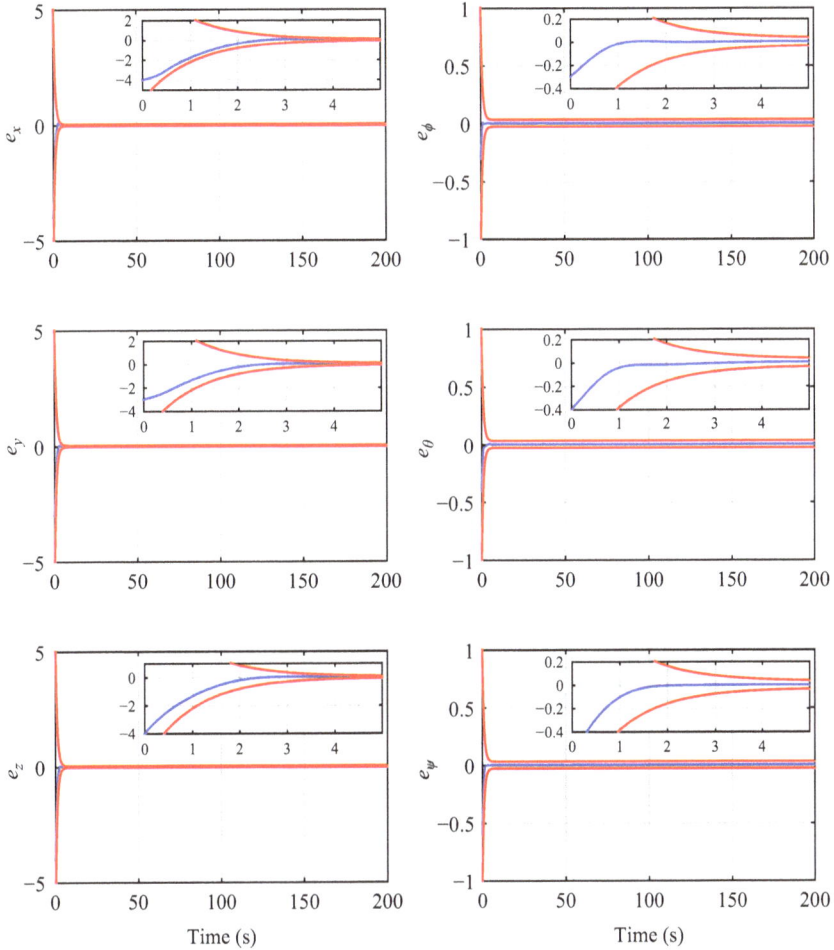

Figure 4.6   Simulation study A: The estimation errors along with the performance
bounds imposed by the proposed method. The estimation errors and
performance bounds are indicated by blue and red colors, respectively,
inside the box for each case

zero. The estimation errors of the proposed estimation scheme are presented in
Figure 4.6. It can be easily seen that the estimation errors smoothly converge to zero
and remain always within the performance envelope defined by the corresponding
performance functions as it was expected from the aforementioned theoretical analysis. The evolution of the Navigation Function potential is presented in Figure 4.7. The
value of Navigation Function remains strictly less than 1 during the simulation study,
which consequently means that obstacles have been successfully avoided. Moreover,
the task space control commands in Figure 4.8.

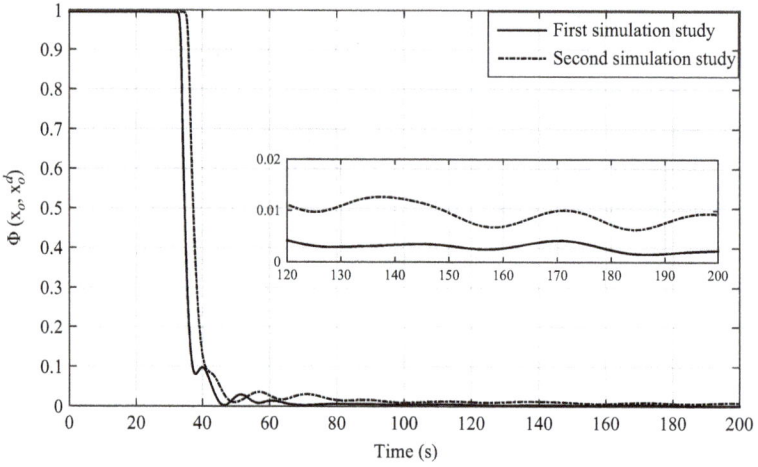

*Figure 4.7   Simulation studies A–B: The evolution of the Navigation Function
potential*

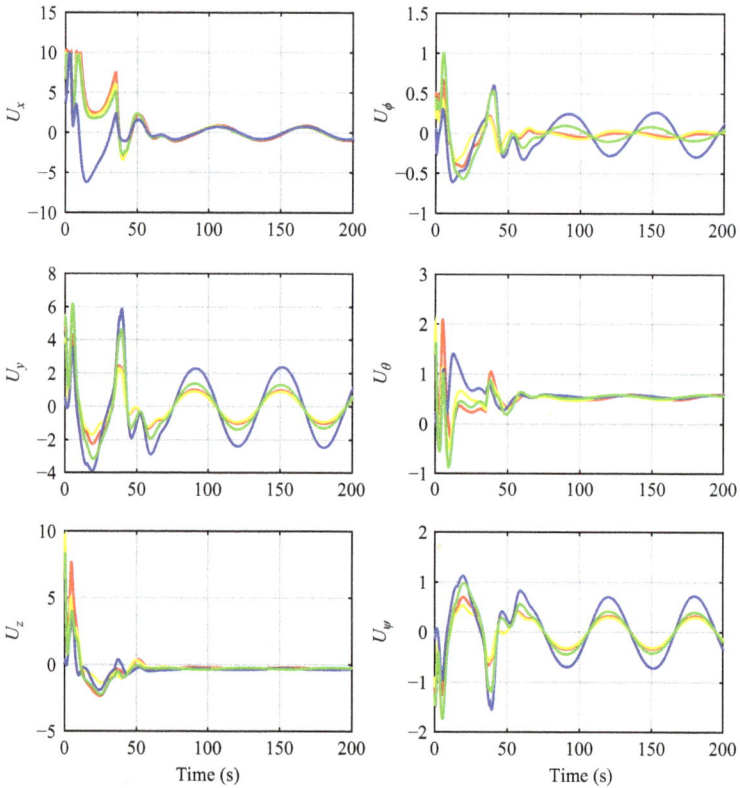

*Figure 4.8   Simulation study A: The evolution of control inputs. The corresponding
control inputs of the leading UVMS is indicated with blue color, while
the following UVMSs are indicated with red, yellow, and green colors,
respectively*

4.3.3.2   Simulation study B: Comparison with a centralized scheme

We considered exactly the same scenario, but instead a centralized control scheme was implemented to compare its response with the proposed scheme. More specifically, we incorporated the Navigation Function concept within the centralized control scheme presented in [9] (i.e., we modified the proportional term of the control scheme) to achieve obstacle avoidance. Observing the error trajectory tracking of Figure 4.9, it can be concluded that the system under the centralized control scheme reached to the desired configuration while avoiding the obstacles within the workspace. However, employing the aforementioned centralized scheme requires 92 variables (i.e., 20 state and velocity variables q_i and ζ_i, $i \in \mathcal{N}$ for each UVMS, as well as 12 state and velocity variables for the object) to be exchanged online among the robots. Considering floating point variables, this implementation requires the transfer of 2944 bytes in every control loop, and it is well known that the achievable

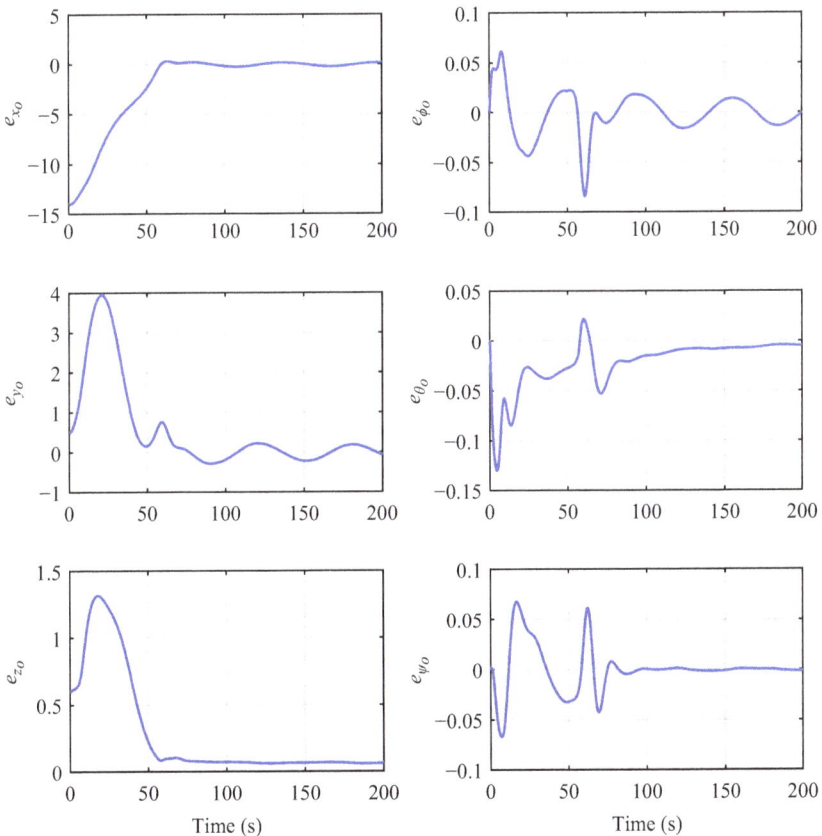

Figure 4.9   Simulation study B: The object tracking errors in all directions employing the centralized control scheme presented in [9]

exchange rate depends directly on the available bandwidth. Therefore, considering a full-duplex communication, to achieve a 10 Hz exchanging rate, which is an ordinary rate in underwater robotics, a modem with at least 30 k bytes s^{-1} bandwidth capability should be available, which is unreasonably demanding based on the current technology in underwater acoustic modems and the small number of cooperating robots. For instance, high-speed acoustic modems provided by *Evologics* company (e.g., S2C 48/78, S2C 40/80, S2C 42/65) supply data transfer rates up to 31 k bits s^{-1}. Although recent advances in underwater acoustic modem technology accomplish continuously higher rate of communication, the number of participating robots remains still small owing to the limited bandwidth. For instance, in the considered scenario, if the number of cooperating robots increases by one (i.e., 5 cooperative UVMSs), the bandwidth requirements will increase respectively up to 35 kb s^{-1}, which would also raise significantly the mission costs. On the contrary, it should be notice that the proposed control strategy imposes no restriction regarding the underwater communication bandwidth.

4.4   Distributed predictive control for multi-UVMS collaborative object transportation under implicit communication

This section presents a distributed control approach for a team of UVMSs to cooperatively transport an object while addressing key challenges, including kinematic and representation singularities, obstacles within the workspace, joint limits, and control input saturation. Specifically, by leveraging the coupled dynamics between the robots and the object and employing appropriate load-sharing coefficients, we design a Nonlinear Model Predictive Control (NMPC) [55] scheme for each UVMS to facilitate cooperative object transportation by steering it along a computed feasible path within the workspace's feasible region. The construction of this feasible path is based on the Navigation Function concept [43], as described in Remark 2, which is employed here to achieve distributed consensus on the object's safe trajectory while ensuring collision avoidance with obstacles and the workspace boundaries. The proposed control scheme distributes the load among UVMSs based on their individual payload capacities. Additionally, the feedback mechanism relies solely on local measurements from each UVMS, eliminating the need for explicit online data exchange between robots. Specifically, each UVMS computes its control signals independently, using only onboard sensor data – such as position and velocity measurements obtained through sensor fusion from various sensors (e.g., IMU, USBL, and DVL), thereby avoiding the complexities of inter-robot explicit communication. This approach not only reduces communication bandwidth requirements but also significantly enhances the robustness of the cooperative control scheme. Moreover, it mitigates limitations imposed by acoustic communication bandwidth, such as constraints on the number of participating UVMSs. In contrast to the approach discussed in the previous section (4.3) and existing cooperative manipulation methods in the literature that utilize decentralized control schemes based solely on local sensory information (e.g., [21,56,57]), most prior studies assume that each robot is equipped with a force/torque sensor at its end-effector to measure interaction forces and torques with the object.

However, such reliance on force/torque sensors can degrade performance due to sensor noise [58–60]. In contrast, the approach proposed in this section circumvents this issue by eliminating the need for force/torque sensors on the agents' end-effectors. Furthermore, many existing studies on cooperative manipulation overlook critical robotic mobile manipulator properties, such as singular kinematic configurations of the Jacobian matrix and joint limit constraints. Our approach explicitly addresses these issues, ensuring improved performance and operational robustness.

4.4.1   Problem formulation

Consider a team of N UVMSs collaboratively manipulating a shared object[¶] within a constrained workspace containing static obstacles. Each UVMS is assumed to be fully actuated at its end-effector frame, enabling it to apply arbitrary forces and torques on the object in any direction. Moreover, each UVMS is equipped with onboard sensors that provide position and velocity measurements. These measurements are obtained through sensor fusion techniques that integrate data from various sources, such as USBL, IMU, DVL, and depth sensors. Additionally, the geometric parameters of both the UVMSs and the commonly grasped object are assumed to be known.

   Problem statement: Given N UVMSs operating in a constrained workspace $\mathcal{W}$ while rigidly holding a common object, and given a desired object configuration x_O^d, the goal is to develop distributed control protocols τ_i, $i \in \mathcal{N}$ that guide the robotic team safely to the target configuration while satisfying the following specifications: (i) ensure obstacle avoidance, preventing collisions with both static obstacles and the workspace boundaries; (ii) operate without imposing strict constraints on underwater communication bandwidth, allowing decentralized execution; (iii) avoid singular configurations to maintain control feasibility and system stability; (iv) achieve distributed consensus on a mutually agreed trajectory for the shared object.

4.4.2   Control methodology

The proposed approach builds on designing an NMPC scheme for the system of UVMSs and the object. Thanks to the novel formulation of the problem, the proposed control strategy relieves the team of robots from intense inter-robot communication during the execution of the collaborative tasks. This, consequently, increases significantly the robustness of the cooperative scheme and furthermore avoids any restrictions imposed by the acoustic communication bandwidth (e.g., the number of participating UVMSs). In this way, we assume that all UVMSs are aware of both the desired configuration of the object and of the obstacles position in the workspace[**]. Thus, the control objective is to navigate the overall formation toward the goal configuration while avoiding collisions with the static obstacles that lie within the workspace. Toward this direction, the concept of Navigation Functions [43] will be

---

[¶]The end-effector frame of each UVMS remains fixed relative to the object's body-fixed frame.

[**]The desired configuration of the object can be transmitted to each UVMS before executing the cooperation task.

incorporated to deal with consensus on a mutually agreed trajectory of the commonly object. That overall dynamics are decoupled next among the object and the robots in N parts accounting individually for each UVMS $i \in \mathscr{N}$ by using certain load sharing coefficients. Each UVMS at each sampling time solves a NMPC subject to its corresponding part of that overall dynamics and a number of inequality constraints that incorporate its internal limitations (e.g., joint limits, kinematic and representation singularities, collision between the arm and the base, and manipulability).

By substituting (4.13) into (4.17) and ignoring the disturbances and uncertainties terms, one obtains the following:

$$\boldsymbol{\lambda} = \boldsymbol{G}^\top \left[\boldsymbol{u} - \boldsymbol{M}(\boldsymbol{q})\dot{\boldsymbol{v}} - \boldsymbol{C}(\dot{\boldsymbol{q}}, \boldsymbol{q})\boldsymbol{v} - \boldsymbol{D}(\dot{\boldsymbol{q}}, \boldsymbol{q})\boldsymbol{v} - \boldsymbol{g}(\boldsymbol{q}) \right] \tag{4.63}$$

which, after substituting (4.7), (4.8), and (4.14) and rearranging terms, yields the overall system coupled dynamics:

$$\widetilde{\boldsymbol{M}}(\tilde{\boldsymbol{q}}_{ov})\dot{\boldsymbol{v}}_O + \widetilde{\boldsymbol{C}}(\tilde{\boldsymbol{q}}_{ov})\boldsymbol{v}_O + \widetilde{\boldsymbol{D}}(\tilde{\boldsymbol{q}}_{ov})\boldsymbol{v}_O + \widetilde{\boldsymbol{g}}(\tilde{\boldsymbol{q}}_{ov}) = \boldsymbol{G}^\top \boldsymbol{u} \tag{4.64}$$

where $\tilde{\boldsymbol{q}}_{ov} = [\boldsymbol{q}^\top, \dot{\boldsymbol{q}}^\top, \boldsymbol{x}_O^\top, \boldsymbol{v}_O^\top]^\top$ and

$$\widetilde{\boldsymbol{M}}(\tilde{\boldsymbol{q}}_{ov}) = \boldsymbol{M}_O(\boldsymbol{x}_O) + \boldsymbol{G}^\top \boldsymbol{M}(\boldsymbol{q})\boldsymbol{G}$$

$$\widetilde{\boldsymbol{C}}(\tilde{\boldsymbol{q}}_{ov}) = \boldsymbol{C}_O(\boldsymbol{v}_O, \boldsymbol{x}_O) + \boldsymbol{G}^\top \boldsymbol{M}(\boldsymbol{q})\dot{\boldsymbol{G}}(\dot{\boldsymbol{q}}, \boldsymbol{q}) + \boldsymbol{G}^\top \boldsymbol{C}(\dot{\boldsymbol{q}}, \boldsymbol{q})\boldsymbol{G}$$

$$\widetilde{\boldsymbol{D}}(\tilde{\boldsymbol{q}}_{ov}) = \boldsymbol{D}_O(\boldsymbol{v}_O, \boldsymbol{x}_O) + \boldsymbol{G}^\top \boldsymbol{D}(\dot{\boldsymbol{q}}, \boldsymbol{q})\boldsymbol{G}$$

$$\widetilde{\boldsymbol{g}}(\tilde{\boldsymbol{q}}_{ov}) = \boldsymbol{g}_O(\boldsymbol{x}_O) + \boldsymbol{G}^\top \boldsymbol{g}(\boldsymbol{q})$$

Now, consider the design constants c_i, $i \in \mathscr{N}$ satisfying

$$c_i \in (0, 1), \forall i \in \mathscr{N} \qquad \text{and} \qquad \sum_{i \in \mathscr{N}} c_i = 1, \tag{4.65}$$

that we introduce to act as the load sharing coefficients for the team of UVMS. In view of (4.65), the object dynamics (4.15) can be rewritten as follows:

$$\sum_{i \in \mathscr{N}} c_i \left\{ \boldsymbol{M}_O(\boldsymbol{x}_O)\dot{\boldsymbol{v}}_O + \boldsymbol{C}_O(\boldsymbol{x}_O, \boldsymbol{v}_O)\boldsymbol{v}_O + \boldsymbol{D}_O(\boldsymbol{x}_O, \boldsymbol{v}_O)\boldsymbol{v}_O + \boldsymbol{g}_O(\boldsymbol{x}_O) \right\} = \sum_{i \in \mathscr{N}} \boldsymbol{J}_{O_i}^\top \boldsymbol{\lambda}_i \tag{4.66}$$

from which, by employing (4.3), (4.10), (4.7), (4.11), (4.5), and (4.8) and after straightforward algebraic manipulations, we obtain the coupled dynamics:

$$\sum_{i \in \mathscr{N}} \left\{ \widetilde{\boldsymbol{M}}_i(\boldsymbol{q}_i)\ddot{\boldsymbol{q}}_i + \widetilde{\boldsymbol{C}}_i(\dot{\boldsymbol{q}}_i, \boldsymbol{q}_i)\dot{\boldsymbol{q}}_i + \widetilde{\boldsymbol{D}}_i(\dot{\boldsymbol{q}}_i, \boldsymbol{q}_i)\dot{\boldsymbol{q}}_i + \widetilde{\boldsymbol{g}}_i(\boldsymbol{q}_i) \right\} = \sum_{i \in \mathscr{N}} \boldsymbol{J}_{O_i}^\top \boldsymbol{u}_i \tag{4.67}$$

where

$$\widetilde{\boldsymbol{M}}_i(\boldsymbol{q}_i) = c_i \boldsymbol{M}_O \boldsymbol{J}_{io} \boldsymbol{J}_i + \boldsymbol{J}_{O_i}^\top \boldsymbol{M}_i \boldsymbol{J}_i$$

$$\widetilde{\boldsymbol{C}}_i(\dot{\boldsymbol{q}}_i, \boldsymbol{q}_i) = c_i \left[\boldsymbol{M}_O \boldsymbol{J}_{io} \dot{\boldsymbol{J}}_i + \boldsymbol{M}_O \dot{\boldsymbol{J}}_{io} \boldsymbol{J}_i + \boldsymbol{C}_O \boldsymbol{J}_{io} \boldsymbol{J}_i \right] + \boldsymbol{J}_{O_i}^\top \left[\boldsymbol{M}_i \dot{\boldsymbol{J}}_i + \boldsymbol{C}_i \boldsymbol{J}_i \right]$$

$$\widetilde{\boldsymbol{D}}_i(\dot{\boldsymbol{q}}_i, \boldsymbol{q}_i) = c_i \boldsymbol{D}_O \boldsymbol{J}_{io} \boldsymbol{J}_i + \boldsymbol{J}_{O_i}^\top \boldsymbol{D}_i \boldsymbol{J}_i$$

$$\widetilde{\boldsymbol{g}}_i(\boldsymbol{q}_i) = c_i \boldsymbol{g}_O + \boldsymbol{J}_{O_i}^\top \boldsymbol{g}_i$$

which is the distributed version of (4.64), since for each UVMS, it is based only individually on its locally measurements (i.e., q_i and $\dot{q}_i$). Now, by using the notation $x_i = [q_i^\top, \dot{q}_i^\top]^\top$, the decentralized dynamics of each UVMS based on (4.67), can be written as the compact form:

$$\dot{x}_i = f_i(x_i, u_i) = \begin{bmatrix} f_{i_1}(x_i) \\ f_{i_2}(x_i, \quad u_i) \end{bmatrix}, i \in \mathcal{N} \tag{4.68}$$

where

$$f_{i_1}(x_i) = \dot{q}_i$$

$$f_{i_2}(x_i, u_i) = \tilde{M}_i^{\#}(q_i)\left(J_{O_i}^\top(q_i)u_i - \tilde{C}_i(\dot{q}_i, q_i)\dot{q}_i - \tilde{D}_i(\dot{q}_i, q_i)\dot{q}_i - \tilde{g}_i(q_i)\right)$$

with

$$\tilde{M}_i^{\#}(q_i) = \tilde{M}_i(q_i)\left[\tilde{M}_i(q_i)\tilde{M}_i^\top(q_i)\right]^{-1}$$

State constraints

Here, we assume that the UVMS must avoid various constraints that can be modified as state constraints of the system. More specifically, joint limits and singularity avoidance should be satisfied by each UVMS. First recall the Jacobian $J_{v,i}$ (see (4.4)). Note that this Jacobian becomes singular at representation singularities, when $\theta_{v_i} = \pm\frac{\pi}{2}$ and $J_i(q_i)$ becomes singular at kinematic singularities defined by the set

$$Q_{s_i} = \{q_i \in \mathbb{R}^{n_i} : \det(J_i(q_i)[J_i(q_i)]^\top) \geq \varepsilon\}, \quad i \in \mathcal{N}. \tag{4.69}$$

with ε to be a small positive number. Moreover, recall the Jacobian $J_O'(\eta_{2,o})$ in (4.16). This Jacobian is singular when $\theta_O = \pm\frac{\pi}{2}$ and the object representation $J_O(\eta_{2,o}) = diag\{I_3, J_O(\eta_{2,o})\}$ becomes singular at kinematic singularities defined by the set

$$O_s = \{x_O \in \mathbb{R}^6 : \det(J_O'(\eta_{2,o})[J_O'(\eta_{2,o})]^\top) \geq \varepsilon\}. \tag{4.70}$$

where ε is a small positive number. These requirements are captured by the state constraint set X_i of the system, given by

$$x_i(t) \in X_i \subset \mathbb{R}^{2n_i} \tag{4.71}$$

which is formed by the following constraints:

$$\theta_O(t) \in \left(-\frac{\pi}{2}, \frac{\pi}{2}\right) \tag{4.72}$$

$$q_i \in \mathbb{R}^{n_i} \backslash \left(Q_{s_i}(q_i) \cup Q_{l_i}(q_i)\right), i \in \mathcal{N} \tag{4.73}$$

$$|\dot{q}_{k_i}| \leq \bar{\dot{q}}_{k_i}, \forall k \in \{1, \dots, n\}, i \in \mathcal{N} \tag{4.74}$$

where $Q_{s_i}(q_i)$ is the set of singular position of the system (4.69), and $Q_{l_i}(q_i)$ is the set of manipulator's joint limits defined as follows:

$$Q_{l_i}(q_i) = \{q_i \in \mathbb{R}^{n_i} : |q_{k_i}| \leq \bar{q}_i\}, \forall k \in \{1, \dots, n_i\}, i \in \mathcal{N} \tag{4.75}$$

where $\bar{q}_{k_i}$ is the limit bound for the corresponding joint q_{k_i}, $k \in \{1,\ldots,n\}, i \in \mathcal{N}$. Moreover, $\dot{\bar{q}}_{k_i}$ is the upper value for the joint velocity $\dot{q}_{k_i}$, $k \in \{1,\ldots,n\}, i \in \mathcal{N}$. Therefore, the set X_i captures all the state constraints of the systems (4.68), i.e., singularity avoidance, as well as joint limits limitations.

Remark 5. *It is important to note that collision avoidance between the entire system (UVMS and the object) and obstacles within the workspace is achieved similarly to the approach described in the previous section (see Figure 4.2). In particular, this is accomplished by computing a safe object trajectory and velocity using the concept of Navigation Functions [43], as detailed in Remark 2.*

Input Constraints
The actuation of the vehicle body and the manipulator is generated by the thrusters and servo, respectively. Hence, the input constraints for τ_{k_i}, $k \in \{1,\ldots,\tau_n\}, i \in \mathcal{N}$, with τ_n being the number of actuated joints, can be given as follows:

$$\|\tau_i\| \leq \bar{\tau}_i \Leftrightarrow \|J_i(q_i)^\top u_i\| \leq \bar{\tau}_i$$

where $\bar{\tau}_i$ is a vector including corresponding limit bound for each actuated joint τ_{k_i}, $k \in \{1,\ldots,\tau_{n_i}\}, i \in \mathcal{N}$. Therefore, we can define the control input set T_i:

$$\tau_i(t) \in T_i \subset \mathbb{R}^{\tau_{n_i}} \tag{4.76}$$

with

$$T_i = \{\tau_i \in \mathbb{R}^{\tau_{n_i}} : \|J_i(q_i)^\top u_i\| \leq \bar{\tau}_i, \forall x_i \in X_i\}$$

4.4.2.1 Control design
Here, by considering the Navigation Function as described in Remark 2, we define the object's desired motion profile similar to (4.22) as follows:

$$v_O^d(t) = -K_{NF}J_O'(\eta_{2,o})\nabla_{x_O}\phi_O(x_O(t), x_O^d) \tag{4.77}$$

where $K_{NF} > 0$ is a positive gain. Now let us define a sequence of sampling time $\{t_j\}_{j\geq 0}$ with a constant sampling time $h > 0$ with $h < T_p$ for the system such that

$$t_{j+1} = t_j + h, \forall j \geq 0 \tag{4.78}$$

Therefore, since all UVMSs $i \in \mathcal{N}$ are aware of both the desired configuration of the object and of the obstacles position in the workspace, given the current position of the object $x_O(t_i)$ and $v_O(t_j)$ at the time t_j, they can propagate for time interval $s \in [t_j, t_j + T_P]$, where T_P is the prediction horizon, a map of desired trajectory and velocity of the object based on (4.21) and (4.77), denoted as $x_O^d(s)$ and $v_O^d(s)$, $s \in [t_j, t_j + T_P]$, which will be used in the subsequent analysis. Therefore, given the current position and velocity of the object at sampling time j denoted by $x_O(t_j)$ and $v_O(t_j)$, respectively, each UVMS $i \in \mathcal{N}$ for a time interval $s \in [t_j, t_j + T_P]$, where T_P is a prediction horizon and based on (4.21), (4.77) and (4.78), can propagate/calculate a map of desired trajectory and velocity for the object denoted by $x_O^d(s)$ and $v_O^d(s)$, respectively. As will be explained in the sequel, at each sampling time, UVMS $i \in \mathcal{N}$ solves its corresponding part of the dynamics (4.67) via an NMPC scheme subject to its dynamics (4.68) and a number of inequality constraints. More specifically, the control objective for each UVMS $i \in \mathcal{N}$ is to follow these desired trajectory and velocity, while respecting the state constraints (4.72)–(4.74), as well as the input constraints (4.76). In particular, in sampled data NMPC, a Finite Horizon Optimal Control Problem (FHOCP) is solved at discrete sampling time instants t_j based on the current state measurements $x_i(t_j)$, $i \in \mathcal{N}$. For UVMS i, $i \in \mathcal{N}$, the open-loop input signal applied between the sampling instants is given by the solution of the following FHOCP:

$$\min_{\hat{\tau}_i(\cdot)} J_i(x(t_j), \hat{\tau}_i(\cdot)) = \tag{4.79}$$

$$\min_{\hat{\tau}_i(\cdot)} \left\{ \int_{t_j}^{t_j + T_p} \left[F_i\big(\hat{x}_O(s), \hat{v}_O(s), \hat{\tau}_i(s)\big) \right] ds + E_i\big(\hat{x}_O(t_j + T_P), \hat{v}_O(t_j + T_P)\big) \right\}$$

subject to:

$$\dot{\hat{x}}_i(s) = f_i(\hat{x}_i(s), \hat{u}_i(s)), \quad \hat{x}_i(t_j) = x_i(t_j), \tag{4.80}$$

$$\hat{\tau}_i(s) = J_i^\top(\hat{q}_i)\hat{u}_i + \tau_{i0}(q_i), \quad s \in [t_j, t_j + T_P] \tag{4.81}$$

$$\hat{x}_O(s) = \mathscr{F}(\hat{q}_i(s)) - \begin{bmatrix} {}^I R_O l_i \\ \alpha_i \end{bmatrix}, \quad s \in [t_j, t_j + T_P], \tag{4.82}$$

$$\hat{v}_O(s) = J_{i_O} J_i(\hat{q}_i(s))\dot{\hat{q}}_i(s), \quad s \in [t_j, t_j + T_P], \tag{4.83}$$

$$\hat{x}_i(s) \in X_i, \quad s \in [t_j, t_j + T_P], \tag{4.84}$$

$$\hat{\tau}_i(s) \in T_i, \quad s \in [t_j, t_j + T_P], \tag{4.85}$$

$$\hat{x}(t_j + T_P) \in \mathscr{E}_f \tag{4.86}$$

where $\mathscr{E}_f$ is a terminal region around the desired trajectory profile, and F and E are the running and terminal cost function, respectively. To distinguish the predicted variables (i.e., internal to the controller), we use the double subscript notation $(\hat{\cdot})$ corresponding to the system (4.80). This means that $\hat{x}_i(s)$, $s \in [t_j, t_j + T_P]$ is the solution of (4.68) based on the measurement of the state at time instance t_j (i.e., $x_i(t_j)$, provided by the on–board navigation system) while applying a trajectory of inputs (i.e., $\hat{u}_i(s)$, $s \in [t_j, t_j + T_P]$). Notice that we use this notation to account for the mismatch between the predicted values of the system and the actual closed-loop values. The cost function $F_i(\cdot)$ and the terminal cost $E(\cdot)$ are both quadratic and given as follows:

$$F_i\big(\hat{x}_O(s), \hat{v}_O(s), \hat{\tau}_f(s)\big) = [\hat{x}_O^\top(s), \hat{v}_O^\top(s)]Q[\hat{x}_O^\top(s), \hat{v}_O^\top(s)]^\top + \hat{\tau}_f^\top(s)R\hat{\tau}_f(s)$$

$$E_i(\hat{x}_O, \hat{v}_O) = [\hat{x}_O^\top(s), \hat{v}_O^\top(s)]P[\hat{x}_O^\top(s), \hat{v}_O^\top(s)]^\top$$

with $P \in \mathbb{R}^{12 \times 12}, Q \in \mathbb{R}^{12 \times 12}$, and $R \in \mathbb{R}^{\tau_{n_i} \times \tau_{n_i}}$ being symmetric and positive definite matrices to be appropriately tuned. The terminal set $\mathscr{E}_f \subset X_i$ is chosen as follows: $\mathscr{E}_f = \{x_i \in X_i : E_i(\cdot) \leq \varepsilon\}$, where ε is an arbitrarily and positive small constant to be appropriately tuned. The solution of FHOCP (4.79)–(4.86) at time t_j provides an optimal control input trajectory denoted by $\hat{\boldsymbol{\tau}}_i^*(s; \boldsymbol{x}(t_j))$, $s \in [t_j, t_j + T_P]$. This control input is then applied to the system until the next sampling time t_{j+1}:

$$\boldsymbol{\tau}_i(s; \boldsymbol{x}(t_j)) = \hat{\boldsymbol{\tau}}_i^*(s; \boldsymbol{x}(t_j)), \ s \in [t_j, t_j + h] \tag{4.87}$$

At time $t_{j+1} = t_j + h$, a new finite horizon optimal control problem is solved in the same manner, leading to a receding horizon approach. Notice that the control input $\boldsymbol{\tau}_i(\cdot)$ is of feedback form, since it is recalculated at each sampling instant based on the then-current state. The pseudo-code description of the proposed real-time control scheme for UVMS i, $i \in \mathcal{N}$ is given in *Algorithm 1*:

Algorithm 1 Real time MPC algorithm:

1: **Triggering time**　　　 ▷ At time instance t_j UVMS i measures its state vector $\boldsymbol{x}_i$
2:　 $\boldsymbol{p}_i(t_j) \leftarrow$ eq.(4.2)　　　　　　　　　　　　　　　 ▷ calculates its EE pose
3:　 $\boldsymbol{v}_i(t_j) \leftarrow$ eq.(4.3)　　　　　　　　　　　　　　　 ▷ calculates its EE velocity
4:　 $\boldsymbol{v}_O(t_j), \boldsymbol{x}_O(t_j) \leftarrow$ eq.(4.6) $-$ (4.7)　　　　 ▷ calculates object pose and velocity
5:　 $\boldsymbol{x}_O^d(s), \boldsymbol{v}_O^d(s), \ s \in [t_j, t_j + T_P] \leftarrow$ eq.(4.21), (4.77)　　　　　 ▷ propagates for the time interval s, $s \in [t_j, t_j + T_P]$ a map of safe/desired trajectory and velocity of the object
6: $\hat{\boldsymbol{\tau}}_i^*(s; \boldsymbol{x}(t_j)), \ s \in [t_j, t_j + T_P] \leftarrow$ FHOCP$(\boldsymbol{x}_i(t_j))$　　　　　　　　　　　 ▷
　 Run FHOCP of (4.79)-(4.86). The solution is a optimal control input trajectory for the time interval $[t_j, t_j + T_P]$.
7: **for** $s \in [t_j, t_j + h]$ **do**
8:　　 Apply the $\boldsymbol{\tau}_i(s; \boldsymbol{x}(t_j)) = \hat{\boldsymbol{\tau}}_i^*(s; \boldsymbol{x}(t_j))$ to the UVMS.
9: **end for**
10:　 $t_{j+1} = t_j + h$　　　　　　　　　　　　　　　 ▷ The next triggering time
11: **goto** *Triggering*.

4.4.3   Results

In this section, the effectiveness of the proposed control strategy is verified through both simulation and experimental studies. The simulation results were conducted using a dynamic simulation environment based on the UwSim simulator [61] running on the Robot Operating System (ROS) [62]. The experimental results were conducted in a test tank employing two small UVMSs with in-house built underwater manipulators.

4.4.3.1   Simulation study

This study examines a scenario involving two UVMSs performing 3D motion while collaboratively transporting a bar-shaped object in a constrained workspace containing static obstacles (see Figure 4.10). Each UVMS follows the same structural design, consisting of an underwater robotic vehicle equipped with a compact four-degree-of-freedom manipulator mounted at the bow (see Figure 4.10). The dynamic properties of the vehicle have been identified using a dedicated identification scheme [63], while the manipulator and object parameters were derived from CAD data. The complete state vector of the vehicle, including 3D position, orientation, and velocity, is

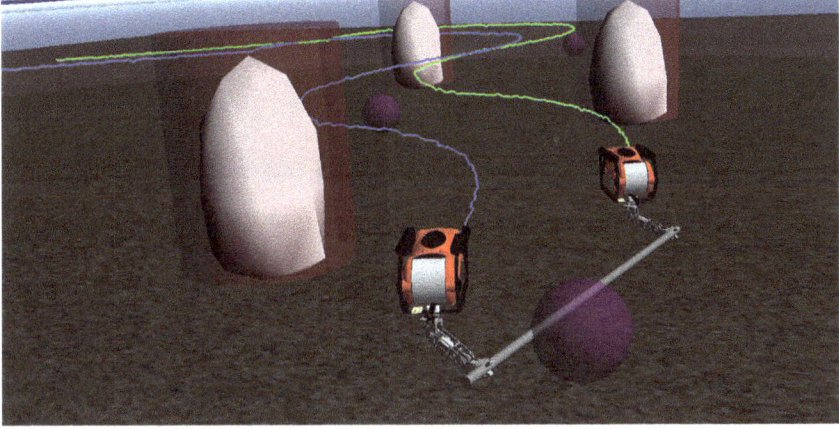

*Figure 4.10   Simulation environment: Cooperatively object transportation using
two UVMSs inside a constrained workspace including obstacles*

obtained through a sensor fusion and state estimation module, as detailed in previous
research [63]. The constrained NMPC utilized in this study has been implemented
using the NLopt Optimization library [64].

In the following simulation, the UVMS team is tasked with tracking a predefined set of waypoints while avoiding obstacles in the workspace. The obstacles'
positions w.r.t. the inertial frame $\mathscr{I}$ in the x–y plane are given as $x_{obs1} = [4, \ -4.5]$,
$x{obs2} = [9, \ -1.5]$, and $x{obs3} = [9, \ 5]$. Each obstacle is modeled as a cylinder
with a radius of $r\pi_i = 0.6m$, $i = 1, 2, 3$ and represented, along with the workspace
boundaries, according to the spherical world model representation. The UVMSs
are enclosed within a bounding sphere $\mathscr{B}(pi, \bar{r})$ of radius $\bar{r} = 1$ m, covering all
possible configurations. The Navigation Function (4.21)–(4.77), as described in
Remark 2, was designed with a gain of $KNF = 0.5$. To ensure operational constraints,
vehicle velocities are limited to $0.5 \, \text{ms}^{-1}$ for translation and $0.1 \, \text{rad} \, \text{s}^{-1}$ for rotation (4.74). Similarly, manipulator joint velocities remain within $(-0.1, 0.1) \, \text{rads}^{-1}$,
while joint positions are constrained to $(-2, 2)$ rad (4.75). Control input saturations for both the vehicle and the manipulator are set as $\bar{\tau}_v = 10N$ and $\bar{\tau}_m = 2N$,
respectively (4.76). The sampling time and prediction horizon are $h = 0.12$ s and
$T_p = 5 \times h = 0.6$ s. The control weight matrices and load-sharing coefficients are
defined as $Px = Qx = 0.8 \cdot I6 \times 6$, $R = 0.3 \cdot I8 \times 8$, $Qv = 0.4 \cdot I6 \times 6$, and $c_1 =
c_2 = 0.5$. The object starts at $xO = [-0.7, 0, 0.72, 0.04, \ -0.07, 0]$ and follows
three waypoints: $xO_1{}^d = [6, \ -6, 0.85, 0, 0, 0], xO_2{}^d = [7.5, 1.5, 0.78, 0, 0, 0]$,
and $xO_3{}^d = [12, 6.5, 0.65, 0, 0, 0]$, presenting a challenging navigation scenario
given the obstacles (see Figures 4.11 and 4.10). The results are depicted in Figures 4.11–4.13. The system trajectory and object coordinates over time are shown
in Figures 4.11 and 4.12, demonstrating that UVMSs successfully transported the
object while avoiding obstacles. The determinant evolution of $det(J(q)[J(q)]^{\top})$ (see
(4.69) and (4.73)) is presented in Figure 4.13, remaining positive throughout the task.
Additionally, Figure 4.14 illustrates the system velocity and joint states, confirming

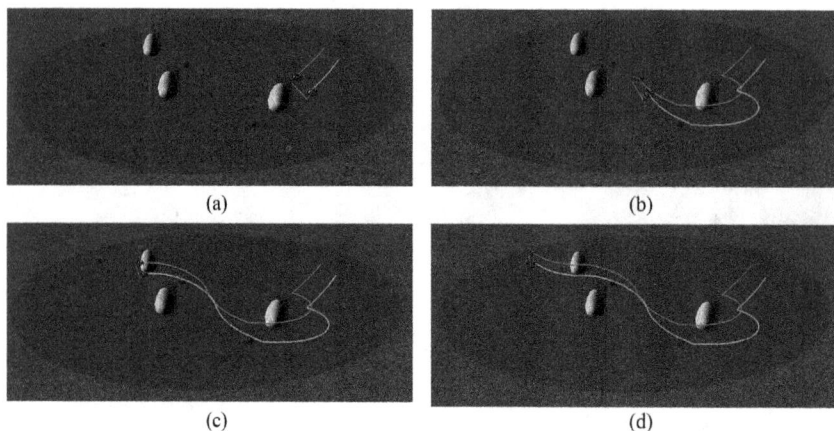

Figure 4.11   Simulation study: the evolution of the proposed methodology in 4 consecutive time instants

that all values remained within their feasible regions, ensuring full compliance with system constraints.

4.4.3.2   Experimental study

This section demonstrates the efficacy of the proposed cooperative control scheme via a set of real-time experiments employing two small UVMSs equipped with in-house built underwater manipulators, carrying a common object (see Figure 4.15). In particular, Section 4.4.3.3 introduces the experimental setup and Section 4.4.3.4 presents the detailed results of two cases of experimental studies employing the proposed controller.

4.4.3.3   Experimental setup

The experiments were carried out inside the *NTUA, Control Systems Lab* test tank, with dimensions $5\,\text{m} \times 3\,\text{m} \times 1.5\,\text{m}$ (Figure 4.15). The bottom of the tank is covered by a custom-made poster with various visual features and markers. In the following experiments, the team of UVMSs consists of two small ROVs equipped with the same custom made small waterproof manipulator (see Figure 4.15). More specifically, a 4 DoF Seabotix LBV (red color), actuated in Surge, Sway, Heave, and Yaw and a 3 DoF VideoRay PRO (yellow color) effective only in Surge, Heave, and Yaw motion were used as the vehicle bases in this work (see Figure 4.15). Notice that the 3 DoF VideoRay robot is under-actuated along the Sway axis. This intuitively means that while the combined vehicle–manipulator system is full-actuated at the end-effector frame, the vehicle base remains under-actuated along the Sway body frame axis. Thanks to the nature of the optimization procedure, the aforementioned difficulty is handled within the FHOCP (4.79)–(4.86), which results in a solution that combines the optimal vehicle and manipulator motion to achieve the desired movement at the end-effector frame.

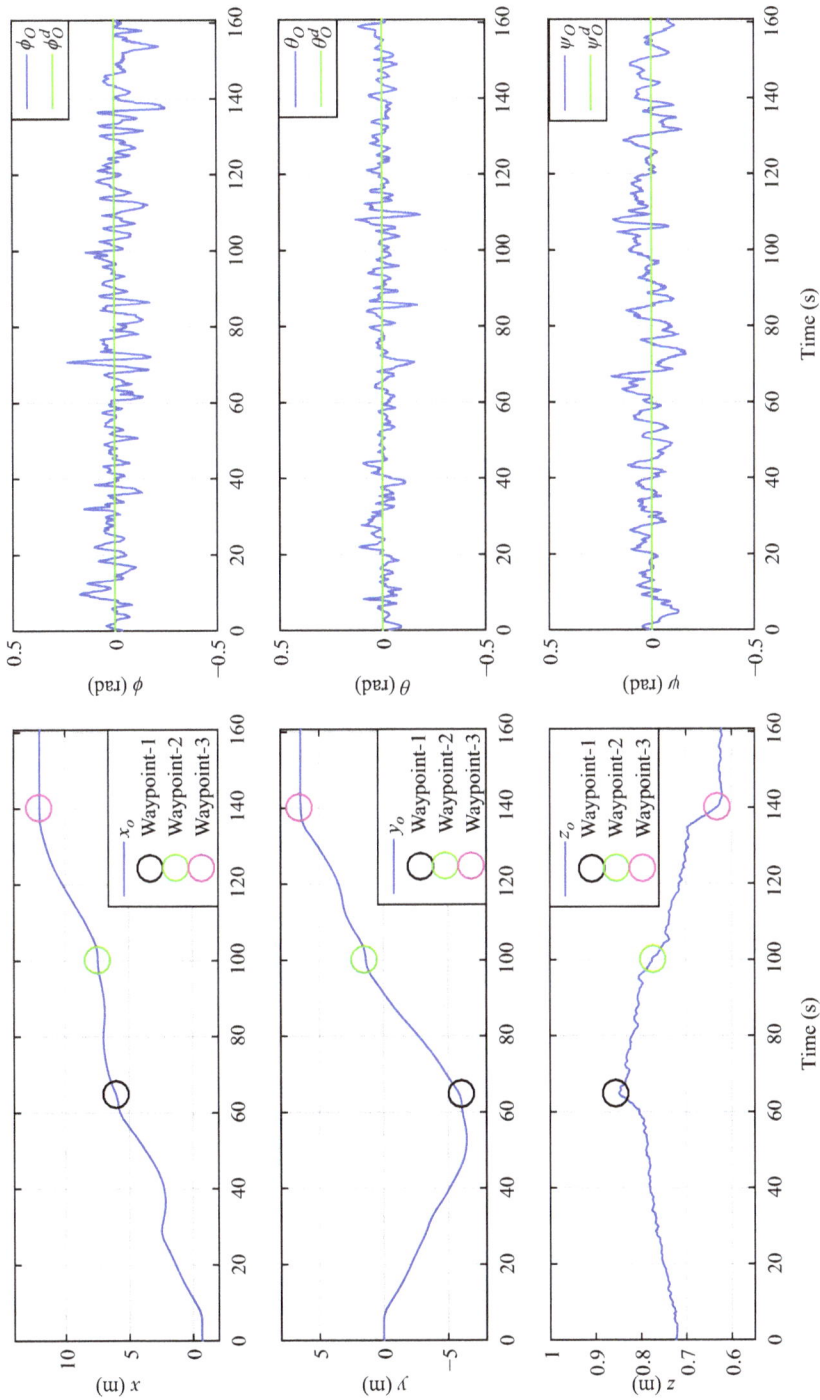

Figure 4.12   Simulation study: object coordinates during the control operation

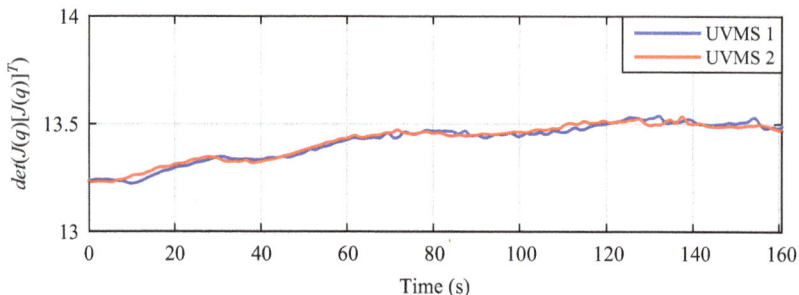

Figure 4.13   Simulation study: $det(\mathbf{J}(\mathbf{q})[\mathbf{J}(\mathbf{q})]^{\top})$ during the control operation

Both underwater vehicles are equipped with identical custom-built four-degree-of-freedom waterproof manipulators (see Figure 4.15). Additionally, each UVMS features a downward-facing Sony PlayStation Eye camera enclosed in a waterproof housing, capturing images at a resolution of 640×480 pixels with a frame rate of 30 fps. The target object in the experiment was a pipe measuring 1.0 m in length and weighing 0.35 kg. To assist with localization, an underwater laser pointer, fixed to the vehicle and aligned with the down-facing camera, projects a green dot onto the bottom of the test tank. The laser dot's position on the image plane, combined with data from the onboard navigation system sensors (e.g., IMU), is processed using a sensor fusion algorithm to estimate the vehicle's state vectors. The Seabotix LBV and VideoRay PRO are also equipped with an *SBG IG $-$ 500A* AHRS and a 10-DOF IMU sensor, respectively, providing temperature-compensated measurements of 3D acceleration, angular velocity, and orientation at a frequency of 100 Hz. For marker localization, the system relies on the ArUco library [65]. In all experimental studies, the complete state vector of each UVMS, including 3D position, orientation, velocity, and dynamic parameters, is derived using a sensor fusion and state estimation module. This module is based on a Complementary Filter approach and an identification scheme detailed in previous research [63]. The specific methodologies behind sensor fusion, state estimation, and parameter identification are beyond the scope of this work but can be found in [63]. The software implementation of the proposed control scheme was carried out in C++ and Python using the Robot Operating System (ROS) [62].

Remark 6 (Real-time implementation). *The Nonlinear Model Predictive Controller employed in this work was implemented using the NLopt Optimization library [64] and was running with 1 ms time step, which is common in a real-time operation with underwater robotic systems. It is worth mentioning that recent advances in technology (i.e., the new generation of very powerful CPUs) motivated engineers and scientists to develop faster and more efficient solvers (e.g., ACADO [66]) that therefore allow the reliable implementation of NMPC controllers in real-time fast applications. In this work, the overall software was running on two conventional laptops (each UVMS was connected to a separate laptop) with 4 cores, 2.80 GHz CPU, and 16 GB of RAM.*

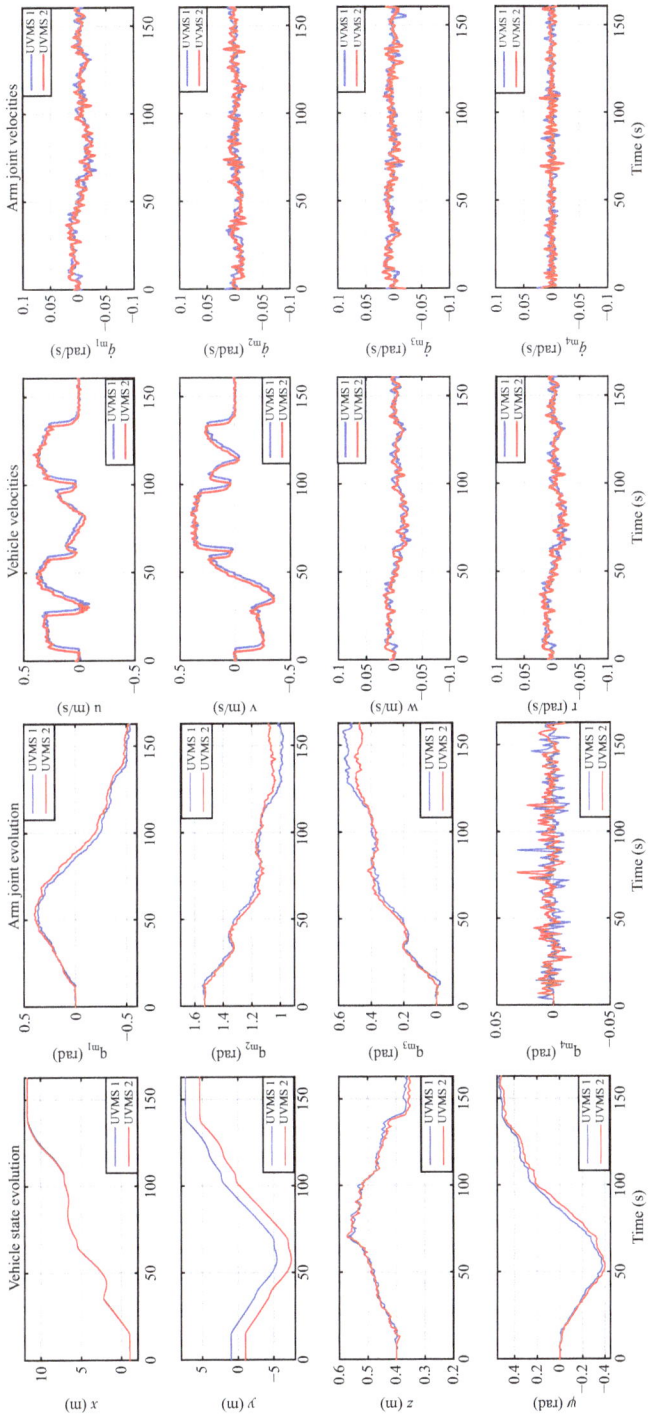

(a) System states

(b) System velocities

Figure 4.14 Simulation study: evolution of the system states and velocities at the joint level

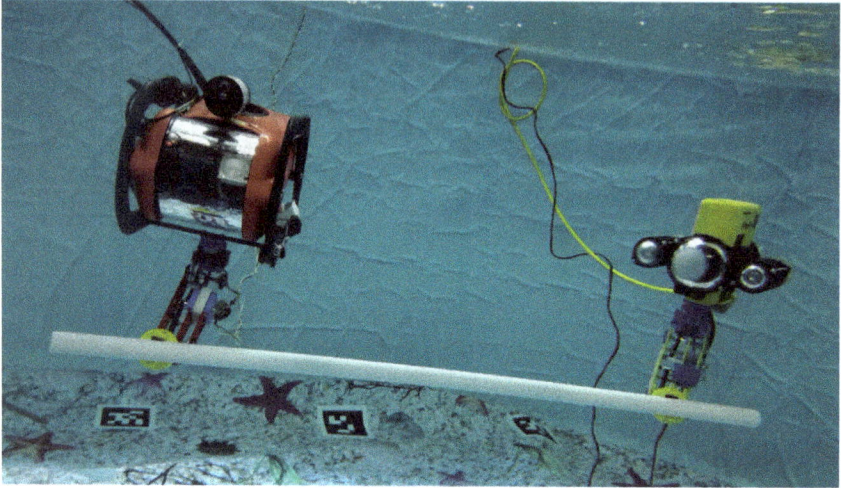

Figure 4.15   Experimental setup: custom made UVMSs during real-time experiment cooperative transportation. The vehicles used in this work were a 4 DoF Seabotix LBV and a 3 DoF VideoRay PRO presented in red and yellow colors, respectively.

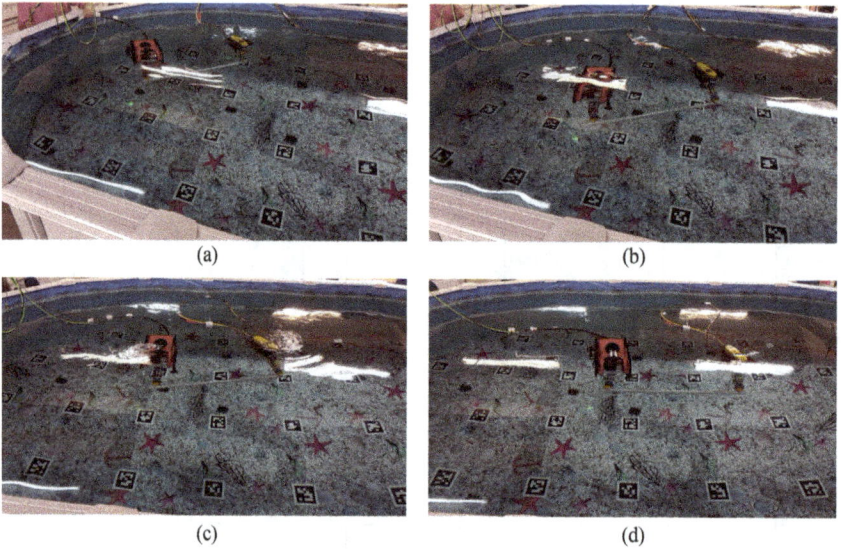

(a)   (b)

(c)   (d)

Figure 4.16   Experimental study - stabilization scenario: The evolution of the proposed methodology in four consecutive time instants

4.4.3.4   Experimental results

In following experimental studies, the objective for the team of UVMSs is to stabilize the object cooperatively in a desired configuration within the test tank. Moreover,

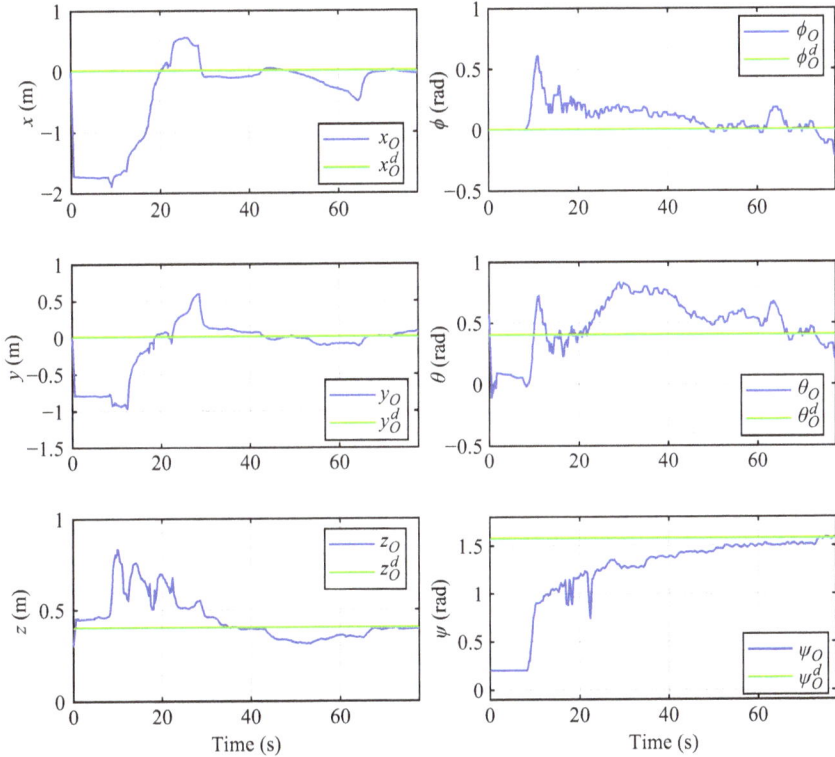

Figure 4.17 *Experimental study – stabilization scenario: Object coordinates during the control operation*

in the following experiments, the team of UVMS should simultaneously avoid the workspace (test tank) boundaries, which were modeled according to the spherical world representation. Notice that the small and limited size of the available test tank did not allow us to consider obstacles inside the workspace. However, collision avoidance with the test tank boundaries demonstrates the efficacy of the proposed scheme for avoiding collisions in real time. The radius of the sphere $\mathcal{B}(\boldsymbol{x}_i, \bar{r})$ which covers all the UVMS volume (i.e., main body of the vehicle, additional equipment and robotic manipulator for all possible configurations) is defined as $\bar{r} = 0.75$ m. Moreover, the radius of the sphere $\mathcal{B}(\boldsymbol{x}_O, r_O)$ that covers the object is defined as $r_O = 0.5$ m. Similar to the simulation part, the Navigation function (4.21)–(4.77) was designed to generate the safe navigation with gain $K_{NF} = 0.1$. Regarding to constraints (4.74), in both experiments, we consider that the vehicle's velocity must not exceed $0.5 \, \text{m s}^{-1}$ for translation and $0.5 \, \text{rad s}^{-1}$ for rotational velocities. In the same vein, the manipulator joint velocities must be retained between $(-0.2, 0.2) \, \text{rad s}^{-1}$. Moreover, the manipulator joint positions (4.75) must be retained between $(-2.5, 2.5)$ rad for the first joint q_{m_1}, $(-1.5, 0.7)$ rad for the second joint q_{m_2}, and $(-0.5, 1.5)$ rad for the third joint q_{m_3}, respectively. Notice that the fourth joint q_{m_4} is limit free and thus no joint limit

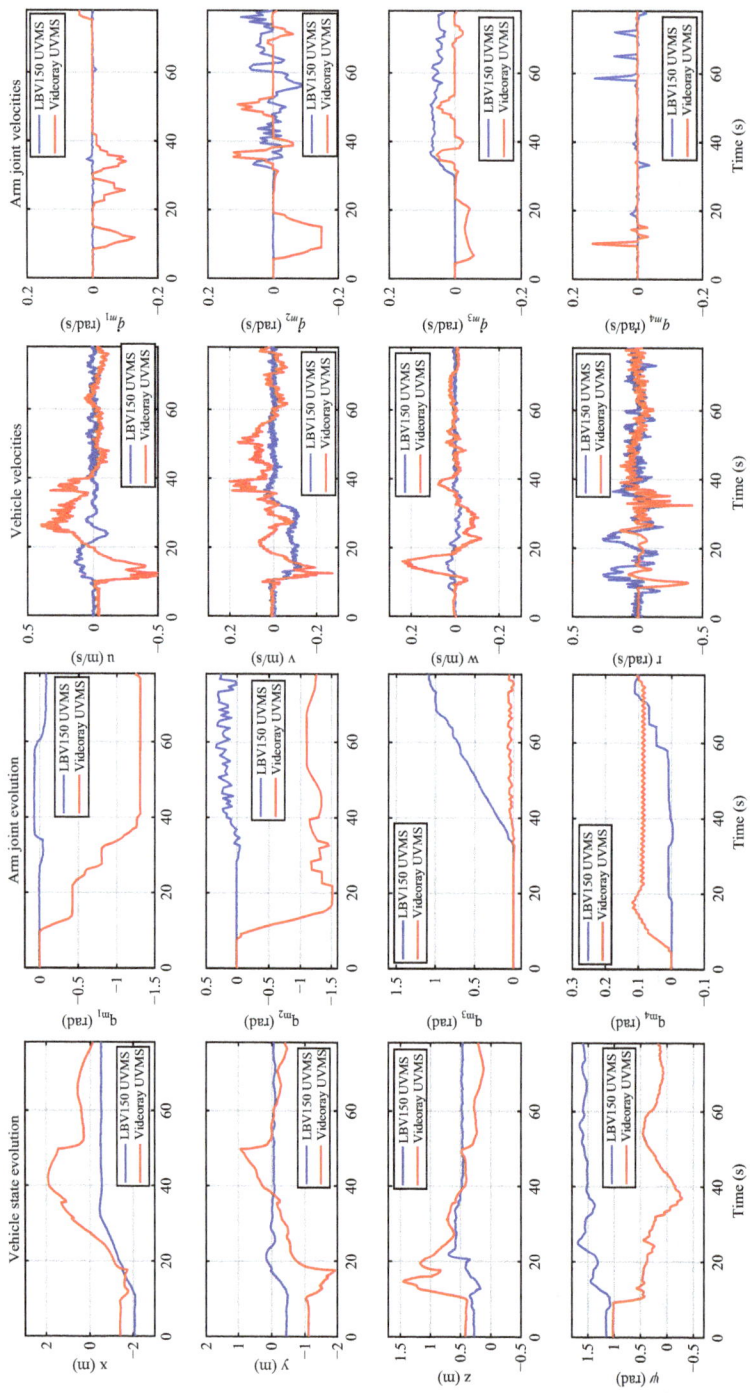

(a) System states

(b) System velocities

Figure 4.18 Experimental study: evolution of the system states and velocities at joint level

constraints are considered to this joint. Furthermore, input saturations (4.76) for the vehicle and manipulator are considered as follows: $\bar{\tau}_v = 2N$ and $\bar{\tau}_m = 0.2N$, respectively. The sampling time (4.78) and the prediction horizon for both of the following experiments are $h = 0.15$ s and $T_p = 5 \times h = 0.75$ s, respectively. Moreover, the matrices P_x, Q_x, Q_v, and R, as well as the load sharing coefficients c_1 and c_2 for both UVMSs are equal and set to $P_x = Q_x = 0.5 \cdot I_{6\times6}, R = 0.15 \cdot I_{8\times8}, Q_v = 0.2 \cdot I_{6\times6}$, and $c_1 = c_2 = 0.5$.

As stated before, the objective for the team of UVMS in the first experiment is to transfer and stabilize the object into a desired configuration inside the small test tank. More specifically, the desired configuration was set as $x_O^d = [0.0, 0.0, 0.4, 0, 0.45, 1.57]$. The results are presented in Figures 4.16–4.18. The trajectory of the system within the workspace and the object coordinates evolution are depicted in Figures 4.16 and 4.17, respectively. It can be seen that the team of UVMSs have successfully transported and stabilized the object to the desired configuration. The evolution of the system states and velocities at the joint level is indicated in Figure 4.18. It can be seen easily that all of the aforementioned values remained in their corresponding constraints sets during the experiment's evolution.

References

[1] Heshmati-Alamdari S., Karras G.C., Marantos P., *et al.*: "A robust model predictive control approach for autonomous underwater vehicles operating in a constrained workspace." *2018 IEEE International Conference on Robotics and Automation (ICRA)*. Piscataway, NJ: IEEE; 2018. pp. 6183–6188.

[2] Ridao P., Carreras M., Ribas D., *et al.*: "Intervention AUVs: The next challenge." *Annual Reviews in Control*. 2015;40:227–241.

[3] Farivarnejad H. and Moosavian S.A.A.: "Multiple impedance control for object manipulation by a dual arm underwater vehicle-manipulator system." *Ocean Engineering*. 2014;89:82–98.

[4] Karras G.C., Fourlas G.K., and Nikou A., *et al.*: "Image based visual servoing for floating base mobile manipulator systems with prescribed performance under operational constraints." *Machines*. 2022;10(7):547.

[5] Londhe P.S., Mohan S., Patre B.M., *et al.*: "Robust task-space control of an autonomous underwater vehicle-manipulator system by PID-like fuzzy control scheme with disturbance estimator." *Ocean Engineering*. 2017;139:1–13.

[6] Sharifi M., and Heshmati-Alamdari S.: "Safe force/position tracking control via control barrier functions for floating base mobile manipulator systems." *2024 European Control Conference (ECC)*. 2024; pp. 3650–3655.

[7] Heshmati-Alamdari S.: *Cooperative and Interaction Control for Underwater Robotic Vehicles* [Ph.D. thesis]. Greece: National Technical University of Athens; 2019.

[8] Padir T. and Koivo A.J.: "Modeling of two underwater vehicles with manipulators on-board." *Proceedings of the IEEE International Conference on Systems, Man and Cybernetics*. 2003;2:1359–1364.

[9]   Sun Y.C. and Cheah C.C.: "Coordinated control of multiple cooperative underwater vehicle-manipulator systems holding a common load." *Ocean '04 – MTS/IEEE Techno-Ocean '04: Bridges across the Oceans – Conference Proceedings.* 2004;3:1542–1547.

[10]   McClamroch N.H.: "Singular systems of differential equations as dynamic models for constrained robot systems." *IEEE International Conference on Robotics and Automation.* 1986; pp. 21–28.

[11]   Padir T.: "Kinematic redundancy resolution for two cooperating underwater vehicles with on-board manipulators." *Conference Proceedings – IEEE International Conference on Systems, Man and Cybernetics.* 2005; pp. 3137–3142.

[12]   Padir T. and Nolff J.D.: "Manipulability and maneuverability ellipsoids for two cooperating underwater vehicles with on-board manipulators." *Conference Proceedings – IEEE International Conference on Systems, Man and Cybernetics.* 2007; pp. 3656–3661.

[13]   Marani G., Choi S.K., and Yuh J.: "Underwater autonomous manipulation for intervention missions AUVs." *Ocean Engineering.* 2009;36(1):15–23.

[14]   Cui R., Ge S.S., How B.V.E., *et al.*: "Leader-follower formation control of underactuated autonomous underwater vehicles." *Ocean Engineering.* 2010;37(17–18):1491–1502.

[15]   Pereira G.A.S., Pimentel B.S., Chaimowicz L., *et al.*: "Coordination of multiple mobile robots in an object carrying task using implicit communication." *Proceedings – IEEE International Conference on Robotics and Automation.* 2002; pp. 281–286.

[16]   Heshmati-Alamdari S., Karras G.C., Sharifi M., *et al.*: "Control barrier function based visual servoing for underwater vehicle manipulator systems under operational constraints." *2023 31st Mediterranean Conference on Control and Automation (MED).* Piscataway, NJ: IEEE; 2023. pp. 710–715.

[17]   Stilwell D.J. and Bishop B.E.: "Framework for decentralized control of autonomous vehicles." *Proceedings of the IEEE International Conference on Robotics and Automation.* 2000;3:2358–2363.

[18]   Uchiyama M. and Dauchez P.: "Symmetric hybrid position/force control scheme for the coordination of two robots." *Proceedings of the IEEE International Conference on Robotics and Automation.* 1988; pp. 350–356.

[19]   Khatib O.: "Object manipulation in a multi-effector robot system." *Proceeding of the 4th International Symposium on Robotic Research.* 1988;4:137–144.

[20]   Tanner H.G., Loizou S.G., and Kyriakopoulos K.J.: "Nonholonomic navigation and control of cooperating mobile manipulators." *IEEE Transactions on Robotics and Automation.* 2003;19(1):53–64.

[21]   Schneider S.A. and Cannon J.R.H.: "Object impedance control for cooperative manipulation: Theory and experimental results." *IEEE Transactions on Robotics and Automation.* 1992;8(3):383–394.

[22]   Luh J.Y.S. and Zheng Y.F.: "Constrained relations between two coordinated industrial robots for motion control." *International Journal of Robotics Research.* 1987;6(3):60–70.

[23]   Sugar T. and Kumar V.: "Decentralized control of cooperating mobile manip-ulators." *Proceedings – IEEE International Conference on Robotics and Automation*. 1998;4:2916–2921.

[24]   Khatib O., Yokoi K., Chang K., *et al.*: "Vehicle/arm coordination and multiple mobile manipulator decentralized cooperation." *IEEE International Conference on Intelligent Robots and Systems*. 1996;2:546–553.

[25]   Dickson W.C., Cannon Jr R.H., and Rock S.M.: "Decentralized object impedance controller for object/robot-team systems: Theory and experi-ments." *Proceedings – IEEE International Conference on Robotics and Automation*. 1997;4:3589–3596.

[26]   Liu Y.H., Arimoto S., and Ogasawara T.: "Decentralized cooperation con-trol: non-communication object handling." *Proceedings – IEEE International Conference on Robotics and Automation*. 1996;3:2414–2419.

[27]   Conti R., Meli E., Ridolfi A., *et al.*: "An innovative decentralized strategy for I-AUVs cooperative manipulation tasks." *Robotics and Autonomous Systems*. 2015;72:261–276.

[28]   Furferi R., Conti R., Meli E., *et al.*: "Optimization of potential field method parameters through networks for swarm cooperative manipulation tasks." *International Journal of Advanced Robotic Systems*. 2016;13(5):1–13.

[29]   Simetti E. and Casalino G.: "Manipulation and transportation with coop-erative underwater vehicle manipulator systems." *IEEE Journal of Oceanic Engineering*. 2016;42:782–799.

[30]   Manerikar N., Casalino G., Simetti E., *et al.*: "On autonomous cooperative Underwater Floating Manipulation Systems." *Proceedings – IEEE Interna-tional Conference on Robotics and Automation, 2015*; 2015.

[31]   Manerikar N., Casalino G., Simetti E., *et al.*: "On cooperation between autonomous underwater floating manipulation systems." *2015 IEEE Under-water Technology*, UT 2015; 2015.

[32]   Simetti E. and Casalino G.: "A novel practical technique to integrate inequal-ity control objectives and task transitions in priority based control." *Jour-nal of Intelligent and Robotic Systems: Theory and Applications*. 2016; 84(1–4):877–902.

[33]   Heshmati-Alamdari S., Bechlioulis C.P., Karras G.C., *et al.*: "Coopera-tive impedance control for multiple underwater vehicle manipulator sys-tems under lean communication." *IEEE Journal of Oceanic Engineering*. 2020;46(2):447–465.

[34]   Heshmati-Alamdari S., Karras G.C., and Kyriakopoulos K.J.: "A dis-tributed predictive control approach for cooperative manipulation of multiple underwater vehicle manipulator systems." *2019 international conference on robotics and automation (ICRA)*. Piscataway, NJ: IEEE; 2019. pp. 4626–4632.

[35]   Heshmati-Alamdari S., Karras G.C., and Kyriakopoulos K.J.: "A predictive control approach for cooperative transportation by multiple underwater vehi-cle manipulator systems." *IEEE Transactions on Control Systems Technology*. 2021;30:917–930.

[36]   Nikou A., Heshmati-Alamdari S., and Dimarogonas D.V.: "Scalable time-constrained planning of multi-robot systems." *Autonomous Robots.* 2020;44(8):1451–1467.

[37]   Nikou A., Heshmati-Alamdari S., Verginis C.K., *et al.*: "Decentralized abstractions and timed constrained planning of a general class of coupled multi-agent systems." *2017 IEEE 56th Annual Conference on Decision and Control (CDC).* Piscataway, NJ: IEEE; 2017. pp. 990–995.

[38]   Nikou A., Verginis C.K., Heshmati-alamdari S., *et al.*: "A robust non-linear MPC framework for control of underwater vehicle manipulator systems under high-level tasks." *IET Control Theory & Applications.* 2021;15(3):323–337.

[39]   Sciavicco L. and Siciliano B.: *Modelling and Control of Robot Manipulators.* Berlin: Springer; 2012.

[40]   Antonelli G.: *Underwater Robots.* Springer Tracts in Advanced Robotics. Berlin: Springer; 2013.

[41]   Siciliano B., Sciavicco L., and Villani L.: "Robotics: modelling, planning and control." in *Advanced Textbooks in Control and Signal Processing.* Berlin: Springer; 2009.

[42]   Soylu S., Buckham B.J., and Podhorodeski R.P.: "Redundancy resolution for underwater mobile manipulators." *Ocean Engineering.* 2010;37(2–3): 325–343.

[43]   Koditschek D.E. and Rimon E.: "Robot navigation functions on manifolds with boundary." *Advances in Applied Mathematics.* 1990;11(4):412–442.

[44]   Sontag E.D.: *Mathematical Control Theory: Deterministic Finite Dimensional Systems*, 2nd ed. New York, NY: Springer-Verlag, 1998.

[45]   Bechlioulis C.P. and Rovithakis G.A.: "Prescribed performance adaptive control for multi-input multi-output affine in the control nonlinear systems." *IEEE Transactions on Automatic Control.* 2010;55(5):1220–1226.

[46]   Kosuge K., Oosumi T., and Seki H.: "Decentralized control of multiple manipulators handling an object in coordination based on impedance control of each arm." *IEEE International Conference on Intelligent Robots and Systems.* 1997;1:17–22.

[47]   Kosuge K., Oosumi T., and Chiba K.: "Load sharing of decentralized-controlled multiple mobile robots handling a single object." *Proceedings – IEEE International Conference on Robotics and Automation.* 1997; p. 4.

[48]   Heshmati-Alamdari S., Sharifi M., Karras G.C., *et al.*: "Control barrier function based visual servoing for Mobile Manipulator Systems under functional limitations." *Robotics and Autonomous Systems.* 2024;182:104813.

[49]   Mohan S. and Kim J.: "Coordinated motion control in task space of an autonomous underwater vehicle-manipulator system." *Ocean Engineering.* 2015;104:155–167.

[50]   Slotine J.J. and Li W.: "Adaptive strategies in constrained manipulation." *Proceedings of 1987 IEEE International Conference on Robotics and Automation.* Vol. 4. Piscataway, NJ: IEEE; 1987. pp. 595–601.

[51] Tatlicioglu E., Braganza D., Burg T.C., *et al.*: "Adaptive control of redundant robot manipulators with sub-task objectives." *Robotica.* 2009;27(6): 873–881.

[52] De Luca A. and Mattone R.: "Sensorless robot collision detection and hybrid force/motion control." *ICRA 2005. Proceedings of the 2005 IEEE International Conference on Robotics and Automation, 2005.* Piscataway, NJ: IEEE; 2005. pp. 999–1004.

[53] Caccavale F., Natale C., Siciliano B., *et al.*: "Resolved-acceleration control of robot manipulators: A critical review with experiments." *Robotica.* 1998;16(5):565–573.

[54] Bechlioulis C.P. and Rovithakis G.A.: "Robust partial-state feedback prescribed performance control of cascade systems with unknown nonlinearities." *IEEE Transactions on Automatic Control.* 2011;56(9):2224–2230.

[55] Allgöwer F., Findeisen R., and Nagy Z.K.: "Nonlinear model predictive control: From theory to application." *The Chinese Institute of Chemical Engineers.* 2004;35(3):299–315.

[56] Gudiño-Lau J., Arteaga M.A., Munoz L.A., *et al.*: "On the control of cooperative robots without velocity measurements." *IEEE Transactions on Control Systems Technology.* 2004;12(4):600–608.

[57] Caccavale F., Chiacchio P., and Chiaverini S.: "Task-space regulation of cooperative manipulators." *Automatica.* 2000;36(6):879–887.

[58] Farivarnejad H. and Moosavian S.A.A.: "Multiple impedance control for object manipulation by a dual arm underwater vehicle–manipulator system." *Ocean Engineering.* 2014;89:82–98.

[59] Moosavian S.A.A. and Papadopoulos E.: "Multiple impedance control for object manipulation." *Proceedings of 1998 IEEE/RSJ International Conference on Intelligent Robots and Systems.* Vol. 1. Piscataway, NJ: IEEE; 1998. pp. 461–466.

[60] Moosavian S.A.A. and Papadopoulos E.: "Cooperative object manipulation with contact impact using multiple impedance control." *International Journal of Control, Automation and Systems.* 2010;8(2):314–327.

[61] Prats M., Perez J., Fernandez J.J., *et al.*: "An open source tool for simulation and supervision of underwater intervention missions." *IEEE/RSJ International Conference on Intelligent Robots and Systems (IROS)*; 2012. pp. 2577–2582.

[62] Quigley M., Gerkey B., Conley K., *et al.*: "ROS: an open-source Robot Operating System." *Proceedings of the IEEE International Conference on Robotics and Automation (ICRA) Workshop on Open Source Robotics.* Kobe, Japan; 2009.

[63] Karras G.C., Bechlioulis C.P., Marantos P., *et al.*: "Sensor-based motion control of autonomous underwater vehicles, part I: modeling and low-complexity state estimation." in *Autonomous Underwater Vehicles: Design and Practice*; 2020. pp. 15–43.

[64] Johnson S.G.: *The NLopt Nonlinear-Optimization Package*; 2007. https://github.com/stevengj/nlopt.

[65]   Garrido-Jurado S., noz Salinas R.M., Madrid-Cuevas F.J., *et al.*: "Automatic generation and detection of highly reliable fiducial markers under occlusion." *Pattern Recognition*. 2014;47(6):2280–2292.

[66]   Houska B., Ferreau H.J., and Diehl M.: "An auto-generated real-time iteration algorithm for nonlinear MPC in the microsecond range." *Automatica*. 2011;47(10):2279–2285.

Chapter 5
Propelling URSULA: bioinspired fins for robotic squids

Berke Gür¹, Alhamzah Ihsan Ali² and Batuhan Akbulut²

Project URSULA aims to develop a next-generation robotic system that will enable dexterous underwater manipulation and seabed intervention in ways that current unmanned underwater vehicles cannot. The robot has a biomimetic design inspired by squids, combining a hydrodynamic and agile body with multifunctional robotic manipulators. The robot also hosts numerous unconventional and innovative technologies such as tendon-driven soft robotic limbs, propellerless underwater propulsion systems, perception-guided autonomous navigation and posture control, high-bandwidth visible light-based wireless underwater communication, model-mediated teleoperation with haptic feedback, and virtual reality enhanced operator support systems. In this work, one of the bioinspired, propellerless locomotion systems of URSULA, namely, the fin propulsion system is introduced and comprehensively evaluated. A theoretical model for the thrust generation mechanism of the fin is presented. The justification and algorithmic details of a pattern generator algorithm implemented together with the fin driver software are presented. The analytic model is first validated through computational fluid dynamics (CFD) simulations and then through laboratory experiments. Both the CFD simulations and the experimental results agree well with the analytic model predictions. These results also provide some new insights that, to the best of the authors' knowledge, have not yet been reported in the literature. These results also indicate areas of further improvement in both the fin design and the theoretical model.

5.1 Introduction

Due to the rapid increase in the world's population in the next 30 years, it is predicted that the available natural resources will be insufficient to meet our daily needs in critical areas such as food and energy [1]. Solutions proposed to alleviate the consequences of this approaching resource crisis include increased resource utilization, reducing waste, and increasing recycling efficiency. In addition to these solutions,

¹Department of Mechanical Engineering, Faculty of Engineering, Marmara University, Türkiye
²Graduate School, Mechatronics Engineering Program, Bahçeşehir University, Türkiye

the effective and sustainable use of the relatively untapped marine environment is considered another viable remedy.

More than 70% of the Earth's surface is covered by marine environments. These aquatic systems are home to more than 230 000 species and play a decisive role in vital oxygen, carbon, water, and many other biogeochemical cycles. Furthermore, the marine environment influences the formation and development of important meteorological and climate systems. Fresh and salty waters are becoming an increasingly important source of food for the world's population. Fish and aquaculture around the world have increased steadily in the last 70 years, mainly due to aquaculture farms.

In the last 30 years, world energy demand has increased by 70% and is expected to increase further (by at least 50%) by 2030. Today, almost all demand is met by fossil fuels such as oil and natural gas. Increased investment has led to significant advances in alternative and sustainable energy technologies such as solar and hydrogen. However, the capacities of existing and projected renewable energy sources are not sufficient to replace fossil fuels in the short to medium term. 25% (40 billion tons out of 160 billion tons) of the total known underground reserves are in known underground reserves. Furthermore, 35% of the total oil production and 30% of the natural gas production are derived from underground wells. It is clear from these figures that underground fossil fuel and hydrocarbon reserves are being depleted much faster than terrestrial reserves. Deep wells are drilled, which were not economically feasible in the past, to replace depleted underground reserves. The transition in energy production from fossil fuels to alternative sources is expected to increase the importance of oceans and seas in energy production in the long term. In 2000, almost all wind turbines in Europe were located on land. At the end of 2024, the installed offshore wind power in Europe had increased to 37 GW, reaching 13% of the total installed wind power [2]. Coastal and underwater tidal and wave energy generators represent an emerging branch of renewable and sustainable marine energy sources [3]. By 2040, the total installed offshore wind energy capacity is expected to increase by a factor of 15 [4].

It is clear from the above discussion that the marine environment, and in particular the underwater and seafloor environment, will be exploited much more effectively for food supply, energy production (from conventional fossil fuels and renewable sources), and infrastructure construction. The list of underwater activities can be expanded to include the mining of precious and rare metals and minerals, gas hydrates and other natural minerals, search and rescue operations, object extraction and recovery, underwater archaeology, geology, biology, ecology, and oceanography studies, biofuel production, and military applications. In addition, these underwater activities must be carried out in deeper waters, far from the coast, in a skillful and safe physical interaction with natural or artificial underwater structures.

However, the underwater environment is unsuitable and even dangerous for the prolonged presence and activities of human divers. Given the depth of operation, the complexity and duration of the task, even commercial divers using specialized equipment can only operate for relatively short periods of time in moderate to shallow waters. These problems make the use of unmanned underwater robots for underwater exploration and exploitation activities highly desirable and often necessary. The fleet of underwater robots used worldwide for commercial and scientific applications has grown significantly since the early 2000s. According to industry research and forecasts, the underwater robot market will grow from a size of 4.5 billion USD in 2023 at an average annual rate of 15% over the next 5 years and reach 12.9 billion USD in 2030 [5]. Despite their growing numbers, market size, and mission requirements, underwater unmanned robotic systems (i.e. remotely operated vehicles (ROVs) and autonomous underwater vehicles (AUVs)) have changed little in terms of their overall design principles and functional capabilities since their introduction in the 1970s. With the significant expansion of application areas and diversification of tasks expected to be performed, current underwater robot designs will not be sufficient to meet operational demands, particularly in terms of underwater intervention [6].

The primary objective of project URSULA (abbreviation for **U**nmanned **R**obotic **S**quid for **U**nderwater and **L**ittoral **A**pplications) is to develop a robotic platform for skillful intervention and dexterous underwater manipulation that combines the benefits of the worker class (ROVs) and traveling class robots (AUVs). The proposed robot will work close to the seabed, around both natural and artificial underwater structures. As a secondary objective, this robotic platform is designed to serve as a test bed to evaluate novel technologies that can be utilized in autonomous underwater systems.

To this end, propellerless propulsion techniques are integrated into URSULA to avoid complications and potential risks that might arise from operating propeller systems close to the seabed and around natural and artificial underwater structures. These risks include, but are not limited to, entanglement with vegetation and limited maneuverability in confined spaces. Two bioinspired propulsion systems, namely undulating fins and jet propulsion systems, are designed and integrated with the robot. This chapter presents the design methods, laboratory experiments, and measurements conducted for the fin propulsion system of project URSULA. The chapter is organized as follows. Section 5.2 provides a general overview of the URSULA robot. Section 5.3 describes the design methodology of the fin propulsion system, followed by Section 5.4 that describes the laboratory experiments carried out to test and validate the design. The chapter ends with Section 5.5, which presents conclusions and planned future developments.

5.2 Overview of URSULA

The URSULA robotic system consists of a cuttlefish-inspired biomimetic underwater robot and an operator station. The main body of URSULA consists of three main waterproof and rigid parts: (1) the head, (2) the pen, and (3) the tail. The body also includes a soft mantle that covers the pen. Four robotic arms are attached to the head, which houses the actuation system for these arms. Four fins (bow and aft, port and starboard, plus two optional central fins) form the robot's main actuation system and are attached to two rails on the port and starboard sides that run the length of the pen, outside the mantle. These side rails are complemented by a pair of identical dorsal and ventral rails used to connect the sensors and visible light communication system. The actuators for these fins, as well as other on-board electronics, computing, power, and battery systems, are housed inside the watertight pen. The jet drive is the secondary propulsion system and is located at the front of the pen, close to the pen–head interface. The jet propulsion system utilizes the mantle and the water-filled cavity inside the mantle. The total body length (excluding the limbs) is approximately 1.2 m, while the limbs are 0.6 m each. The largest cross section is at the head, with a diameter of 0.25 m. The robot has a total dry weight of 30 kg and is designed to be slightly buoyant in water. The current prototype is rated for a maximum dive depth of 50 m and an endurance of minimum 45 min while operating on batteries in the intervention mode. However, endurance is dependent on the battery configuration and payload. A 3D model of URSULA is presented in Figure 5.1.

The four initial robotic limbs are made of soft, hyper-elastic material and form the robot's manipulation interfaces. These four limbs are grouped into two pairs as tentacles and arms. The two arms are the main manipulation interfaces capable of tactile sensing due to pressure sensors embedded in their tips. The longer tentacles are used for support tasks, such as fixing the robot in place. These tentacle tips also house an RGB underwater camera and an underwater light system to assist underwater manipulation activities. A microcontroller hosted on a custom-designed electronic board is used to move and control the limbs.

The primary motion system of the robot is the four fins located on the bow and aft, port, and starboard sides of the main body. A pair of optional port- and starboard-central fins can also be used to support these main fins. Inspired by the fleeing mechanism of common squid, a jet propulsion system is incorporated as a secondary motion system. This jet propulsion system generates propulsion by squirting water compressed inside the mantle through a steerable funnel. The fin and jet

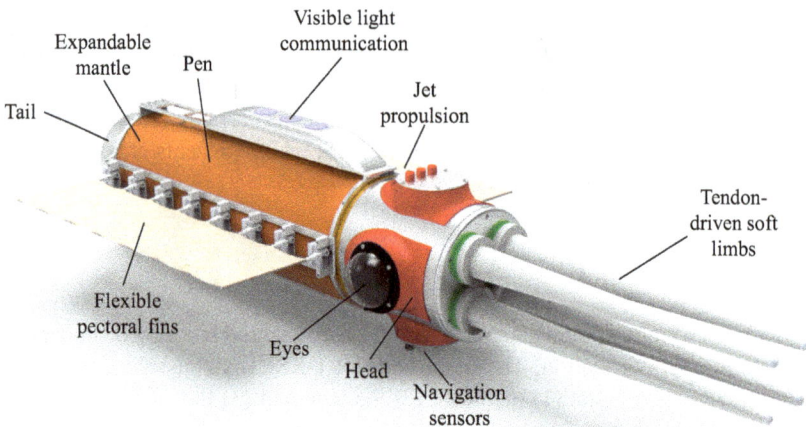

Figure 5.1   A 3D model of URSULA with major systems labeled

propulsion systems are controlled by separate microcontrollers, housed on custom designed electronic boards. The robot can be powered from either a surface plat-form via an umbilical cable in the tethered (TET) mode or from onboard batteries in the untethered (NOWIRE) mode. Onboard Li-ion batteries can be charged using the umbilical cable without the need to disassemble the robot to remove the batteries after each deployment.

URSULA's mission computer hosts the ROS1-based [7] mission software, including navigation, control, guidance, communication, situational awareness, system health monitoring, and fault detection. Process manager software ensures reliable operation and, if necessary, seamless reboot of the mission computer. The robot's on-board sensing systems can be divided into three separate functional groups. Although each group of sensing systems can operate independently of the other sensing systems, sensor fusion helps increase the dexterity of the robot and the situational awareness of the operator. The first group of sensors includes basic navigational sensors, such as inertial measurement unit (IMU), depth sensors, and global navigation satellite system (GNSS) sensors, as well as more sophisticated sensors, such as a Doppler velocimeter log (DVL). The second group of perception sensors includes a sonar and a 3D vision system that require image processing. The sonar system is used primarily for autonomous navigation in Exploration (EXP) mode, while the 3D vision system is used for the teleoperation of the soft limbs in Intervention (INT) mode. The complexity of these image processing algorithms requires a separate, specialized GPU-based sensing computer.

In the TET mode, communication with the robot is established using a 4-wire Ethernet connection through the umbilical cable. The visible light system is designed to operate in the NOWIRE mode as the main communication system between the robot and the operator station. The system includes a dome-shaped surface module suspended in the air just above the robot and a fin-shaped subsurface module attached to the dorsal part of the robot. The high bandwidth provided by the visible light communication system is used to provide a video feed to the operator in the INT mode.

The compact and mobile operator station of URSULA consists mainly of the operator computer and the physical interfaces used by the operator to control the robotic limbs during the INT mode. Haptic devices, touchless [8] or virtual reality (VR) interfaces [9] can be utilized to command the robot and interact with the under-water environment. In addition to the computer, the high voltage DC power supply and the surface module of the visible communication system are also connected to the operator computer.

5.3   Methods and design

Bioinspired, deformable, and flexible undulating fins are the primary locomotion mechanism of URSULA. This section presents a concise review of the fin mechanics and the models developed in the design of these fins. These models take time-dependent fin motion as the input and calculate the thrust and moment generated by the fin. The thrust forces and moments can be used to predict and control the motion of the robot using dynamic motion models, enabling underwater navigation. In addition, models that accurately predict the loads generated by the fin support the design and optimization of the fins. Detailed descriptions of the numerical simulations utilized in the development of the fin system for project URSULA are provided. The simulation results are analyzed and utilized to optimize the fins for efficient operation.

5.3.1   *Working principles*

The simplest fin design consists of flexible beams with membranes stretched in between. The oscillatory motion of the beams is controlled through actuators attached to their stem (see Figure 5.2).

Given this single, simple fin, it is envisioned that the fin can operate in three different modes.

1.   Standing wave (SW) or flapping mode,
2.   Traveling wave (TW) or undulating mode,
3.   Fixed fin (FF) or wing mode.

Depending on the mode of operation, a single fin can generate forces in all three directions (surge, sway, and heave). In standing wave mode, the beams oscillate in phase and the fin moves as a rigid body, generating a flap-like motion. This results in a net thrust in the sway direction. However, the fins are designed to swivel $\pm 90°$ from the horizontal around the surge axis, allowing one to control the direction of the generated thrust in the heave-sway plane. The fin beams undulate rhythmically according to a certain pattern in the traveling wave mode, forming a forward or backward traveling wave. This type of undulation generates thrust along the surge axis. As a first step, the standing and traveling wave modes are studied. The fixed fin mode is not capable of generating thrust on its own and therefore was not considered in this work. However, in the future, the use of the fixed fin mode in combination with alternative propulsion methods such as jet propulsion will be studied.

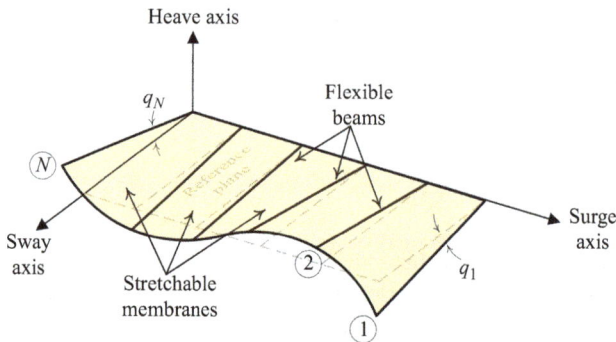

Figure 5.2   A depiction of an undulating fin with N beams

5.3.2 Modeling

Modeling of undulating fins in an underwater environment is challenging due to the fluid–structure interaction (FSI) involved. In addition, modeling becomes more difficult because of the complexities introduced by the flexibility of the beams and the elasticity of the membrane. When the fins undulate in water, drag forces are induced on the fin surface. The total thrust force produced by the fin is obtained by calculating this drag force per unit surface area and summing these forces for the entire fin surface. The total moment is also calculated on the basis of the force generated per unit area. By moving the total force and moment obtained to the center of mass of the vehicle, the total thrust that propels the robot is obtained. In this mathematical dynamic model, we adopted the approach presented by Sfakiotakis *et al.* [10]. Accordingly, the fin geometry assumed in this study is shown in Figure 5.3.

The motion of a single beam can be described as

$$q_i(t) = A \sin[2\pi ft - (i - 1)\phi] + X, \qquad i = 1, 2, \ldots, N \qquad (5.1)$$

where $q_i(t)$ is the fin beam angle at time t, A and f are the frequency and amplitude of the wave, respectively, ϕ and X represent the inter-beam phase shift and angular offset, respectively, and N is the number of fin rays (see Figure 5.2). The inter-beam phase angle ϕ determines actuation mode. When $\phi > 0$, the undulation spreads from the first beam toward beam N, resulting in forward thrust along the positive surge axis. Alternatively, if $\phi < 0$, the undulation of the fin spreads from the last link N toward the first beam, resulting in the generation of a backward thrust. Finally, if $\phi = 0$, the fin oscillates as a single rigid surface in the flapping mode.

Based on the simplified fluid drag model, the distributed drag force generated on the infinitesimal surface element $ds = dw \times dh$ (or hydrodynamic force intensity) is perpendicular to this surface and is given as follows:

$$\vec{f}_n = -\frac{1}{2}\rho C_D v_n^2 \cdot \text{sgn}(v_n) \cdot \hat{e}_n \qquad (5.2)$$

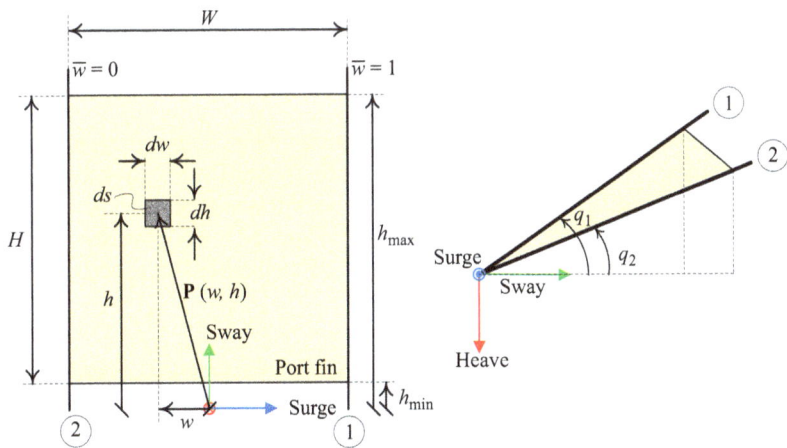

Figure 5.3   The assumed fin geometry depicting model parameters

where ρ is the fluid mass density, C_D is the drag coefficient, v_n is the normal component of the local velocity vector $\vec{p}(w,h)$ of the fin surface and is defined as follows:

$$v_n = \frac{Wh}{\delta}\{(1-2\overline{w})\dot{q}_2 + [\overline{w}^2 + \overline{w}(1-\overline{w}\cos(\Delta q))] \cdot (\dot{q}_1 + \dot{q}_2)\} \tag{5.3}$$

where $\Delta q = q_1 - q_2$ is the difference between the joint angles, $\dot{q}_i$ is beam i's undulation velocity, W is the total width of the membrane, w and h are the width and height coordinates, respectively, of the point on the membrane for which the velocity is computed, $\overline{w} = w/w_{max}$ is the normalized width coordinate, with the maximum length of the stretched membrane w_{max} defined as follows:

$$w_{max}(h, \Delta q) = [2h^2(1 - \cos(\Delta q) + W^2]^{1/2} \tag{5.4}$$

and δ is defined as follows:

$$\delta = [W^2 - 2W^2 \cdot \overline{w}(1-\overline{w})(1-\cos(\Delta q)) + h^2\sin^2(\Delta q)]^{1/2} \tag{5.5}$$

Finally, $\hat{e}_n$ is the unit vector normal to the membrane surface and is computed as follows:

$$\hat{e}_n = \hat{e}_w \times \hat{e}_h = \frac{W}{\delta} \cdot \begin{bmatrix} -\overline{w}\cos q_1 - (1-\overline{w})\cos q_2 \\ -\overline{w}\sin q_1 - (1-\overline{w})\sin q_2 \\ -(h/W) \cdot \sin\Delta q \end{bmatrix} \tag{5.6}$$

where $\hat{e}_w$ and $\hat{e}_h$ are the unit vectors along the width and height directions, respectively, at the coordinate (w, h).

Integrating the force intensity defined in (5.2) over the surface area yields the total hydrodynamic force developed by the fin

$$\vec{F}_h = [F_x \quad F_y \quad F_z]^T = \int_{h_{min}}^{h_{max}} \int_0^1 \vec{f}_n \cdot d\overline{w}dh \tag{5.7}$$

Note also that the resultant couple generated by the hydrodynamic fin forces at the fin coordinate system origin can be computed as follows:

$$\vec{M}_h = [M_x \quad M_y \quad M_z]^T = \int_{h_{min}}^{h_{max}} \int_0^1 (\vec{p}(\overline{w}, h) \times \vec{f}_n) \cdot d\overline{w}dh \tag{5.8}$$

A final note should be made that the resultant forces and moments given in (5.7) and (5.8), respectively, are defined with respect to the fin frame. The origin and orientation of the fin frame are different from the standard body frame attached to the center of mass of the robot. Therefore, a correction must be applied to both the resultant force and the moment values when they are transferred to the body-attached coordinate system.

5.3.3   Central pattern generator

Locomotion and maneuvering are not achieved by the actuation of a single fin, but rather through the coordinated actuation of all of the fins that make up the fin-based locomotion system. Therefore, actuating the fins individually and independently based on prescribed trajectories may result in uncoordinated motion of the robot. Furthermore, joint-specific trajectories may result in abrupt and uncoordinated transitions between different modes of locomotion, which hinders smooth locomotion. To overcome this problem, central pattern generator (CPG) algorithms are employed in the URSULA fin system. CPGs in invertebrate and vertebrate animals are self-organizing biological neural networks that generate synchronous and rhythmic motor outputs for stereotyped locomotion behavior such as walking, flying, and swimming without the need for any feedback input [11].

In this work, the CPG originally proposed by Sproewitz *et al.* [12] and utilized by Gliva *et al.* [13] is used. Accordingly, the joint trajectories are computed based on the sinusoidal waveform:

$$q_i(t) = a_i \sin(\xi_i), \quad i = 1, 2, \ldots, N \tag{5.9}$$

where N is the number of beams in the fin. The oscillation amplitude a_i and phase ξ_i are derived from the following coupled linear oscillator equations

$$\ddot{a}_i = k_a^2(A - a_i) - 2k_a\dot{a}_i \tag{5.10}$$

and

$$\ddot{\xi}_i = 2c(N-1)(2\pi f - \dot{\xi}_i) - c^2 \sum_{j=1}^{N} [\xi_i - \xi_j - (i-j)\phi], \text{ for all } i \neq j \tag{5.11}$$

where k_a is a weight that determines the convergence rate of a_i and c is the coupling weight of the oscillator.

5.3.4 Dynamic simulations

The analytic fin model described above was first implemented and evaluated in MATLAB®. A basic two-beam and single-membrane model was initially utilized. Various hydrodynamic forces and moments were computed for different undulation parameters. Simulations were performed for both the flapping (standing wave) and undulation (traveling wave) modes.

In standing wave mode, the beams of the fin oscillate in phase and the fin moves as a rigid body. The forces generated by a single membrane port-side fin are shown in Figure 5.4. A relatively large amplitude oscillating force with zero mean is generated in the *x*-direction, a small amplitude oscillating force with zero mean is generated in the *y*-direction, and no force is generated in the *z*-direction. Relatively small-amplitude oscillatory moments are generated about all three axes. With a pair of fins, due to the symmetry of the port and starboard fins, only the *x*-direction forces and the *y*-direction moments interfere constructively, resulting in a heave and pitch motion. However, because of the zero-mean oscillatory nature of this force and moment, zero (or very small) oscillatory motion is obtained in heave and pitch. It should be noted that, similar to the swimming flipper, a net positive *y*-direction (sway) force is expected to be generated in this mode of operation. However, the simulation results refute this expectation. More is discussed on this matter in the latter part of this chapter.

In the traveling wave mode, the rays of the fin oscillate out of phase, generating a traveling wave. The forces and moments generated by an individual single-membrane

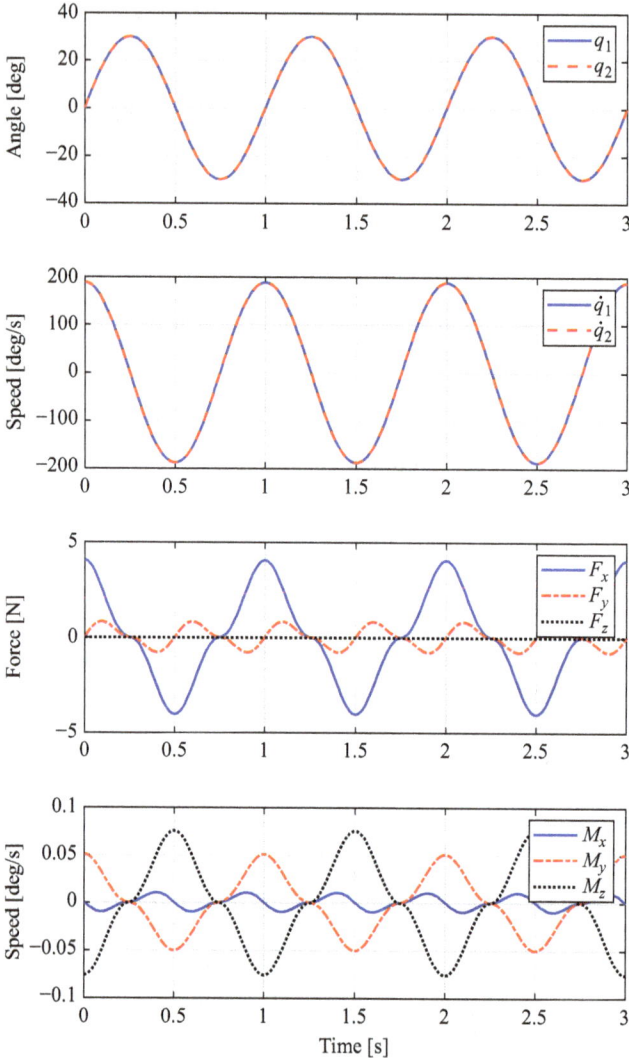

Figure 5.4   The beam kinematics and hydrodynamic loads generated by a single membrane simple port-side fin in flapping mode. Hydrodynamic loads are computed with respect to the fin coordinate system. The fin geometric and kinematic parameters are $h_{max} = 0.2$ m, $h_{min} = 0$ m, $W = 0.1$ m, $A = 30°$, $\phi = 0°$, $f = 1$ Hz, and $X = 0°$

port and starboard fin are shown below in Figure 5.5. Again, a relatively large amplitude oscillating force with zero mean is generated in the x-direction, whereas a smaller amplitude oscillating force with zero mean is generated in the y-direction. In contrast to the flapping mode, an oscillatory force with a non-zero mean is generated along the longitudinal z-direction, resulting in a surge motion for the robot. Similarly to the flapping mode, moments are generated about all three axes of the fin. With

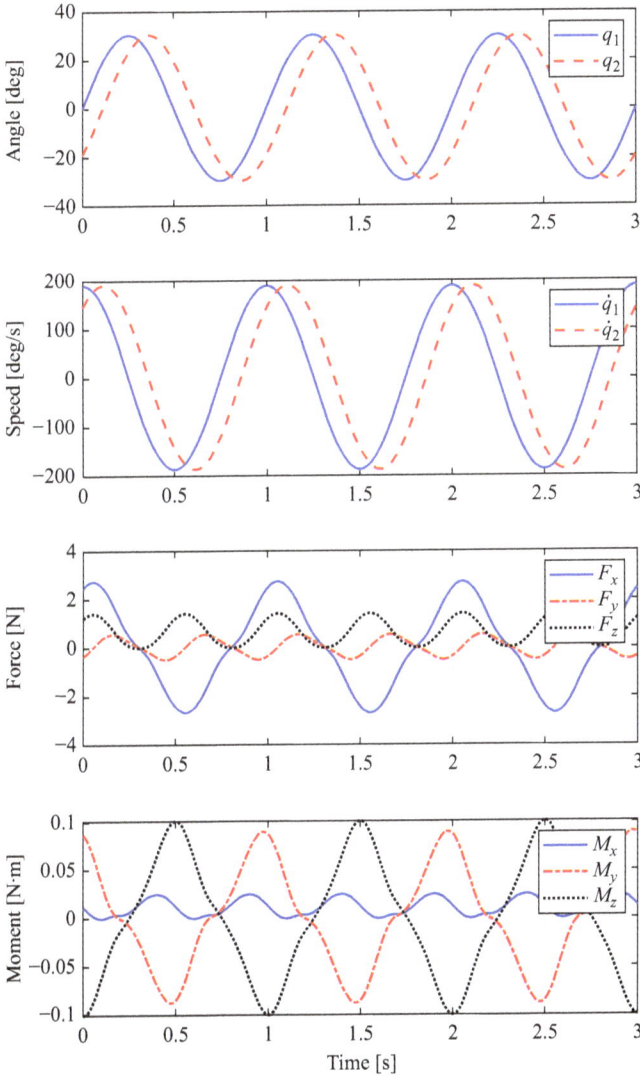

Figure 5.5   *The beam kinematics and hydrodynamic loads generated by a single membrane simple port-side fin in undulation mode. Hydrodynamic loads are computed with respect to the fin coordinate system. The fin geometric and kinematic parameters are $h_{max} = 0.2$ m, $h_{min} = 0$ m, $W = 0.1$ m, $A = 30°$, $\phi = 40°$, $f = 1$ Hz, and $X = 0°$*

a pair of fins, due to the symmetry of the port and starboard fin pairs, the x- and z-direction forces constructively interfere, while the y-direction forces cancel out. The x-moments of the two fins cancel out, resulting in no net yaw moment. The out-of-phase y- and z-moments constructively interfere, resulting in oscillatory pitch and roll motions.

5.3.5   Computational fluid dynamics

To validate the analytical model of the fin system (before the manufacture of the fin prototypes and laboratory testing), a CFD study was performed. Since the fin system consists of rigid and elastic components that move underwater, a fully coupled FSI approach was utilized in this study. The CFD model is designed as closely as possible to the geometry and operating conditions of the mechanical fin. COMSOL Multiphysics [14] software was used to simulate the hydrodynamics of the simple fin. The fluid flow is assumed to be laminar. The rigid parts of the fin are modeled using the multibody dynamics interface. The flexible fin membrane geometry is solved using the arbitrary Lagrangian–Eulerian (ALE) method. The ALE method models the dynamics of the deforming geometry and moving boundaries with a moving grid. A time-dependent analysis is used to simulate the motion of the fin model in fluid. An elastic material is assumed for the fin membrane, while other fin mechanisms such as connections and joints are assumed to be made of structural steel. Fresh water is considered the fluid medium. Free tetrahedral elements are used for the mesh. This mesh consists of 142 040 volume elements, 14 894 boundary elements, and 1 188 edge elements (see Figure 5.6). The parameters applied in the simulations are summarized in Table 5.1.

A comparison of the typical analytic model and the CFD results is presented in Figure 5.7 for the flapping and undulation modes. These results indicate that there is general agreement between the analytic and CFD computations. In particular, the calculated force magnitudes are comparable for both methods. Furthermore, both

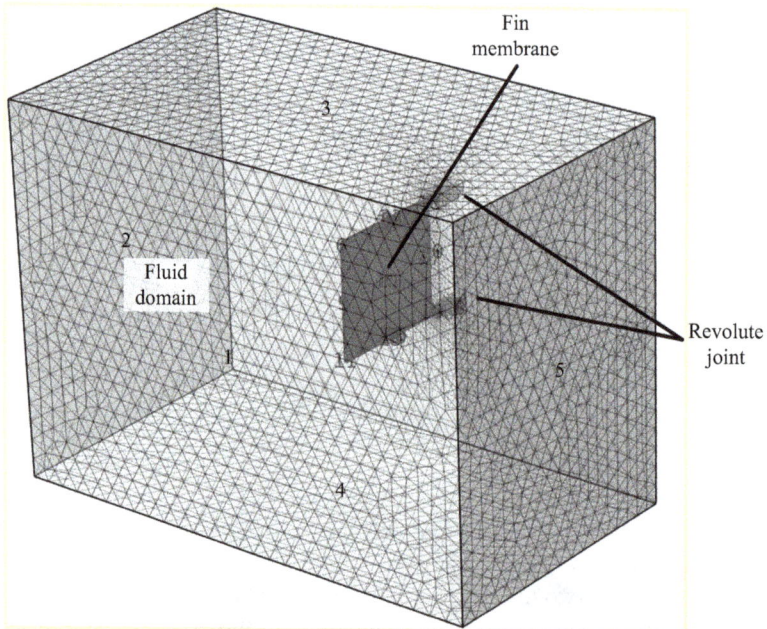

Figure 5.6   The model used in the CFD simulations for the basic fin

Table 5.1   Parameters used in the CFD simulations for the fin

Parameter	Value
Geometric properties	
Height of the membrane (H)	140 mm
Beam length (L)	144 mm
Membrane width (W)	90 mm
Membrane thickness (t)	3 mm
Beam material properties (Structural steel)	
Mass density (ρ)	7850 kg/m^3
Modulus of elasticity (E)	200 GPa
Poisson's ratio (ν)	0.3
Membrane material properties (Silicon rubber)	
Mass density (ρ)	750 kg/m^3
Modulus of elasticity (E)	1.3 GPa
Poisson's ratio (ν)	0.499
Fluid material properties (Fresh water)	
Mass density (ρ)	1000 kg/m^3

methods accurately predict that there will be no net force generated in the surge direction in the flapping mode, whereas a force fluctuating with twice the undulation frequency and with a net positive surge component is generated in the undulating mode. Likewise, the generated thrust also oscillates at twice the undulation frequency in the sway direction. Finally, both methods accurately predict that the zero mean heave force oscillates slower than the previous two force components, at the undulation frequency. Also, the effect of the CPG is evident in the form of a smooth and gradual increase in the amplitudes of the undulation forces from zero at the beginning of the simulation compared to Figures 5.4 and 5.5.

However, unlike the analytical model that failed to predict the generation of a net sway force in the flapping mode, CFD simulations indicate its existence (see Section 5.4). Furthermore, small variations in the hydrodynamic forces between analytical and CFD results also exist. The fin force computations in the analytic model are based on drag only (the so-called 'quasi-static' approach) and do not take into account the effects of fluid interaction [15]. In contrast, the CFD results come from a time-dependent (transient) study that more closely represents the actual fin moving in fluid. Moreover, there is a complete coupling between the fluid and the fin structure in the CFD analysis. That is, pressure and viscous fluid forces affect the motion of the structure, while structural velocity acts as a boundary condition for the fluid. The CFD model uses an elastic material for the membrane, while such deformations are ignored in the analytical model (more specifically, w_{max} is assumed to be constant in the derivation of (5.3)). Finally, the apparent phase shift between the analytical and CFD model results is also observed in other studies. The uncertainties in CFD modeling and the overprediction of the added mass term by CFD are cited among

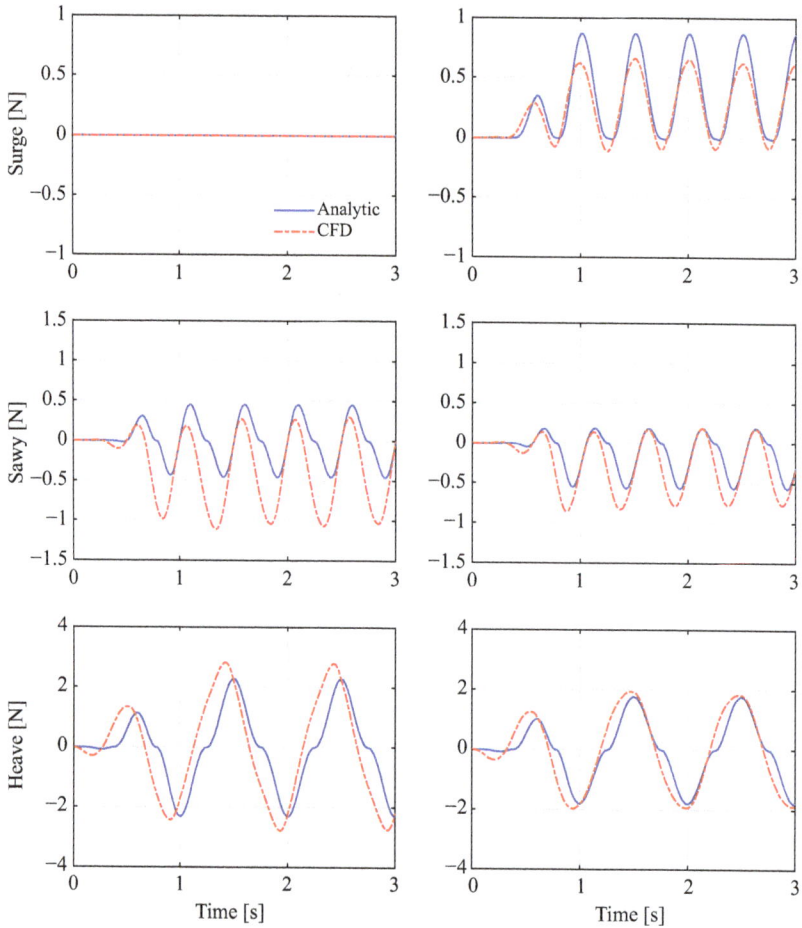

Figure 5.7 *A comparison between the analytic model and CFD results for the fin forces. Left column represents the flapping mode (A = 30°, f = 1 Hz, φ = 0°, X = 0°), while results for undulation with A = 30°, f = 1 Hz, φ = 40°, X = 0° is presented on the right column.*

potential causes of this phase difference [16]. A similar CFD analysis was performed for the moments generated by the fins.

The basic fin, which consists of a membrane stretched between a pair of beams, was extended to more elaborate designs with multiple beams and membranes. An optimization analysis was performed to determine the most suitable parameters for fin undulations. Various central pattern generators (CPG) are implemented for the smooth startup of the fins and transition between operation modes. The reader is referred to the thesis of the coauthor of this chapter for further details [17].

5.4   Measurements and results

This section explains the manufacture of the physical fin prototype and the laboratory experiments conducted to validate the modeling and simulation results presented in the previous section for the fin.

5.4.1   Prototype development and manufacture

A prototype of a two-membrane, three-beam ($N = 3$) fin was constructed to test and validate the analytical results discussed in Section 5.3. Parachute fabric was chosen as the membrane material. A membrane of size 200×200 mm was sewn onto 200 mm-long, 3 mm-diameter carbon fiber rods. The fin membranes are designed to be removable and replaceable. Although analytical modeling was based on rigid beams, highly flexible carbon fiber rods are preferred in the prototype because of their low cost, ease of procurement, strength, durability, resistance to corrosion, and light weight. These carbon fiber rods are connected to the gearboxes of the waterproof servomotors with aluminum brackets. The prototype fin is shown in Figure 5.8.

5.4.2   Laboratory tests and measurements

A test tank is used to perform the laboratory test of the fin system. The details of this experimental setup are shown in Figure 5.9. The fin is placed vertically (with the sway direction facing downward) inside the test tank. An inverted "L"-shaped bracket is used to mount the fin to the test tank structure. The forces and moments generated by the fin are measured using a high-resolution force/torque sensor. To remove the weight and other effects of the bracket and other components, the sensor was recalibrated prior to each experiment.

Several experiments are carried out in both the traveling and standing wave modes to measure the total thrust and moment produced by the fin and compare them

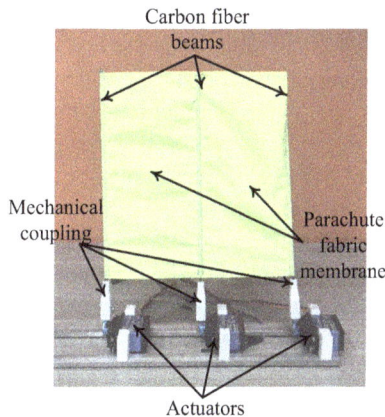

Figure 5.8   The two-membrane fin prototype

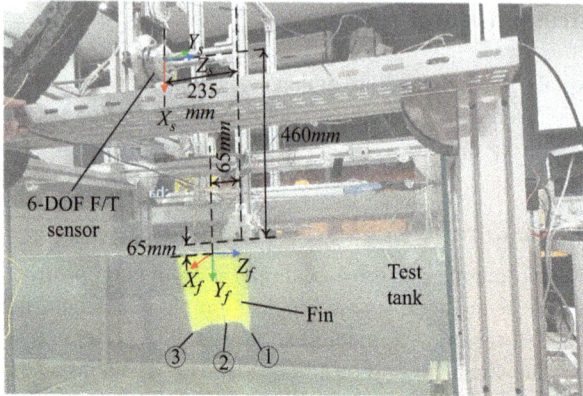

Figure 5.9   The test tank and bench used in laboratory experiments of the fin system

with analytical model results. The beam undulation parameters are generated in software at a rate of 100 Hz and passed to the CPG before being sent to the servomotor drivers. The CPG output is also fed to the analytical model. A camera is placed outside the tank to track the motion of the first beam and check whether the low-cost servos are capable of tracking the commanded trajectories in the presence of hydrodynamic loads. A simple image processing algorithm is used to track the angle of the coupling elements, as shown in Figure 5.10. A typical measurement obtained from this video camera and image-processing software overlaid with the CPG commanded trajectory is presented in Figure 5.11, indicating fairly reasonable agreement between the two. Such comparison analyses are performed with all experiments to ensure that the desired fin undulation is achieved.

The force/torque sensor also samples measurements at 100 Hz. Experimentally measured forces and torques are passed through a 10th-order Butterworth low-pass digital filter (LPF) with a cut-off frequency of 10 Hz to remove high-frequency mechanical vibration and electronic noise. Synchronization of experimental hardware and simulation software is accomplished using ROS timestamps. Representative force and moment comparisons obtained for the flapping mode are given in Figure 5.12. Immediately evident from the figures is that the analytic model underestimates the hydrodynamic force and moments generated by the fins. Furthermore, although the analytic model and CFD simulations agree that no forces should be generated in the surge direction, the experiment reveals a non-negligible surge force. The main cause of this unexpected surge force and the exaggerated hydrodynamic loads in general is attributed to the inevitable oscillatory motion of the "L" bracket due to hydrodynamic loads acting on the fin. This motion of the base of the fin is further exacerbated by complex flow patterns that emerge in the fluid due to the limited volume of the test tank. In the alternative experimental setup discussed by Sfakiotakis *et al.* [10], the fins are mounted on a rail system, enabling the system to make very stable and noise-free surge force measurements. However, the rail setup prevents the measurement of sway and heave forces, as well as any other moment generated by the

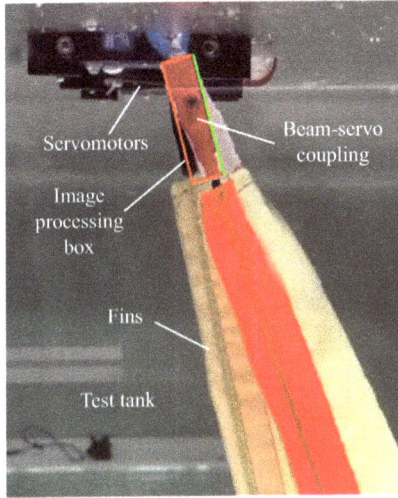

Figure 5.10   A photo captured from the video camera showing the image processing box around the first beam coupling, which is used to ensure the servomotors are providing the desired undulation.

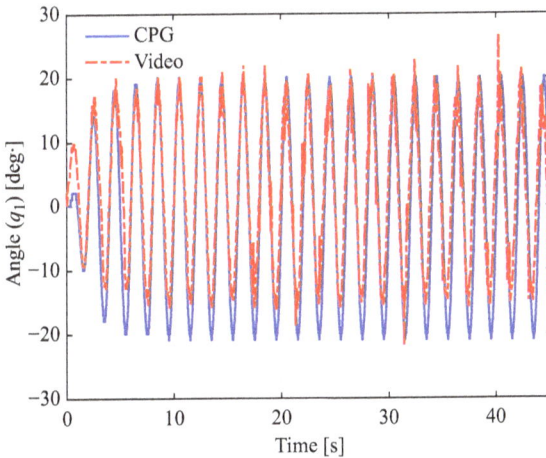

Figure 5.11   The CPG-commanded and video-measured actual joint angles for the first beam for an $A = 20°$ amplitude commanded oscillation

fin. In this work, a holistic approach was taken to measure all of the hydrodynamic loads generated by the fin simultaneously. Since the actual robot will also experience such oscillatory loads underwater, it is believed that the results provided in this work are more representative of the actual loads that the fins will generate. However, because the actual robot has a higher inertia compared to the bracket, despite being

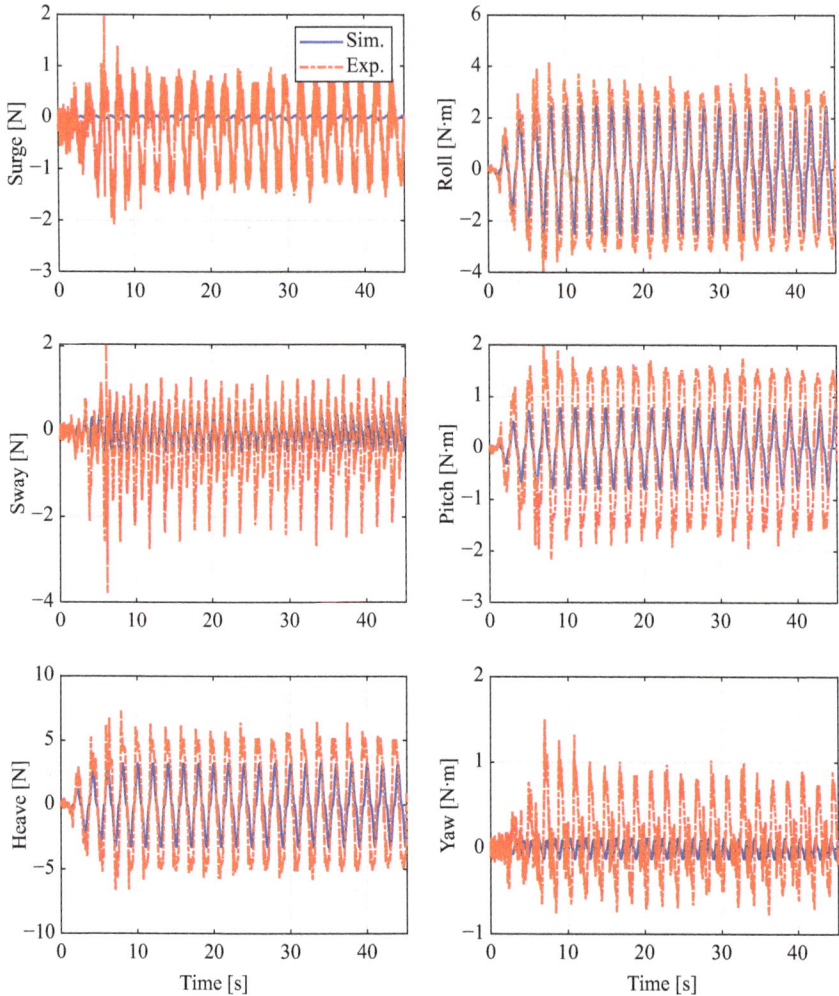

Figure 5.12   *A typical comparison between the analytic model hydrodynamic load predictions and experimentally measured force generated by a fin in flapping mode. The results presented here are for a two membrane, three beam (N = 3) fin with $h_{max} = 0.2$ m, $h_{min} = 0$ m, $W = 0.1$ m (per membrane), $A = 20°$, $\phi = 0°$ (flapping mode), $f = 0.5$ Hz, and $X = 0°$*

present, these coupling effects are expected to be less pronounced. In accordance with the CFD simulations, experiments also prove the existence of a net sway force generated in the opposite direction of the fin sway axis. Finally, the experiments agree well with the prediction that the heave force as well as the roll and pitch moments oscillate with the flapping frequency while the rest of the loads oscillate twice as fast.

Representative results for the undulation mode are also presented in Figure 5.13. Similar arguments can be made for the experimental results in the case of undulating

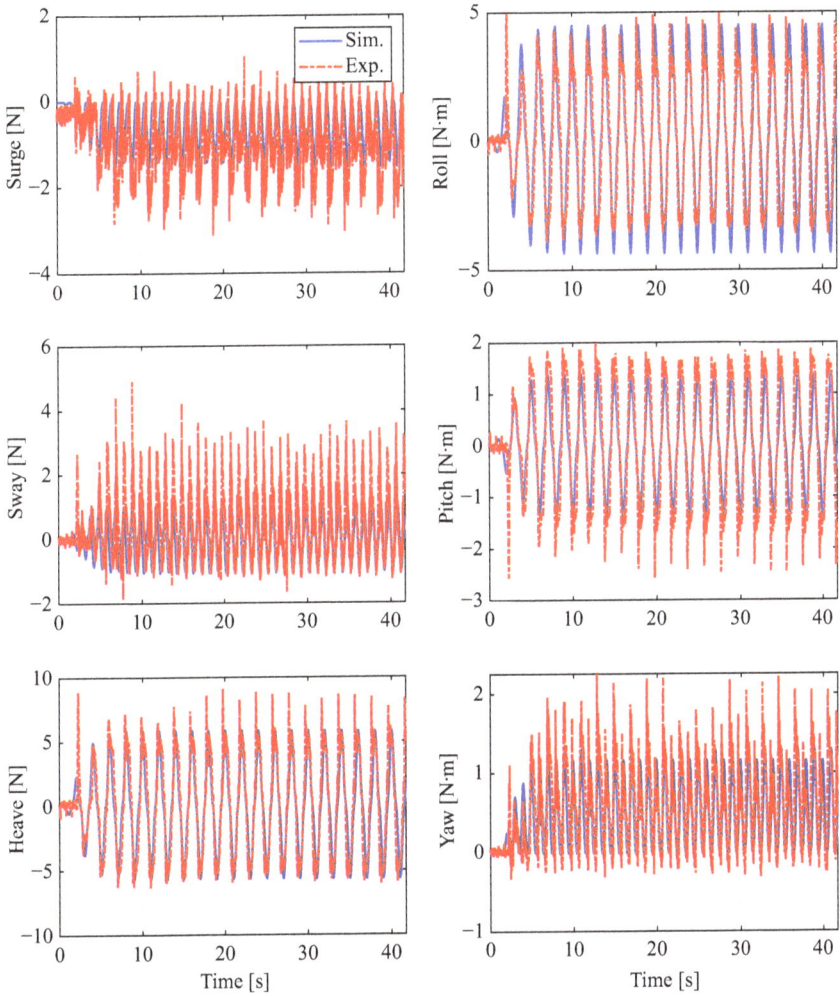

Figure 5.13   *A typical comparison between the analytic model hydrodynamic load predictions and experimentally measured force generated by a fin in flapping mode. The results presented here are for a two-membrane, three-beam (N = 3) fin with $h_{max} = 0.2$ m, $h_{min} = 0$ m, $W = 0.1$ m (per membrane), $A = 20°$, $\phi = 30°$ (undulation mode), $f = 0.5$ Hz, and $X = 0°$*

mode operation. Hydrodynamic loads are generally higher than those predicted by the analytic model. An oscillating surge force with a nonzero mean is observed in the surge direction, which is used to propel the robot forward and backward. Despite some mismatch in the amplitudes, there is again a good agreement with theoretical results in the oscillation frequencies of the hydrodynamic forces and moments.

There are several other reasons for the discrepancy between the analytical and experimental results. As elaborated in Section 5.3, the quasi-static assumption utilized in the analytic model may not be very accurate. In addition, it is very difficult to determine the effective drag coefficient of the fin membrane. The analytic model assumes that the beams are rigid and the membranes are inelastic. Both assumptions are grossly inaccurate, in particular when highly flexible carbon-fiber rods and parachute fabric are used for the beams and membranes, respectively. Finally, albeit limited, a discrepancy does exist between the CPG commanded beam motions and the servo motions. Nevertheless, the analytic model is validated through both CFD simulations and tank experiments. Furthermore, the CFD simulations and the experiments highlight areas in the analytic model that need improvement. In particular, the drag model must be modified, and the restrictions on the nonelasticity of the beams and membranes must be relaxed.

5.5 Conclusions

URSULA is a robotic squid designed specifically for dexterous underwater manipulation and seabed intervention tests. Due to potential risks such as entanglement that propeller-based thrusters possess, URSULA deploys two novel propellerless locomotion systems. One of these systems, namely the fin system is presented in this chapter. A fin consists of elastic membranes stretched between multiple rigid rods (termed beams) protruding from the body of the robot. The fins are actuated through servomotors attached to the base of the beams and can swivel about the longitudinal axis of the robot to any offset angle between $-90°$ (pointing down vertically) and $90°$ (pointing up vertically) angles. The robot possesses four such fins (bow, aft, port, and starboard), providing it with omnidirectional motion capability. The operating modes (flapping and undulation) of the fin are explained. A detailed theoretical derivation of the thrust generation mechanism and the analytic model of the fin is provided. Unlike the existing literature, which reports only hydrodynamic forces generated by fin systems, both fin forces and moments are computed by the model presented in this work. The model can also compute the loads on the servomotors directly or indirectly through the Jacobian. However, this aspect of the analytic model is not discussed here due to space concerns. The CPG algorithm that synchronizes the motion of all four fins as well as makes transitions between different modes of operation is presented.

The analytic model was first evaluated by comparing its fin hydrodynamic load predictions with numerical CFD simulations. The analytic results align well with the CFD computations, except in the flapping mode, where contrary to expectations, the analytic model predicts no net force in the sway direction. Following CFD simulations, an actual prototype of the fin is built and evaluated in a test tank. Similar to CFD simulations, the experimental results indicate the presence of a net sway force in the flapping mode. The fins are shown to generate net sway and surge force when operating in the flapping and undulation modes, respectively. Coupled with the quartet arrangement of the fins, the possibility to swivel the fins fully up or down about the robot's longitudinal axis provides URSULA with omnidirectional motion capabilities. A detailed discussion on the shortcomings of the analytic model, fin design, and experiment setup is also provided. Furthermore, an optimization analysis based on the geometric and operational parameters of the fin, as well as the size of the fin (in terms of the number of membranes), is not discussed here because of space constraints. The latter is of great interest as it makes possible the superposition of multiple undulation waves on a single fin. The reader is referred to the thesis of the coauthors of this chapter for further details.

Several improvements to the fin model and the actual fin design are planned for the future. The analytic model will be modified to incorporate the elasticity of the membranes and the flexibility of the beams. A more accurate model for the fin hydrodynamic force generation mechanism that can predict the sway forces in flapping mode is necessary. Smaller, rigid fins that are actuated by a single servomotor may represent a viable alternative to the elastic designs discussed in this chapter. Such rigid fins can also help improve reliability, consistency, and durability, which appear to be problematic with elastic fins.

References

[1] Williams N. "Environmental credit crunch." *Current Biology.* 2008;**18**(21):R979–80.

[2] Costanzo G., Brindley B., Tardieu P.: *Wind Energy in Europe 2024*. WindEurope asbl, Brussels, Belgium: European Wind Energy Association (EWEA); 2024. 64 pp. [Accessed date: 6 June 2025].

[3] Dhavle J. and Pirttimaa L.: *Scaling Up Investments in Ocean Energy Technologies 2023*. Abu Dhabi, UAE: International Renewable Agency (IRENA); 2023. 31 pp. [Access date: 6 June 2025]

[4] Lehmköster J., Löschke S., Ladischensky D., *et al.*: *World Ocean Review 2021*. Hamburg, Germany: Maribus gGmbH; 2021. 7. 169 pp. [Accessed date: 6 June 2025]

[5] Offshore AUV and ROV – Global Strategic Business Report. CA, USA: Global Industry Analysts, Inc; 2025. 284 pp. [Accessed date: 6 June 2025].

[6] Bleicher A.: "The gulf spill's lessons for robotics." *IEEE Spectrum.* 2010;**47**(8):9–11.

[7] ROS – Robot Operating System [Website on the Internet]. *Open Robotics.* San Jose, CA: Open Robotics; 2025 Available from: https://ros.org/ [updated 2025 Mar 31; accessed 2025 Mar 31]

[8] Kapicioglu K., Getmez E., Akbulut B.E., *et al.*: "A touchless control interface for low-cost ROVs." *IEEE Proc. OCEANS Conf. San Diego–Porto*; San Diego, CA, USA, 2021 Jul. 20–23. Piscataway: IEEE; 2021. pp. 1–6.

[9] Sümey E.M. and Gür B.: "A robot simulation environment for virtual reality enhanced underwater manipulation and seabed intervention tasks." *IEEE Proc. Int. Conf. Robot. Autom.*; Atlanta, GA, USA, 2025 May 19-2-3. Piscataway: IEEE; 2025. pp. 1–6.

[10] Sfakiotakis M., Fasoulas J., and Gliva R.: "Dynamic modeling and experimental analysis of a two-ray undulatory fin robot." *IEEE/RSJ Int. Conf. Intell.*

Robot. Sys. (IROS), Hamburg, Germany, 2015 Sep. 28–Oct. 2. Piscataway: IEEE; 2015. pp. 339–46.

[11]   Guertin P.A.: "Central pattern generator for locomotion: Anatomical, physiological, and pathophysiological considerations." *Frontiers in Neurology.* 2013;**3**(183):1–15.

[12]   Sproewitz A., Moeckel R., Maye J., *et al.*: "Learning to move in modular robots using central pattern generators and online optimization." *The International Journal of Robotics Research.* 2008;**27**(3–4):423–43.

[13]   Gliva R., Mountoufaris M., Spyridakis N., *et al.*: "Development of a bio-inspired underwater robot prototype with undulatory fin propulsion." *Proc. 9th Int. Conf. New Horizons in Industry, Business and Education (NHIBE'15)*; Skiathos, Greece, 2015 Aug. 2015. p. 1–6.

[14]   COMSOL Multiphysics [Website on the Internet]. COMSOL, Inc., 100 District Avenue, Burlington, MA 01803, USA: Comsol Inc.; 2025 Available from: https://www.comsol.com/ [updated 2025 Mar 31; accessed 2025 Mar 31]

[15]   Sefati S., Neveln I., MacIver M.A., *et al.*: "Counter-propagating waves enhance maneuverability and stability: A bio-inspired strategy for robotic ribbon-fin propulsion." *Proc. IEEE RAS-EMBS Int. Conf. Biomed. Robot. Biomechatronics*; 2025 May 19–23; Atlanta, GA, USA. Piscataway: IEEE; 2012. pp. 1620–5.

[16]   Randeni S.A.T.P., Leong Z.Q., Ranmuthugala D., *et al.*: "Numerical investigation of the hydrodynamic interaction between two underwater bodies in relative motion." *Applied Ocean Research.* 2015;**51**:14–24.

[17]   Ali A.I.: *Dynamic Modeling and Control of Bioinspired Undulating Fins for Squid-like Biomimetic Robots* (thesis). Beşiktaş, İstanbul, Türkiye: Bahçeşehir University; 2021.

Chapter 6

Application of the swarm-capable micro-AUV MONSUN for monitoring of vegetation and water quality in lakes

Erik Maehle[1], Gaurav Kapoor[1] and Sören Nienaber[1]

This chapter introduces the concept and experimental findings related to the MONSUN micro-Autonomous Underwater Vehicle (micro-AUV) created at the University of Lübeck for monitoring environmental conditions in lakes. In addition to assessing physical–chemical parameters, the evaluation of macrophyte vegetation is carried out to determine water quality. The configuration of MONSUN's sensors and the navigation techniques employed for capturing images of macrophytes are detailed, along with the recognition process for their automatic evaluation using AI methods.

6.1 Introduction

Water bodies such as lakes, rivers, transitional, or coastal waters represent delicate ecological systems that are vulnerable to human activities and climate change. Achieving and maintaining a favorable status presents a significant challenge that is addressed, for example, by the European Water Framework Directive [1]. This directive mandates, among other things, the consistent monitoring of both the ecological and chemical conditions. In addition to the physical and chemical metrics, macrophytes serve as a critical indicator of water quality. Currently, their monitoring is predominantly conducted manually by divers. The research initiative "*MO*nitoring of *VE*getation and Water Quality of Lakes with Underwater Robot Swarms *(MOVE)*" aims to automate this process through the use of cooperating MONSUN micro-AUVs, which have been developed at the University of Lübeck. The MONSUN micro-AUV is already described in detail in a previous IET book on the design and practice of AUVs [2]. For MOVE, several mission-specific extensions were required, which are the subject of this chapter.

In this chapter, contemporary methods and associated research are introduced initially, focusing particularly on the approach to observing macrophytes in lakes as per the European Water Framework Directive, and the application of micro-AUVs for monitoring tasks. Additionally, modern AI techniques for underwater

[1] Section of Computer Science and Technology, Institute of Computer Engineering, University of Lübeck, Germany

image identification are overviewed. Following that, the hardware and software of the MONSUN micro-AUVs in the configuration used for MOVE, along with the monitoring strategies employed, are outlined. Subsequently, the framework for the AI image identification of macrophytes is illustrated. Finally, the experimental findings related to navigation, hovering, vegetation boundary identification, and AI assessment of macrophyte images are presented. The chapter is partly based on a previous conference paper about the MOVE project [3].

6.2   Current methods and associated research

6.2.1   Monitoring of water quality in lakes

Healthy rivers, lakes, coastal regions, and groundwater are vital resources for individuals, industries, and ecosystems. In Europe, the Water Framework Directive (WFD) serves as the primary legislation for water conservation established by the European Union and adopted by its member countries [1,4]. It pertains to inland, transitional, coastal surface waters, and groundwater. Its aim is the sustainable management of water resources and the maintenance of the ecological integrity of water bodies. The primary objective is to attain a "good status," which signifies a "good ecological and chemical status" for surface waters. Additionally, the degradation of water bodies must be avoided. To achieve and maintain a good status, monitoring is utilized to evaluate specific data regarding biology, hydro-morphology, and chemistry. The reference status signifies the largely unaltered condition of the corresponding water body, from which only minimal variations are permitted. The ecological status is then categorized on a 5-level scale that ranges from "very good" (no variation from reference status) through "good" (minor variation) to "bad" (significant variation).

This chapter will focus on the assessment of lakes. In accordance with the WFD in Germany, their status must be evaluated at regular intervals (typically every 3 years), following a standardized method (called *Phylib* [4]). A key aspect here is the monitoring of submerged macrophytes as a crucial indicator of water condition. Since complete monitoring is not practical, representative portions of the lake are chosen. These so-called transects begin at the shoreline and extend straight into the lake until reaching the vegetation boundary, i.e. the point at which the vegetation on the lakebed ceases entirely due to insufficient sunlight. The number of transects relies on the lake's size and the shape of its shoreline. Typical figures range from 1 to 5 for very small lakes, escalating to 30–50 for larger ones. The usual width of a transect spans 20–30 m, while the typical length varies from several tens of meters to a few hundred meters, depending on the specific lake. For these transects, the macrophyte vegetation is mapped. To accomplish this, the occurrence frequency of each macrophyte species is assessed. This analysis is performed at four depth levels (0–1 m, 1–2 m, 2–4 m, and 4 m up to the vegetation boundary). The occurrence frequency is evaluated using a 5-level scale that ranges from "very rare" to "plentiful." Currently, the monitoring and mapping of macrophytes must primarily be conducted manually by human divers. They carry out the assessment while diving along the transect, necessitating their possession of biological expertise regarding macrophytes. A suitable method utilizing GPS at the surface is detailed in [5]. Alternatively, samples of the macro-phytes can be collected from a boat using a rake, which are then analyzed by human specialists. In very shallow waters at the depth level 1 (0 m to 1 m), viewing boxes may also be utilized. All these methods are quite labor-intensive and costly. Additionally, it would be highly beneficial to conduct mapping not just for transects but for entire areas, which, however, is often not practical due to financial constraints. More frequent monitoring over shorter periods would also be advantageous. To achieve this, the automation of monitoring and mapping processes through the use of autonomous underwater robots and AI image recognition is suggested in this chapter. Other types of remote sensing from the air (e.g. satellites, airplanes, or drones) or acoustic methods have also been explored in the literature (see [6] for a review) that, nonetheless, do not permit the resolution and precision needed here.

6.2.2   Underwater robots for environmental monitoring

Autonomous underwater robots are already widely used. Nevertheless, a majority of them are quite big, heavy, and costly devices intended for use in seas and oceans. Additionally, they demand expensive support facilities such as vessels and cranes. They are thus not ideally tailored for shallow environments, which is necessary for observing lakes. Recently, the so-called micro-AUVs have also become available in the market. These are rather small (less than ca 1 m length) and light (less than ca 10 kg weight), less expensive, and easy to manage by one person without further support. Examples are Vertex [7], YUCO [8], NemoSens [9], ecoSubμ5 [10], or Hydrus [11]. They can function in shallow waters and also partially assist in swarming activities. A major challenge is underwater localization because GPS is not obtainable beneath the surface. Nevertheless, precise geo-referencing of the observed data is crucial for monitoring purposes (see [12] for an overview). There are acoustic systems available, such as USBL (Ultra Short Baseline), which are, however, costly and necessitate a complex setup.

The advantage of the MONSUN micro-AUVs utilized in MOVE is their ease of handling and modularity. With their six motors, they are highly maneuverable and capable of hovering. Being swarm-enabled, they can operate concurrently and utilize cooperative localization and navigation. Consequently, they are highly appropriate for monitoring tasks in lakes as intended here.

6.2.3   AI techniques for macrophyte image recognition

At present, the recognition of macrophytes for monitoring and mapping water quality in lakes is conducted by human specialists. Recent advancements in AI techniques, especially deep neural networks, have enabled the creation of effective applications for identifying terrestrial plants. Examples include PlantNet [13] or Flora Incognita [14]. These applications primarily focus on assisting humans in recognizing plants in common scenarios. Additionally, there are numerous studies that address deep learning for identifying underwater entities such as fish, plankton, or corals.

Here, impressive outcomes have already been attained (refer to [15,16]). With regard to aquatic vegetation, the imaging and classification of seagrass in coastal regions have captured the interest of researchers [17]. Primarily, the scope and density of seagrass meadows have been evaluated, while individual plants have also been recognized using neural networks [18]. To the best of the authors' knowledge, there is still a lack of research on AI-driven classification and detection of various macrophytes as needed here. Most methods in computer vision concentrate on enhancing the image quality of underwater photos to boost visibility for human observation, rather than targeting AI applications. Specifically, color correction techniques (see [19–21]) are being investigated, as they enable enhanced visibility of the flora and fauna underwater.

6.3 MONSUN micro-AUV

6.3.1 Hardware

MONSUN is a compact (under 1 m in length), lightweight (under 10 kg in weight), adaptable, and modular micro-AUV that can be easily customized for specific applications. It provides exceptional maneuverability, thanks to four vertical and two horizontal thrusters, and can be fitted with a variety of sensor equipment. A comprehensive description can be found in chapter 12 of the prior IET publication "Autonomous Underwater Vehicles: Design and Practice" [2]. In previous projects, it has been utilized for tasks such as monitoring eddies in the Baltic Sea [2], detecting water contamination in harbors, or inspecting quay walls [22].

The configuration utilized for MOVE is displayed in Figure 6.1. An oriented downward camera equipped with LED lights (1) is included to capture footage of

Figure 6.1   MONSUN micro-AUV configuration utilized for monitoring vegetation and water quality in lakes

the macrophytes as the robot maintains a steady elevation of 1–1.5 m above the lake ground. To achieve this, a downward-facing ping sonar (3) from *BlueRobotics* is utilized. To identify the vegetation boundary, a second online camera is installed (2). Communication underwater with other MONSUNs is accomplished by *ahoi-acoustic mini modems* [23] (depicted at the bottom section of the robot (4)), communication and localization at the surface by WiFi and GPS, respectively (refer to antenna mast at the top (5)). Intermediate rings can optionally be incorporated within the hull. Figure 6.1 shows, for example, a sensor ring developed by the MOVE project partner *Sea & Sun Technologies* (SST sensor ring). It can be flexibly configured to accommodate high-precision physical–chemical sensors, for example for temperature, pressure, conductivity, oxygen, pH value, turbidity, or chlorophyll. Additionally, the light spectrum can be measured (6).

6.3.2   Software

The program responsible for the robot's autonomous actions is built upon the widely recognized *Robot Operating System (ROS) Kinetic* [24] in conjunction with Ubuntu 16.04, running on a Raspberry Pi 3 single-board computer. ROS utilizes a framework of nodes that interact with one another via publishers and subscribers over so-called topics. The software of the robot is structured into three hierarchical layers.

Drivers for sensors and actuators represent the most fundamental layer. Communications with the sensors and actuators are realized via an I2C bus or USB and distributed across designated ROS topics. The sensors and actuators consist of the following:

- *GPS:* The information from the GPS comprises the present time and the coordinates of the GPS fix.
- *IMU:* The information obtained from the Inertial Measurement Unit (IMU) includes the roll, pitch, and yaw positioning of the robot.
- *Ping Sonar:* The information obtained from the Ping Sonar encompasses the distance from the lakebed along with its associated confidence value.
- *Battery:* The information from the battery monitor includes the charge status of the LiPo batteries utilized to operate the robot.
- *Acoustic modem:* Besides being used for communication, the acoustic modem supplies roundtrip duration of the data packets transmitted to the other modems for measuring the distance between the robots.
- *Thrusters:* The thrusters node is responsible for transmitting a specific speed value to both the vertical and horizontal motors of the robot.

Controllers layer governs the fundamental operations of the robot. The information obtained from the sensors is accessed by the controller nodes through their designated ROS topics. This encompasses the following:

- *Hovering:* While hovering, the robot is directed to keep a steady distance above the ground by utilizing the depth measurements from the pressure sensor, D_p and the distance to the ground measured by the ping sonar, S. Therefore, for a desired hovering distance, *H*, the commanded depth value for MONSUN's depth controller, D_r is determined by the subsequent formula:

$$D_r = D_p + S - H$$

- *Depth, Roll, and Pitch:* The depth controller is utilized to command a given depth value to the robot. The actual depth indicated by the pressure sensor and the target depth are input into a controller to determine the necessary thrust for the four vertical motors of the robot to reach the designated depth. By employing the roll and pitch data from the IMU, another controller is applied to maintain stability for the robot when operating underwater.
- *Heading:* This controller is utilized to command the intended compass course for the robot. It incorporates a controller that determines the necessary force exerted by the robot's two horizontal motors to reach the target heading by utilizing the present heading (yaw) readings from the IMU.

Behavior layer instructs the robot to execute particular actions such as GPS Navigation, V-Formation, Point and Shoot, and Obstacle Avoidance. These actions can be initiated or halted using ROS services, as well as through a timeout in their individual ROS nodes when the robot is underwater. The data flow for both the hovering and GPS navigation actions is illustrated in Figure 6.2. The description of the actions is outlined as follows:

- *GPS Navigation:* This functionality is utilized to guide the robot to specific GPS coordinates on the water's surface. The GPS navigation functionality is activated through a ROS service, which simultaneously enables the Heading control node. The necessary yaw orientation of the robot toward a GPS waypoint is derived from the current GPS coordinates and those of the waypoint. The current GPS

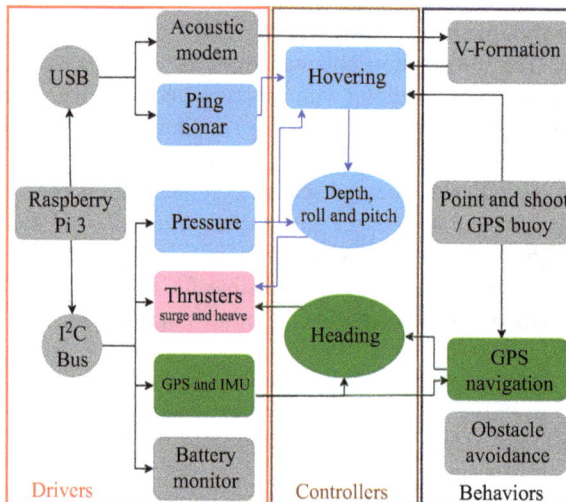

Figure 6.2   Layers of abstraction in MONSUN software

coordinates are supplied by the GPS sensor data node. Once the heading is determined, it is sent to the Heading control node, which directs the robot's horizontal motors via the motor node to attain it, utilizing the yaw data from the IMU. When the target heading is achieved, it is sustained by the heading controller, while the horizontal motors are instructed to achieve a predetermined linear speed toward the GPS waypoint. Upon the robot reaching the GPS waypoint, the Heading control node instructs the motor node to halt the robot.

- *V-Formation:* Given that the radio signals utilized by GPS are unable to pass through water, collaborative swarm localization in a V-formation is utilized as outlined in earlier studies referenced in [25]. Two MONSUN AUVs remain on the surface, maintaining their locations via GPS, while a third unit is underwater and hovers above the macrophytes to capture videos of them. In addition to communication, the acoustic modems support distance measurements through roundtrip delays. By utilizing trilateration, the submerged robot can calculate its relative position to the surfaced units, enabling it to geo-reference its monitoring data.

- *Point and Shoot/GPS Buoy:* These behaviors serve as a substitute for V-Formation to maneuver beneath the surface for observation aims (refer to the following chapter below).

- *Obstacle Avoidance:* Typically, one does not anticipate any obstacles when navigating along a transect in a lake. Nonetheless, for safety purposes, a basic obstacle avoidance behavior is made available as an option. This utilizes a forward-facing ping sonar (not depicted in Figure 6.1). In the most straightforward scenario, it halts the mission by shutting down the thrusters and surfacing. Additionally, more intricate responses, such as attempting to navigate around the obstacle, could certainly be developed, but this is outside the focus of this chapter.

6.4   Monitoring strategies for MONSUN

6.4.1   Navigation

Three different navigation approaches were examined:

(1)   *V-Formation:* In this behavior, a minimum of two extra MONSUNs remain at the surface, referred to as S-AUVs (surfaced AUVs). The S-AUVs determine their location via GPS and exchange information among themselves and with a control station located on a boat or at the shoreline using WiFi. The submerged D-AUVs (dived AUVs) interact with the S-AUVs through acoustic modems. The two-dimensional positioning of the D-AUVs is established by assessing their distance from the S-AUVs using the acoustic modems. Three-dimensional localization of the D-AUVs is unnecessary for the acoustic communication since the depth information is provided by the pressure sensor. This swarm behavior enables the robot to achieve autonomous navigation underwater. The D- and S-AUVs can also switch roles in the event of a malfunction. For instance, if one of the D-AUVs experiences a failure in the pressure

sensor or vertical motors, an S-AUV can function as a D-AUV and vice versa, owing to their comparable sensor configurations.

(2)   *Point and Shoot:* This navigation method serves as an alternative to the V-formation technique and requires only one robot, unlike the V-formation, which needs a minimum of three. In this behavior, the robot determines the necessary direction toward a GPS waypoint with respect to its current GPS location while it is on the surface. It then turns itself toward that GPS waypoint (*point*) on the surface using the yaw heading from the IMU at a linear speed of zero. Once the desired heading is achieved, the robot initiates Hovering at a defined hovering distance underwater and begins to move toward the GPS waypoint (*shoot*), maintaining the calculated constant heading using merely the current IMU yaw readings. This hovering continues for the calculated duration based on the distance from the waypoint and the estimated speed over ground, or for a predetermined time, such as two minutes, whichever is shorter. The speed over ground of the robot is derived from its pace during the preceding dive. After the allotted time, the robot surfaces and realigns its heading by pointing toward the waypoint, then resumes hovering underwater as it progresses toward it. This cycle repeats until the waypoint is attained. If the robot surfaces and realizes that it has exceeded the waypoint, it will bypass that waypoint and initiate the Point and Shoot action toward the subsequent waypoint until all GPS waypoints are accomplished.

(3)   *GPS buoy:* This navigation method serves as an alternative to V-formation navigation, requiring only one robot. In this approach, a GPS buoy that floats on the water's surface is linked to the MONSUN through a cable. The MONSUN tows the GPS buoy behind it while navigating underwater using the GPS Navigation behavior together with Hovering. This method addresses the limitations of the Point and Shoot behavior, as the robot continuously receives a GPS signal while submerged. It facilitates hovering underwater with ongoing adjustments to the heading based on the robot's GPS positions. The robot's GPS position relative to the GPS buoy is determined by considering the GPS coordinates of the buoy, the robot's yaw heading obtained from the IMU, the cable length of the GPS buoy, and the robot's depth from the pressure sensor. However, a significant concern here is the limited length of the cable.

6.5   Image recognition of macrophytes using artificial intelligence

6.5.1   *Approach for AI visual recognition*

For AI image recognition, a workflow was created. It comprises two primary elements (see Figure 6.3). The initial section focuses on the development of a dataset for the training process, while the second element outlines the production pipeline. In the initial section (Figure 6.3(a)), images (frames) are extracted from the MONSUN videos and other sources, which are subsequently labeled using CVAT (Computer Vision Annotation Tool) [26]. The resulting dataset will provide the foundation for

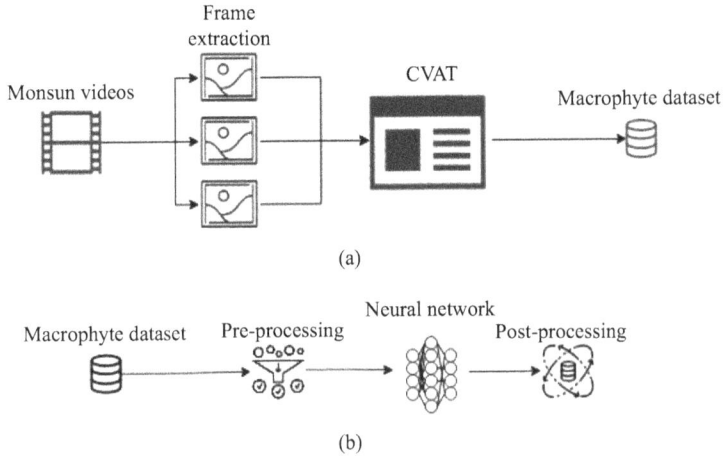

Figure 6.3   AI image recognition process. (a) Dataset pipeline and (b) production pipeline

training the neural network. During the production stage (Figure 6.3(b)), the trained neural network will be utilized to recognize images from the MONSUN videos. In this phase, pre- and post-processing occur before and after the analysis, to enhance the outcomes.

6.5.2   Development of a labeled collection of macrophyte images

To train the neural network, a dataset with labeled macrophyte images needed to be generated initially. For this reason, a software application was created to extract images from videos of macrophytes. Within this application, the images from the videos are sampled by assessing the video frames through a similarity measure. When the images show significant differences, a compilation of the prior images is formed. From this compilation, the image featuring the most characteristics is chosen and added to the dataset as the optimal image. This process is carried out throughout the entire video. The goal of this application is to gather as many images as possible from the recordings while ensuring a high level of variety among the different images. In addition to the information obtained from MONSUN videos, manually recorded videos were employed that concentrate more on specific species. The dataset was further expanded to encompass specific plant images from the GBIF database (Global Biodiversity Information Facility [27]). Nearly 30,000 macrophyte images were thus gathered throughout the duration of the project.

The images were submitted to a biological specialist at the MOVE project partner *Landesamt für Umwelt Schleswig-Holstein* for pixel-by-pixel annotation of the macrophytes depicted in them. To assist the specialist, the computer vision annotation tool (CVAT) [26] was utilized (Figure 6.4). In this way, a cumulative count of 2105 images representing 46 distinct species, along with other structural components was labeled throughout the duration of the project.

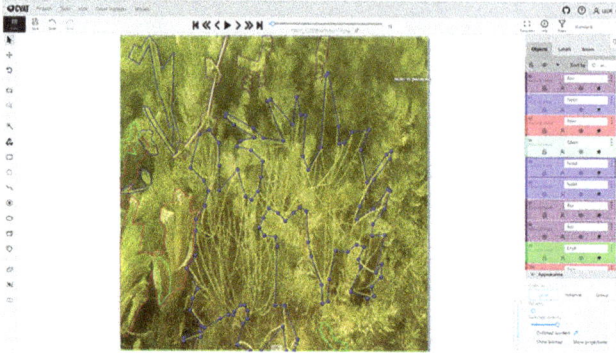

Figure 6.4   Screenshot of the editing interface in the "Computer Vision Annotation Tool (CVAT)." Shown on the right are the various colored categories (distinct growth patterns) that were annotated by hand in the image. (Source: Landesamt für Umwelt Schleswig-Holstein, Germany.)

6.5.3   Training of neural networks

With an initial modest dataset that was extracted from the CVAT tool, numerous training sessions utilizing neural networks for semantic segmentation were conducted. Specifically, DeepLab v3+ [28] and Unet [29] were evaluated. Both were executed with a ResNet [30] as a classification network. It was also examined whether better results could be obtained using a classification network that had been pretrained on plant data.

Nonetheless, the training sessions revealed that semantic segmentation is not an appropriate approach for the challenges faced in the MOVE project. This is partly due to the complexity of both the technique and the issue at hand, including inadequate lighting and color conditions underwater. Additionally, the resemblance among the aquatic plants adversely affected the effectiveness of the approach. Another contributing factor is the significant class imbalance present in the dataset. On one side, this arises because certain plants are more commonly found in the lakes and thus appear more often in the videos. On the other side, categories such as sediment are frequently observed in images, regardless of the plant depicted. Though the semantic segmentation network was capable of predicting the outlines and forms of the plants in our tests, it failed to differentiate between various species or vegetation categories, as it primarily predicted the most prominent plant class within the training set [3].

In order to tackle this issue, multiple methods were explored including image augmentation. Regrettably, none of these methods succeeded in enhancing the segmentation outcomes. On one side, this is still attributed to the significant complexity of segmenting underwater plants on a pixel-by-pixel basis. On the other side, the imbalance could only be somewhat addressed, as certain classes, particularly sediment, persisted in appearing abundantly in both the new and augmented images.

Ultimately, the intricacy of the issue was simplified by shifting the technique from semantic segmentation to classification. In this approach, a classification network is intended to acquire a vector that represents the percentage distribution of vegetation in the image. This offered the benefit that, among other things, methods and results from the prior dataset could be applied without modifications. Nevertheless, this no longer allows for the identification of individual plants, as needed by the *Phylib* method [4]. Vegetation categories and their limits, nonetheless, are distinguishable and can be utilized for vegetation mapping.

To execute the classification approach, the pixels for each vegetation class were initially tallied in the masks generated from the annotations and recorded as a percentage in a vector. Following this, various classification networks of differing complexities were evaluated (e.g. ViT [31], ResNet [30] or EfficientNet [32]), and finally, ResNet was selected.

6.5.4   Pre- and post-processing

To enhance the outcomes, two measures were adopted. First, *pre-processing* was carried out. This pre-processing initially included four components.

(1)   *Color correction* was performed by adjusting the red channel within the HSV (Hue, Saturation, and Value) color space. Initially, the image was converted to the HSV color space. Subsequently, the hue values of the red channel were altered until the distribution of these values matched that of the blue and green values [33].

(2)   *Brightness adjustment* utilizing a maximum-suppressed histogram equalization. The histogram was generated from the intensity values of an image. Next, to prevent excessive adjustment, especially the high values were truncated to ensure that these bright values stayed elevated. Following this, the histogram was uniformly spread across the values [34].

(3)   *Dehazing*, designed to eliminate hazy distortions from an image. Utilizing the dark channel prior to the image, the pixel that was least impacted by the hazy distortions was identified. Subsequently, the image was adjusted by a guided filter [35].

(4)   *Marine snow removal* was conducted as a concluding step. Marine snow refers to tiny, white, analog artifacts that frequently appear in underwater images. The processing method utilizes the distinct characteristics of marine snow to categorize pixels. For instance, when a cluster of pixels significantly contrasts with its environment, appears white, and remains below a specific size, it is presumed to be marine snow. Subsequently, the pixels are replaced with the median of their neighbors [36].

An illustration of an image that has undergone preprocessing is displayed in Figure 6.5.

Given that the vegetation zones are to be forecasted, it can be taken for granted that the outcomes should align with seamless zones. Consequently, *post-processing* was carried out to identify the zones and, particularly, the transitions with greater

Figure 6.5   Illustration of a preprocessed image: initial image (left). (Copyright: Landesamt für Umwelt Schleswig-Holstein, Germany.) Modified image after color correction (right).

accuracy. The fundamental concept is to adjust the individual elements to align with those of neighboring elements to eradicate anomalies.

6.5.5   Detection of vegetation boundary

While carrying out the monitoring tasks along a transect, the MONSUN AUV shall also have the capability to identify the vegetation boundary online, as mentioned earlier. To achieve this, a method utilizing basic computer vision for classifying the lakebed is introduced. In this case, two techniques, Nu-Support Vector Classifier (Nu-SVC) and Multi-Layer Perceptron (MLP), have been employed.

Nu-SVC represents a specific variation of the Support Vector Machine algorithm. The primary distinction in Nu-SVC is that it incorporates a hyperparameter (ν parameter) utilized to manage the quantity of support vectors and margin inaccuracies [37].

MLP represents a category of neural network comprising an input layer, various hidden layers, and an output layer [38], which leads to a minimal number of adjustable parameters.

The inputs of the classification models consist of histograms derived from Local Binary Patterns (LBP). The LBP serve as an image descriptor capable of extracting features from a grayscale image by leveraging texture information through a windowed methodology [39]. Both techniques, Nu-SVC and MLP, do not necessitate substantial computational resources, enabling them to operate online during the mission. To facilitate this, a separate vision processor (Raspberry Pi 3) is optionally incorporated into an intermediate ring of MONSUN, communicating with the primary processor via Ethernet.

6.6   Experimental results

To practically test the use of the MONSUN underwater robots for monitoring lakes, numerous test drives were conducted on Lake Ratzeburg in the summer months of 2022, 2023, and 2024, and their results were evaluated. The aim was to record videos of the underwater vegetation following the standard procedure according to the EU Water Framework Directive along transects from the shore vertically into the lake up to the vegetation boundary (lower macrophyte limit), which would then be evaluated using AI image recognition. For this, an area on the northwestern shore was selected for the test, which had already been monitored and mapped in previous tests using divers (transect 1). The test drives were usually carried out from the shore out to the lake and accompanied by a boat. The drives were completely autonomous. The robots were simply put into the water, started, and then recovered again after surfacing at the end of the drive. Some of the results of these test drives are presented and discussed below.

6.6.1   Navigation along a transect

The V-formation had already been tested in a previous project, showing good results (see [25]). A disadvantage, however, is that three MONSUN AUVs are required, which complicates handling and makes this approach more expensive, with a higher CO_2 footprint compared to a single AUV solution. Therefore, the tests reported here concentrate on the Point and Shoot strategy, as well as the use of a GPS buoy.

6.6.1.1   Point and Shoot (P&S) strategy

The GPS track of a test run with P&S on the surface is shown in Figure 6.6(a). The time interval between the track points shown is 10 seconds. It can be seen that the method works in principle, although a fairly large wind drift can be observed. Possible countermeasures include waypoints at shorter intervals, holding ahead on course and stronger thrusters.

In further tests, runs involving hovering over the lakebed were conducted. The diving depth, measured with the pressure sensor, and the hovering distance to the lakebed, measured with the ping sonar, for one of the test runs are shown as an example in Figure 6.6(b). The cycle duration is 2 minutes, during which a distance of approx. 20 m is covered. The surfacing for the "point" phases can be clearly seen from the increase in the sonar distance and the simultaneous decrease in the diving depth (see dotted vertical lines, which indicate the start of the descent and ascent phases). During the "shoot" phase, the hovering distance remains almost constant at approx. 1.1 m, as desired, with only slight fluctuations. The diving depth follows the lake bottom and increases slightly toward the end of the 2nd cycle. The water depth (sum of diving depth and hovering distance) is approx. 2.1 m at the beginning and increases to approx. 2.2 m. The peaks in the diving depth are incorrect measurements of the pressure sensor, which can easily be filtered out.

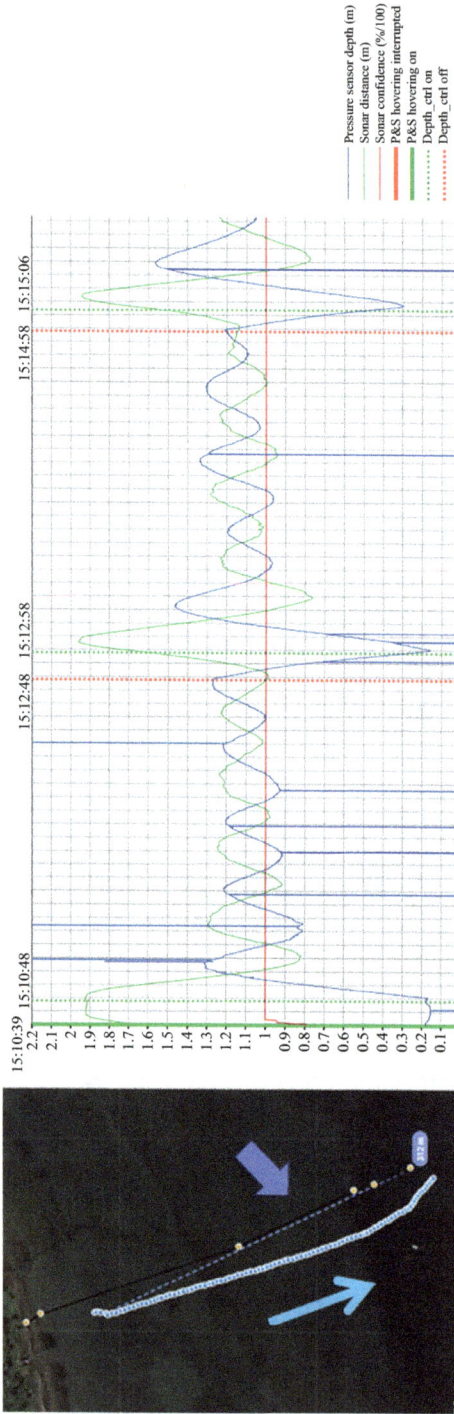

Figure 6.6 Test results with Point and Shoot strategy: (a) P&S trip on the surface along transect 1 toward vegetation boundary with northeast wind of Beaufort 2–3 (cycle duration: 2 minutes). (b) Diving depth and hovering distance (in meters) for two P&S cycles over time (cycle duration: 2 minutes).

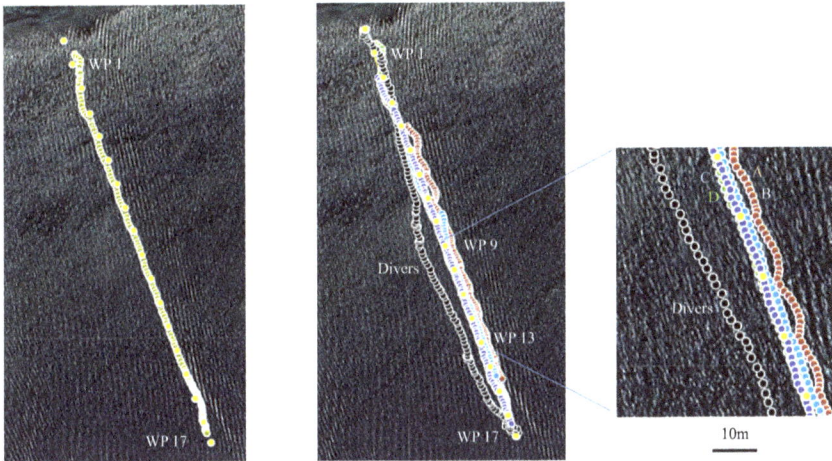

Figure 6.7   GPS tracks of test trips along transect 1.02 with GPS buoy and track of the divers for comparison (distance between the track points 10 s respectively). The yellow points are the given waypoints of the transect.

6.6.1.2   GPS buoy

Several test runs were conducted with a GPS buoy. Figure 6.7 on the left shows one of the recorded test tracks in light wind, and on the right, this track is shown together with three further tracks from test runs on other days.

During trip A, a slightly larger drift to the east can be seen. On this day, there was a westerly wind of Beaufort 2 to 3. The other three trips showed only very slight deviations from the target course. Here, there were only light winds of Beaufort 1 to 2. It should also be noted that the GPS points shown represent the position of the GPS buoy, to which the robot is connected via a 6 m cable. A mathematical correction would still be possible here, but this has been omitted for the sake of simplicity.

For comparison, the GPS track of the divers is also shown, which shows a significantly higher deviation of the transect's waypoints. The speed is about the same. Overall, it has been shown that the GPS buoy method achieves sufficient accuracy of a few meters for georeferencing, which is within the range of GPS accuracy. However, this requires that the wind is not too strong.

6.6.2   Hovering

The hovering distance in the test drives described above was set between 1.3 m and 1.5 m. If the hovering distance is too short, there is a risk that the robot will get stuck in very high submerged macrophytes and be slowed down. If the distance is too high, the image quality will deteriorate.

Figure 6.8 shows the diving data from test run C in the area from WP 9 to WP 17. Two areas can be clearly distinguished. In the first area up to time point 39, the hovering distance (pinger.dist) is very constant at 1.4 m, although the water depth

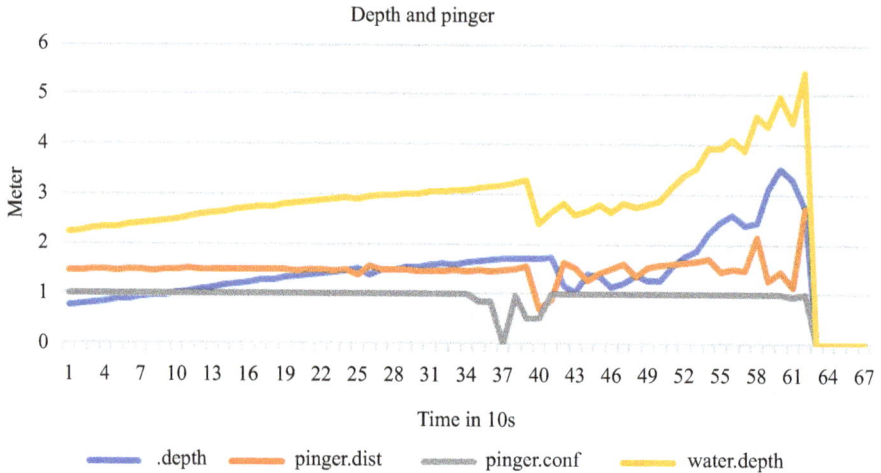

Figure 6.8   *Dive data from MONSUN test trip C from WP 9 to WP 17 in transect 1.02: water depth (water.depth), diving depth (.depth), hovering distance (pinger.dist), and sonar confidence (pinger.conf in %/100). The x-axis indicates the time in 10 s steps.*

increases from 2.1 m to 3.2 m. This is compensated very well by the diving depth. In the second area from time point 39 on, the hovering distance and thus also the diving depth fluctuate significantly by a few centimeters. The water depth increases in this area from approx. 3 m to over 5 m.

These results can be explained by the fact that the first area is the stonewort zone, where the lakebed is covered with a dense meadow of aquatic plants (see Figure 6.14 top). In the second area, the vegetation changes to higher submerged macrophytes with irregular vegetation, some of which include quite tall plants (see Figure 6.14 bottom). The sonar measurement therefore fluctuates, and the robot adjusts its diving depth accordingly. This is also clearly visible in the macrophyte videos recorded by MONSUN. At times, the confidence of the sonar measurement even drops briefly. Fortunately, these fluctuations are not a problem for image quality – in fact, they are positive, as the taller plants become more visible. In addition, the risk of the robot getting stuck in the plants is reduced. Since the sonar not only provides a distance but also a profile of the echoes, strategies for hovering that automatically adapt to the vegetation are also conceivable. However, this was not pursued further in the MOVE project.

In summary, it was observed during the test dives that the hovering distance was maintained with good accuracy in the centimeter range and good-quality video recordings were achieved.

6.6.3   Recording of sensor values

The following examples present the results of recorded sensor values from a test trip, during which the SST sensor ring (see Section 6.3.1) was tested in particular.

Transect 1.02 was selected for the test trip, which was also mapped by divers from the company *lanaplan GbR* by order of the MOVE project partner Landesamt für Umwelt Schleswig-Holstein (Figure 6.9). The transect begins near the shore with the reed bed zone, which MONSUN cannot navigate. This is followed by a small zone of higher submerged macrophytes, which is followed by a larger zone of stonewort. Toward the end of the transect at the vegetation boundary, there is another zone of higher submerged macrophytes. Approximately in the middle of transect 1.02, there is a small shallow area with a water depth of 1.8–2 m.

The divers made five dives at a distance of about 50 m from each other. Their tracks were recorded with a GPS buoy and are shown in Figure 6.9 as lines. During three dives (transects 1.02, 1.03, and 1.05 from right to left), the vegetation zones were assessed manually; during the two remaining ones (transects 1.01 and 1.04), the divers made a classification of the individual macrophytes according to the *Phylib* method.

Figure 6.10 shows the GPS track of a test run by MONSUN with a GPS buoy along part of transect 1.02. The waypoints of the transect are again depicted in yellow, being about 20 m apart from each other. The time interval between MONSUN's trajectory points, shown (blue), is 10 seconds. As can be seen, the course was maintained quite well. The deviation from the target course was max. 6 m, which was also the stopping radius, which was set for reaching a waypoint. Due to the westerly wind of Beaufort 2 with gusts of 3, a light drift to the east can be observed. The average speed was 0.19 m s^{-1}.

The recorded diving data is shown in Figure 6.11. At the beginning of the trip, the water depth was 2.3 m, increasing to 3.5 m by the end. The shallow area with a water depth of 1.9 m is clearly visible. The hovering distance was set at 1.4 m, which was kept fairly constant. The diving depth, therefore, followed the course of the water depth. The confidence value of the ping sonar is also shown, which was 100% throughout.

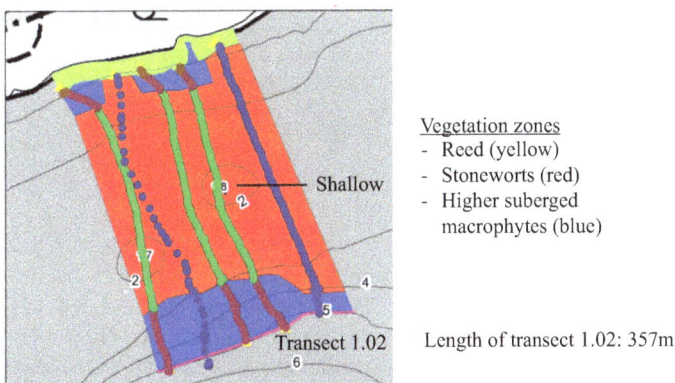

Figure 6.9   Results of vegetation mapping of transect 1.02 by divers. (Source: Landesamt für Umwelt Schleswig-Holstein, Germany.)

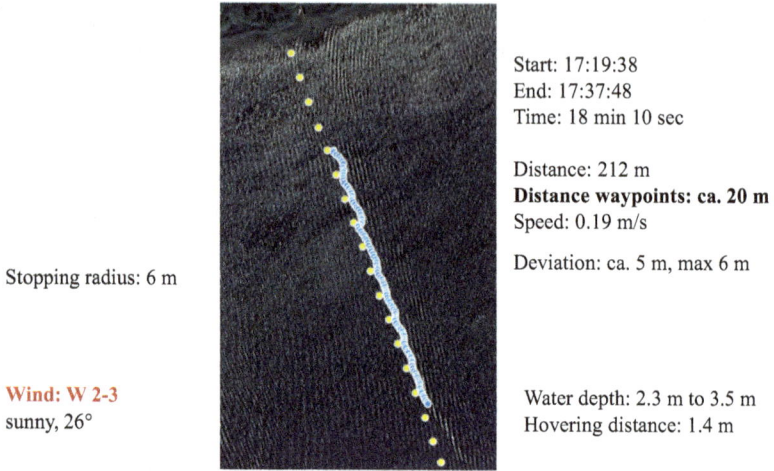

Start: 17:19:38
End: 17:37:48
Time: 18 min 10 sec

Distance: 212 m
Distance waypoints: ca. 20 m
Speed: 0.19 m/s

Deviation: ca. 5 m, max 6 m

Stopping radius: 6 m

Wind: W 2-3
sunny, 26°

Water depth: 2.3 m to 3.5 m
Hovering distance: 1.4 m

Figure 6.10   Test drive of MONSUN along transect 1.02

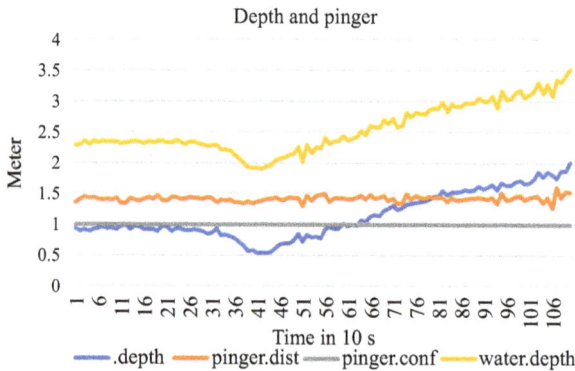

Figure 6.11   Diving data from the MONSUN test trip: water depth (water.depth), diving depth (.depth), hovering distance (pinger.dist), and sonar confidence (pinger.conf in %/100). The x-axis indicates the time in 10 s intervals.

During the test trip, some physical data were also recorded using the SST sensor ring. In addition to the usual temperature and pressure sensor (Figure 6.12), a newly developed light sensor was in operation (Figure 6.13). The influence of the shallow area is clearly visible around time point 41 in all sensor values.

MONSUN's GoPro camera was used to create videos during the dives, from which images were extracted for the AI recognition of the macrophytes. Furthermore, divers recorded videos of the transects, which served as a basis for a comparison test with the AI image recognition (see below).

Two typical image pairs from the divers and MONSUN, respectively, are shown in Figure 6.14. The images from MONSUN (Figure 6.14 right) come from the test

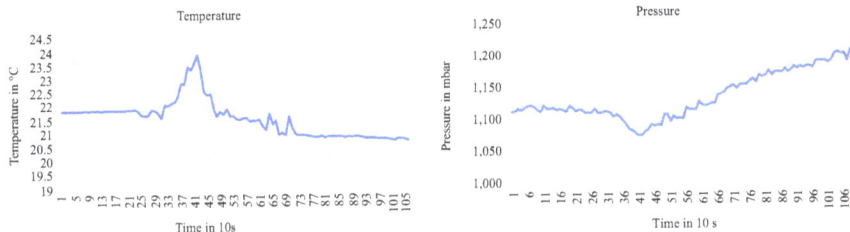

Figure 6.12   SST sensor ring values for water temperature and absolute pressure

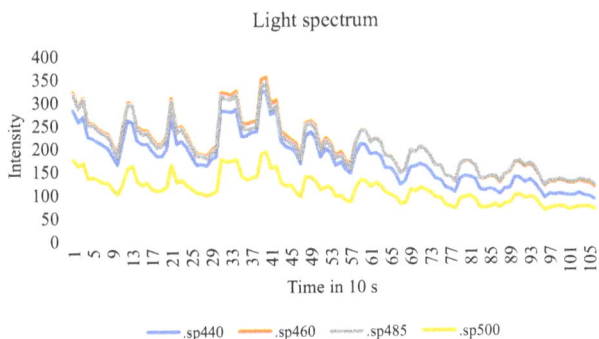

*Figure 6.13   SST sensor ring values for light spectrum for wavelengths 440, 460,
485, and 500 nm*

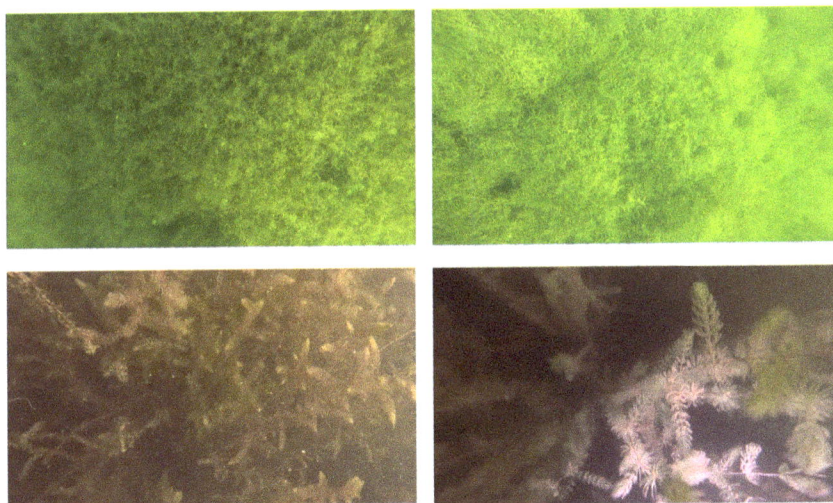

*Figure 6.14   Comparison of images from diver videos (left) (copyright: Landesamt
für Umwelt Schleswig-Holstein, Germany) with MONSUN video
images (right) for stoneworts (top) and higher submerged
macrophytes (bottom)*

dive described above. Based on georeferencing, images were selected from approximately the same location – one pair from the stonewort zone at a water depth of approx. 2 m and the other one from the zone of the higher submerged macrophytes at a water depth of approx. 5 m. As can be seen, the images are very similar, which was also confirmed for other MONSUN recordings. This justifies the approach of using the diver videos for the AI image recognition instead of the original MONSUN videos, which were not available for all transects.

6.7   AI image recognition

6.7.1   *Vegetation zones*

In order to evaluate the AI image recognition for mapping vegetation zones, a comparison test was made with the vegetation mapping done by human divers. For this, the divers investigated four selected areas at Lake Ratzeburg, diving along five transects for each one (see Figure 6.9). Three transects were monitored for vegetation mapping, the other two according to the *Phylib* method. So, a vegetation mapping was available for a total of 12 transects.

For each transect, a video was taken by the divers. From these videos, images were extracted using a slightly modified version of the image extraction program described above. These images were then used for testing the AI image recognition. The results of AI image recognition were evaluated using conventional neural network metrics, which are shown in Table 6.1.

As stated above, the final tests had to be restricted to vegetation zones, since recognizing individual plants as required by the *Phylib* method was not possible. The AI tests were carried out on a ResNet, which was pretrained using the ImageNet dataset. Then, a fine-tuning was trained for 200 epochs without augmentation and with adapted classes. A categorical cross-entropy with a learning rate of 0.0001 was used as the loss. The network was trained and tested on a Nvidia GeForce RTX 3070 Mobile. An epoch with 2105 training images of size 512×512 pixels took an average of 22 s to train. Figure 6.15 shows the metrics during training. The graph on the left shows the development of the loss over the epochs. Here, it can be observed that the network converges after just 50 epochs. However, a strong improvement can still be

Table 6.1   Metrics for evaluating AI image recognition

Metric	Description
Accuracy	$\dfrac{\text{correct elements}}{\text{all elements}}$
Precision	$\dfrac{\text{true positives}}{\text{true positives + false positives}}$
Recall	$\dfrac{\text{true positives}}{\text{true positives + false negatives}}$

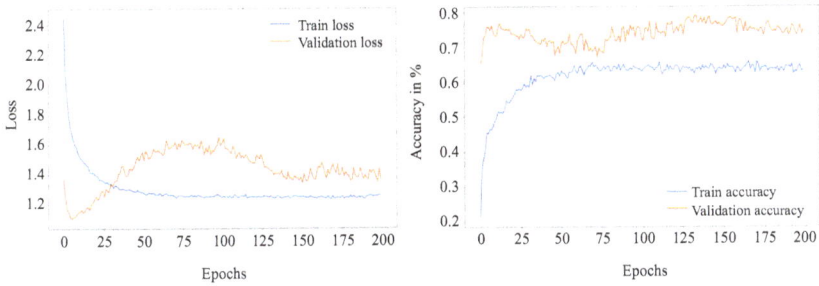

Figure 6.15   Metrics from training: loss (left) and accuracy (right) during training

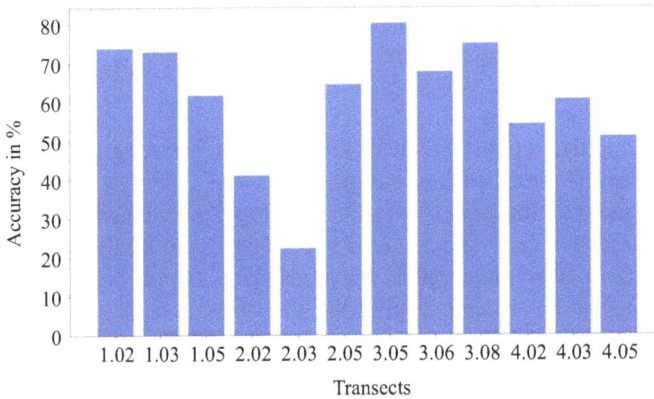

Figure 6.16   Accuracies in percent over the 12 tested transects

seen in the validation loss after the 100th epoch. A similar development can be seen in the accuracies. The strongest increase in training accuracy is up to epoch 50, while validation accuracy also shows an increase after epoch 100.

The trained network was then tested on images from the 12 transects using the modified image extraction program. The average time per image in the tests is 14 s, but most of this time is spent sampling and selecting the image. The time for classifying with the network is only 0.05 s per image. For the entire algorithm, this results in an average execution time of 36 min for a 20-min video.

Figure 6.16 shows the accuracy of the network in the respective transects. In the evaluation metrics, parts of the video that show, for example, the descent are ignored because they do not contain any important information for the vegetation zones. The median accuracy across all transects is 63.19%. While 8 of the 12 transects have an accuracy of over 60%, transect 2.03 is particularly noticeable with only 24%. The low accuracy in this transect can be explained by the fact that, in contrast to the analysis of the divers, the neural network considers the vegetation class "*Dreissena*" to be a separate class. The divers, on the other hand, continue the surrounding vegetation zone and list the coverage of "*Dreissena*" separately. Figure 6.17 shows an example image from transect 2.03, which was labeled "*Zone of higher submerged macrophytes*" by

Figure 6.17   Example image from transect 2.03 showing Dreissena (copyright: Landesamt für Umwelt Schleswig-Holstein, Germany)

the divers, while the network predicts "*Dreissena*." Correcting the labels here and adding "*Dreissena*" is expected to improve the metrics.

Another factor influencing the results is that the zone transitions are fluid. It is possible that the change to another zone is predicted later by the network than the divers. This fact is also influenced by the adjustments in post-processing. However, if you compare the water depth level of the zone transitions between the network's classification and the divers' labels in the results, it is at the same depth level in 85% of cases and at the same depth in 61% of cases.

As a final factor, it is important to mention that the tests on the transects do not specifically filter out disturbances caused by divers. These are, for example, turbulence in the sediment or the divers themselves, which can be seen in the image. These disturbances were not filtered out because they were eliminated in individual cases by post-processing. On the other hand, it can be assumed that such disturbances are not present in the MONSUN images.

Figure 6.18 shows the receiver operating characteristic (ROC) of the five learned vegetation zones across all transects. The area under curve (AUC) is also shown. The ROC represents the true positive rate of a class against the false positive rate of all other classes. The more the true positives are compared to false positives, the better, as there are more correctly classified than incorrectly classified data points. This fact corresponds to a strong early rise in the graph, which then levels off later. This can also be observed in the AUC: the closer this is to 1.0, the better the ROC. In Figure 6.18, the results for "*stonewort zone*" are particularly good with an AUC of 0.87. This is because the stonewort is correctly assigned within the zone and is not labeled by the network outside the zones. When looking at the "*zone of higher submerged macrophytes*" with an AUC of 0.72, the problem of the zone transition is particularly evident here. Within the zone, the classification is mostly correct, but there are images in this zone in particular that show a few tall plants, which the divers classified as a "*zone with little vegetation*."

The lack of ROC for "*Dreissena*" shows that these do not appear in the divers' labels. This supports the theory that the metrics improve when the "*Dreissena*" are

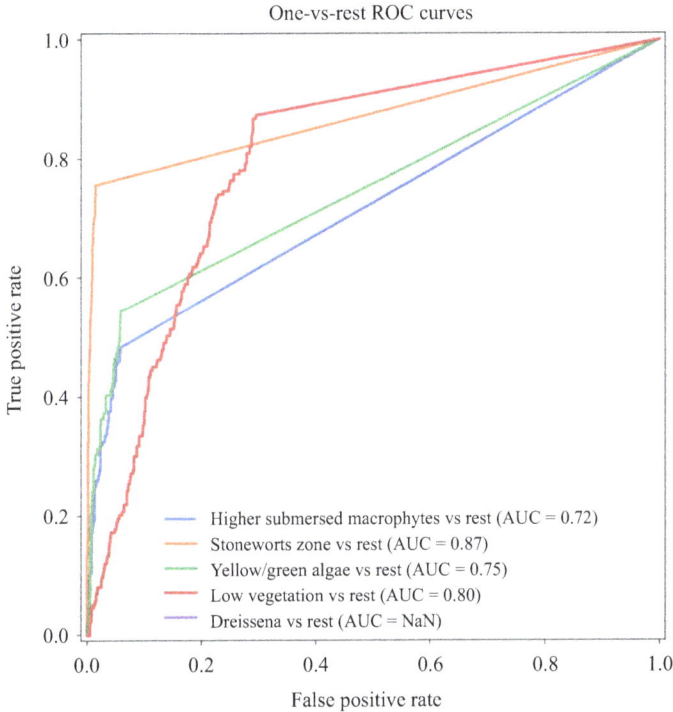

Figure 6.18   ROC curves of the five different vegetation zones across all transects

Table 6.2   Precision and recall for the vegetation zones across all tested transects

Class	Precision	Recall
Higher submerged macrophytes	0.915	0.485
Stoneworts	0.936	0.756
Yellow/green algae	0.403	0.545
Low vegetation	0.299	0.873
Dreissena	0.0	0.0

included in the labels. This is because they appear as false positives in the ROC curves of the other classes but not as true positives.

Table 6.2 shows precision and recall for the five vegetation zones across all transects. Here, too, the values for "*Dreissena*" show that predicting "*Dreissena*" as classes has a negative impact on the results. Both metrics are 0.0 because there are no true positives, false positives, or false negatives. This would result in the following for all equations $\frac{0}{0}$. The problem with the zone transitions is also evident in the metrics. The true positives are very high and the false positives very low, particularly in the "*zone of higher submerged macrophytes*" but also for the "*stoneworts.*" This suggests that the classes within the zones are often correctly identified. The low recall of the *"zone of higher submerged macrophytes"* suggests that many elements of this zone are classified as *"stoneworts"* and *"zone with little vegetation"* at the zone transition. This is especially true since the metrics for the *"low vegetation zone," for example,* are inverted, so that the recall is high and the precision is low. These assumptions are mainly since continuous zones were created through post-processing, so that few false elements appear within the zones.

Preprocessing was tested exemplarily on transects 1.05, 2.02, and 3.05. It was found that it did not improve the results. In two of the three transects tested, the accuracies were even significantly worsened. It should be noted that only the test data was preprocessed, not the training data. It can therefore be assumed that the dataset is sufficiently diverse to achieve good results in the tests and that the strong changes deteriorate the network. How the preprocessing of the training data influences the results cannot be said with the available results but requires further investigations.

6.7.2   Vegetation boundary

To test the approaches for the detection of the vegetation boundary, 800 images were taken from the macrophyte dataset and manually classified by their vegetation density. The images contain two balanced classes as sparse and dense vegetation. The dataset is split into 75% and 25% for training and testing, respectively. The images are all set to a fixed input size of 512×512 pixels. Training the models requires several hyperparameters. For Nu-SVC, the ν-parameter is set as 0.5. For the MLP, two hidden layers are created with 32 neurons and 16 neurons, respectively. Table 6.3 shows the accuracy of the two models on the test images. The accuracy of the MLP was 78.5%, while Nu-SVC performed with 81.5%. In both precision and recall, the Nu-SVC achieves better performances with 0.832 and 0.79, respectively, while the MLP reaches performances of 0.828 and 0.72 in these metrics.

Figure 6.19 shows the results of the computer vision-based vegetation boundary classification. The plot shows the ROC curve, which is the true positive rate against the false positive rate. The area under the curve describes the relation between the true positives of the classification against the false positives, i.e. a larger area under the curve means more correctly predicted samples. The area under the ROC curve is 0.906 for Nu-SVC and 0.878 for MLP, which can be seen in the plot. This shows that both approaches perform similarly, with a slight edge to the Nu-SVC approach.

Table 6.3   Accuracy, precision, and recall of the two trained models Nu-SVC and MLP for the vegetation boundary detection

Model	Accuracy	Precision	Recall
Nu-SVC	0.815	0.832	0.79
MLP	0.785	0.828	0.72

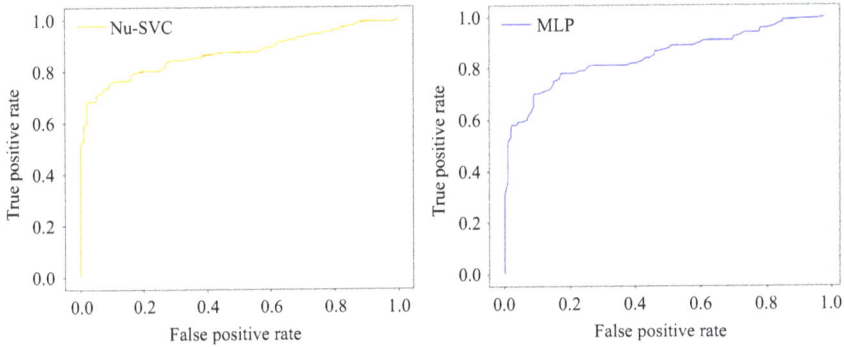

Figure 6.19   ROC curve for the two vegetation boundary classification models: Nu-SVC and MLP

6.8   Concluding remarks

This paper has presented an approach and first results for monitoring vegetation and water quality with MONSUN micro-AUVs. The micro-AUVs hover above the lake ground at a constant distance of about 1–1.5 m. Various experiments have shown that MONSUN can do this with an accuracy of a few centimeters with a low-cost ping sonar and a pressure sensor for water depth measurements. While hovering, the robot is taking videos of the macrophyte vegetation using a downward-looking video camera with LED lights. The quality of these videos proved to be sufficient for the intended automatic recognition of the macrophyte images with AI methods. For mapping the macrophyte vegetation, MONSUN must dive along a transect from the shore perpendicularly until the vegetation boundary is reached. The images taken on its way must be geo-referenced with an accuracy of a few meters. For the necessary localization, various methods were considered. One is cooperative localization using three MONSUNs in a V-formation. As alternatives, a Point and Shoot strategy or the usage of a GPS buoy, which both only need a single AUV, is considered. Various test drives have been made at Lake Ratzeburg that showed that all three methods are suitable and allow for the required georeferencing accuracy.

For the offline evaluation of the macrophyte videos, an AI image recognition pipeline was developed, including an automatic frame extraction from the videos and a subsequent pre-processing. The frame extraction was used successfully for building a macrophyte dataset for training a neural network. However, the required annotation of the images still had to be done by human experts, which was very time-consuming. The current dataset comprises about 30,000 images, from which a total of 2105 images for 46 different species plus other structural elements were annotated. The recognition of single macrophytes as required by the standard *Phylib* method turned out to be too challenging. The mapping of vegetation zones using a classification network and post-processing was, however, possible. Also, the vegetation boundary could be detected online by simpler machine learning techniques with good results. Using an optional sensor ring chemical–physical values and the underwater light spectrum could also be recorded.

Acknowledgments

The MOVE project has been funded by the German Federal Ministry of Education and Research (BMBF) within the Digital GreenTech funding measure under grant number 02WDG1640A. The authors want to thank Ulrich Behrje, Cedric Isokeit, Benjamin Meyer, and Julian Petzold for their contributions to the MONSUN hardware and software as well as Onurcan Köken for his contributions to the video frame extraction and vegetation boundary detection.

Our thanks go also to Inga Kostelnik and her colleagues from Landesamt für Umwelt Schleswig-Holstein for the work on the labelling of the macrophyte images and on the diver's tests, as well as to the sensor ring team around Gerd Seidel at Sea & Sun Technology, Trappenkamp.

The maps in Figures 6.6(a), 6.7, and 6.10 were created with Google Maps (https://www.google.com/maps).

References

[1]   European Commission. *Water Framework Directive* [online]. 2000. Available from https://environment.ec.europa.eu/topics/water/water-framework-directive_en [Accessed 09 Apr 2025].

[2]   Maehle E., Meyer B., Isokeit, C., and Behrje U.: "MONSUN: a swarm AUV for environmental monitoring and inspection" in Ehlers F. (ed.). *Autonomous Underwater Vehicles: Design and Practice.* London: IET– The Institution of Engineering and Technology; 2020. pp. 329–55.

[3]   Kapoor G., Nienaber S., Köken O., and Maehle E.: "Monitoring of Vegetation and Water Quality with MONSUN Micro-AUV Swarms". *Proceedings of OCEANS 2023 – MTS/IEEE U. S. Gulf Coast*; Biloxi, MS, USA, Sep 2023. Piscataway, NJ: IEEE; 2023.

[4]   Schaumburg J., Stelzer D., Schranz C., Vogel A., and van de Weyer K.: *Verfahrensanleitung für die ökologische Bewertung von Seen zur Umsetzung der EG-Wasserrahmenrichtlinie: Makrophyten & Phytobenthos – Phylib* [online]. 2021. Available from https://www.gewaesser-bewertung. de/media/o_2.20_verfahrensanleitung_phylib_seen_stand_august_2021.pdf [Accessed 09 Apr 2025].

[5]   van de Weyer K., Nienhaus I., Tigges P., Hussner A., and Hamann U.: "A simple and cost-efficient method for the planar collection of submerse plant existence in lakes." *Wasser und Abfall.* 2007, vol. 9(1), pp. 20–22.

[6]   Rowan, G. S. L. and Kalacska, M.: "A Review of remote sensing of submerged aquatic vegetation for non-specialists." *Remote Sensing*, 2021, vol. 13(4), p. 623.

[7]   Quraishi A., Bahr A., Schill F., and Martinoli, A.: "Autonomous feature trac-
        ing and adaptive sampling in real-world underwater environments". *IEEE
        Int. Conf. on Robotics and Automation (ICRA)*. Piscataway, NJ: IEEE; 2018.
        pp. 1–6.

[8]   Seaber SAS. *YUCO Micro-AUV* [online]. Available from https://seaber.fr
        [Accessed 09 Apr 2025].

[9]   RTSYS. *NemoSens µAUV* [online]. Available from https://rtsys.eu/nemosens-
        micro-auv [Accessed 09 Apr 2025].

[10]   ecoSUB Robotics. *ecoSUBµ5* [online]. Available from [Accessed 09 Apr
        2025].

[11]   Advanced Navigation. *Hydrus – The Drone Revolution Underwater* [online].
        Available from https://www.advancednavigation.com/robotics/micro-auv/h
        ydrus/ [Accessed 09 Apr 2025].

[12]   Gonzalez-Garcia, J., Gomez-Espinosa, A., Cuan-Urquizo E., *et al.*:
        "Autonomous underwater vehicles: localization, navigation, and communi-
        cation for collaborative missions." *Applied Sciences*. 2020, vol. 10(4), p. 1256.

[13]   Barthelemy D., Boujemaa N., Mathieu D., *et al.*: "The Pl@ntNet project:
        Plant computational identification and collaborative information system."
        *Biosystematics 2011 - Digital Identification Methods in the Digital Age -
        Morphological and Molecular, Computer-Aided and Fully Automated,
        Research and Citizen science*; Berlin, Germany, 2011.

[14]   Mäder P., Boho D., Rzanny M., *et al.*: "The flora incognita app – interactive
        plant species identification." *Methods in Ecology and Evolution*. 2021, vol.
        12(7), pp. 1335–1342.

[15]   Moniruzzaman M., Islam S. M. S., Bennamoun M., and Lavery P.: "Deep
        learning on underwater marine object detection: A survey" *Advanced
        Concepts for Intelligent Vision Systems;* Antwerp, Belgium, 2017. Berlin:
        Springer; 2017. pp. 150–160.

[16]   Gomes D., Saif A. F. M. S., and Nandi D.: "Robust underwater object
        detection with autonomous underwater vehicle: a comprehensive study"
        *Proceedings of the International Conference on Computing Advancements,
        ICCA 2020;* New York, NY, USA, 2020. New York: Association for
        Computing Machinery; 2020.

[17]   Moniruzzaman M., Islam S. M. S., Lavery P., Bennamoun M., and Lam
        C. P.: *Imaging and classification techniques for seagrass mapping and
        monitoring: A comprehensive survey* [online]. 2019. Available from
        https://arxiv.org/abs/1902.11114 [Accessed 09 Apr 2025].

[18]   Moniruzzaman M., Islam S. M. S., Lavery P., and Bennamoun M.: "Faster
        R-CNN based deep learning for seagrass detection from underwater digital
        images." *Digital Image Computing: Techniques and Applications (DICTA)*;
        Perth, Australia, 2019. Piscataway, NJ: IEEE; 2019. pp. 1–7.

[19]   Akkaynak D. and Treibitz T.: "Sea-thru: A method for removing water
        from underwater images." *IEEE/CVF Conf. on Computer Vision and Pattern
        Recognition (CVPR)*; Long Beach, CA, USA, 2019. Piscataway, NJ: IEEE;
        2019. pp. 1682–1691.

[20] Li C., Guo J., and Guo C.: "Emerging from water: Underwater image color correction based on weakly supervised color transfer." *IEEE Signal Processing Letters*. 2018, vol. 25(3), pp. 323–327.

[21] Wang Y., Song W., Fortino G., Qi L.-Z., Zhang W., and Liotta A.: "An experimental-based review of image enhancement and image restoration methods for underwater imaging." *IEEE Access*. 2019, vol. 7, pp. 140233–140251.

[22] Behrje U., Isokeit C., Meyer B., Ehlers K., and Maehle E.: "AUV-based quay wall inspection using a scanning sonar-based wall following algorithm." *Proceedings of OCEANS 2022 – MTS/IEEE*; Chennai, India, 2022, Piscataway, NJ: IEEE; 2022.

[23] Renner B.-C., Heitmann J., and Steinmetz F.: "Ahoi: Inexpensive, low-power communication and localization for underwater sensor networks and μAUVs." *ACM Transactions on Sensor Networks*. 2020, vol. 16(2), pp. 1–46.

[24] Quigley M., Gerkey B., Conley K., *et al.*: "ROS: An open-source robot operating system." *Proceedings of the IEEE International Conference on Robotics and Automation (ICRA) Workshop on Open Source Robotics*; Kobe, Japan, 2009. Piscataway, NJ: IEEE; 2009.

[25] Behrje U., Isokeit C., Meyer B., and Maehle E.: "A robust acoustic based communication principle for the navigation of an underwater robot swarm." *Proceedings of 2018 MTS/IEEE Kobe Techno-Oceans (OTO)*; Kobe, Japan, 2018. Piscataway, NJ: IEEE; 2018.

[26] Sekachev B., Manovich N., Zhiltsov M., *et al.*: *opencv/cvat: v1.1.0.* [online]. 2020. Available from https://zenodo.org/records/4009388 [Accessed 09 Apr 2025].

[27] *GBIF Global Biodiversity Information Facility* [online]. Available from https://www.gbi [Accessed 09 Apr 2025].

[28] Chen L.-C., Zhu Y., Papandreou G., Schroff F., and Adam H.: *Encoder-Decoder with Atrous Separable Convolution for Semantic Image Segmentation* [online]. 2018. Available from https://arxiv.org/abs/1802.02611 [Accessed 09 Apr 2025].

[29] Ronneberger O., Fischer P., and Brox T.: "U-net: Convolutional networks for biomedical image segmentation." *Conference on Medical Image Computing and Computer Assisted Intervention – MICCAI;* Munich, Germany, 2015. LNCS vol 9351, Cham: Springer; 2015.

[30] He K., Zhang X., Ren S., and Sun, J.: "Deep residual learning for image recognition". *IEEE Conference on Computer Vision and Pattern Recognition (CVPR)*; Las Vegas, NV, USA, June 2016. Piscataway, NJ: IEEE; 2016. pp. 770–778.

[31] Dosovitskiy A., Beyer L., Kolesnikov A., *et al.*: *An Image is Worth 16×16 Words: Transformers for Image Recognition at Scale* [online]. 2021. Available from https://doi.org/10.48550/arXiv.2010.11929 [Accessed 09 Apr 2025].

[32] Tan M., Le Q.V.: "EfficientNet: Rethinking model scaling for convolutional neural networks." *Proceedings of the 36th International Conference on Machine Learning*; Long Beach, CA, USA, May 2019 [online]. 2019.

Available from https://doi.org/10.48550/arXiv.1905.11946 [Accessed 09 Apr 2025].

[33] Andersen N. B.: *Underwater-image-color-correction* [online]. Available from https://github.com/nikolajbech/underwater-image-color-correction [Accessed 09 Apr 2025].

[34] Zuiderveld K. J.: *Contrast limited adaptive histogram equalization* [online]. 1994. Available from https://doi.org/10.1016/b978-0-12-336156-1.50061-6 [Accessed 09 Apr 2025].

[35] He K., Sun J., and Tang X.: "Single image haze removal using dark channel prior." *IEEE Conference on Computer Vision and Pattern Recognition*; Miami, FL, USA, 2009, pp. 1956–1963.

[36] Farhadifard F.: *Underwater image restoration: super-resolution and deblurring via sparse representation and denoising by means of marine snow removal* [online]. 2017. Available from https://api.semanticscholar.org/CorpusID:51881629 [Accessed 09 Apr 2025].

[37] Schölkopf B., Smola A. J., Williamson R. C., and Bartlett P. L.: "New support vector algorithms." *Neural Computation*. 2000, vol. 12(5), pp. 1207–1245.

[38] Ruck D. W., Rogers S. K., and Kabrisky M.: "Feature selection using a multilayer perceptron." *Journal of Neural Network Computing*. 1990, vol. 2(2), pp. 40–48.

[39] Ojala T., Pietikäinen M., and Harwood D.: "Performance evaluation of texture measures with classification based on Kullback discrimination of distributions." *Proceedings of 12th International Conference on Pattern Recognition*; Jerusalem, Israel, 1994. Los Alamitos, CA: IEEE Computer Society Press; 1994, pp. 582–585.

Chapter 7

Reference modeling for automatic target recognition

Lovis Justin Immanuel Zenz[1], Erik Heiland[1],
Peter Hillmann[1] and Andreas Karcher[1]

Automatic Target Recognition (ATR) constitutes a fundamental capability of *Maritime Autonomous Vehicles*, with corresponding services manifesting in a variety of forms, while relying on common core concepts – such as an overarching process and the main tasks forming it. Expanding beyond maritime contexts, such services find further application across different domains. For instance, radar systems serve both the classification of items in maritime environments and the identification of flying objects in air traffic control [1,2].

A *Reference Model for ATR (ATR-RM)* facilitates the communication of these core concepts, enabling diverse *ATR* applications to build upon it. Subsequently, solution models can be derived from the *ATR-RM* and knowledge acquired during that process can be reintegrated into the *ATR-RM*. Altogether, this fosters the cross-domain sharing of knowledge.

In this chapter, we address the obtainment, maintenance, and application of such a reference model. First, we examine the domain of *ATR* in a generalized way in Section 7.1. Subsequently, we present our approach for obtaining and maintaining reference models in Section 7.2. Finally, we exemplarily demonstrate the result of applying our approach in Section 7.3.

7.1 Common core concepts of automatic target recognition

Before the generation of the *ATR-RM* can be conceptualized and undertaken, understanding of *ATR* and especially its common core concepts across different applications needs to be acquired.

7.1.1 Generalization of the automatic target recognition process

One such communality is the overarching *ATR Process* with its main tasks. Figure 7.1 provides a generalized overview of this process in the form of a *Business Process Model and Notation (BPMN)* diagram [3]. It contains the following seven tasks [4–6]:

[1]Department of Computer Science, Institute for Applied Computer Science, Chair for Software Tools and Methods for Integrated Applications, University of the Bundeswehr Munich, Germany

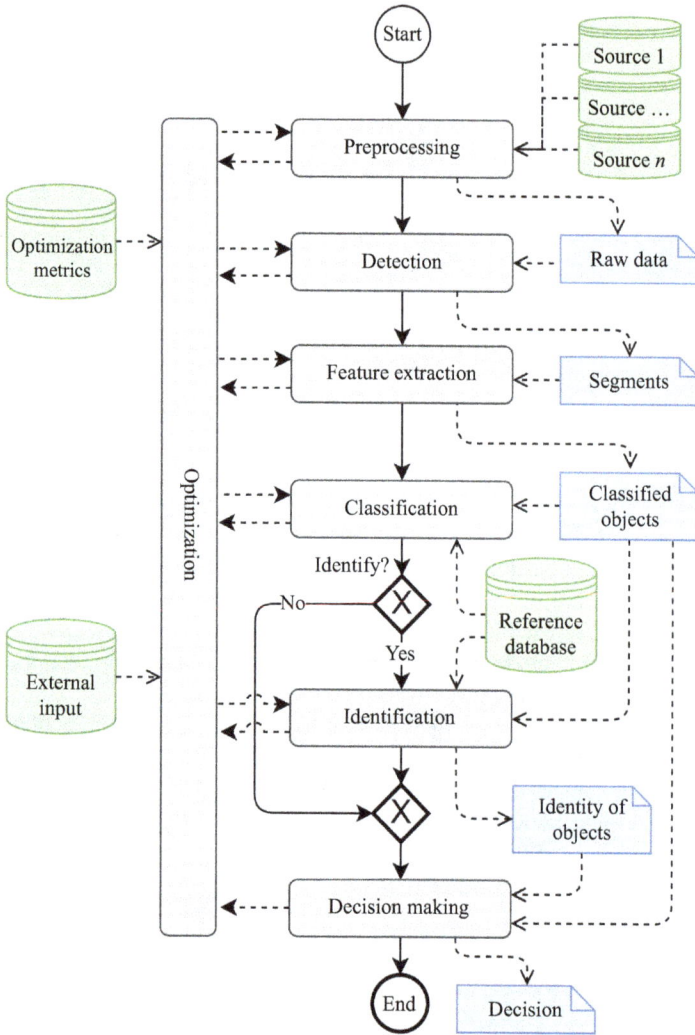

Figure 7.1   ATR process

Preprocessing: Obtaining the input and improving its quality – such as through cal-ibration, cleaning, normalization, and/or transformation – such that the raw data is prepared for and can therefore be utilized in the following tasks.

Detection: Identifying interesting parts of the prepared data, depending on its type and quality to obtain smaller observation segments that optimally only contain a single object to be classified.

Feature extraction: Extracting relevant features from each detected segment and reducing the dimensions – i.e., the amount of different features observed – to ensure that the following classification and identification tasks are provided with sufficient information on one side and can be finished in a reasonable time on the other side.

Classification: Choosing an appropriate classifier and employing it in order to determine the general kind of object represented by a segment of raw data based on the extracted feature.

Identification: Discerning the identity of a classified object based on its general kind and further extracted features to distinguish individual entities within a class of entities.

Decision-making: Choosing decision structures and employing them to decide whether a classified (and identified) object is a wanted or an unwanted target and how to proceed from there.

Optimization: Iteratively refining the above tasks regarding approaches and their parameterization based on previous results as well as updated or additional information – e.g., user data or feedback – with respect to specified optimization metrics – e.g., minimizing the amount of false negatives – to continuously improve the *ATR Process*.

The following four channels provide data to these tasks:

Source: Source of raw data – like sensor recordings – that provides input to the task PREPROCESSING.

Reference database: Source of reference data – like training data – that provides input to the tasks CLASSIFICATION and IDENTIFICATION.

Optimization metrics: Source of conditional data – like requirements and demands – that provides input to the task OPTIMIZATION.

External input: Source of on-demand data – like user input and feedback – that provides input to the task OPTIMIZATION.

Within the *ATR Process*, at the beginning of each subsequent task's execution, it may be necessary to group the results of one or more previously executed tasks. Furthermore, the task IDENTIFICATION is depicted as optional because, in certain cases, knowledge about the kind of an object is sufficient for making a decision. Finally, the in- and outgoing dashed lines of the task OPTIMIZATION represent in- and outgoing control flow structures as exemplified for the paired task PREPROCESSING in Figure 7.2.

With its in- and outgoing control flows in their expanded form, the task OPTIMIZATION creates an interactive feedback loop. This feedback loop generally only leads backward – e.g., a control flow ingoing from the task CLASSIFICATION can only result in a control flow outgoing to either one of the tasks PREPROCESSING, DETECTION, or FEATURE EXTRACTION.

7.1.2 Realization of automatic target recognition

The actual benefit of a reference model is highly dependent upon recurrence within the corresponding domain [7,8]. For the *ATR-RM*, this means that its beneficialness is directly tied to recurrence within the domain of *ATR* – e.g., similarities between different applications of *ATR*.

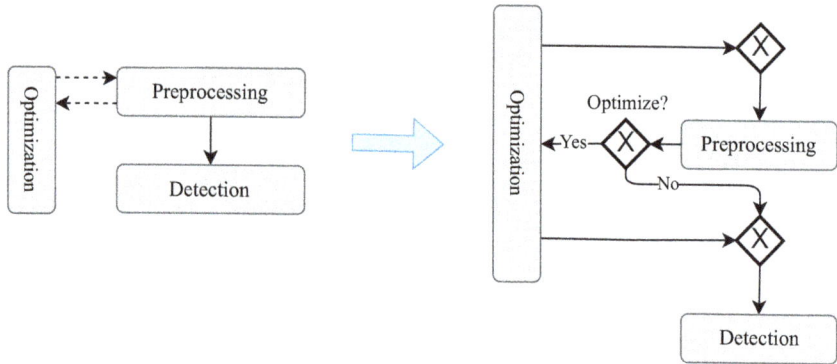

Figure 7.2   Expanded control flow structure for optimization paired with preprocessing

One such similarity is the reoccurring employment of the same technologies. For instance, when we look at the obtainment of raw data for the *ATR Process* – see Figure 7.1 – from Section 7.1.1, radar technology is frequently utilized [1,2]. This can be found, for example, in aircraft with onboard radar, in the area of air traffic control (ATC) and in modern vehicles to regulate the distance to vehicles in front [9,10]. Different sensors are often used in combination to have a sufficiently reliable data basis for classifying objects and to be able to react to different environmental conditions. Imaging methods via cameras or thermal imaging sensors, LiDAR, sonar, sound recordings, and so on are also suitable for *ATR* [11–14].

Depending on the technologies used, various characteristics of the object to be identified can be derived. Examples include color, shape, size, weight, distance, position, temperature, speed and direction of movement, noise, and material. The most relevant characteristics in the respective application context are used to narrow down the data set to potential targets within the detection phase. For example, in the area of ATC, very small objects – such as individual birds – and immovable objects close to the ground – such as large buildings – are ignored, while wind turbines still have too many similarities with a flying object in this phase and cannot be reliably excluded. Additional features from the feature extraction phase are used to classify the object. Various methods and combinations of rule-based decision systems and machine learning algorithms are frequently employed to realize *ATR* applications [4–6].

The classification of an object is not always sufficient to declare it as a target in the decision-making phase or to exclude it as such. Consequently, it is necessary to identify the object of a class if this is possible with the help of other data sources. In the context of determining individuals, this can be done by comparing with existing data, for example. In the field of aviation, special identification friend or foe (IFF) systems are used, which ensure identification via a transmitted transponder code [15].

Our *ATR Process* represents a generally valid procedure for the examples mentioned. Depending on the specific application context, there will be different configurations of the *ATR* service with regard to the technologies and procedures used and whether an identification phase is necessary or possible. The ongoing further development of technologies and methods, as well as their use in domains not pre-viously considered, requires continuous further development of the reference model and regular checks of its validity.

7.2 Obtainment and maintenance of reference models

The examination of *ATR* described above, in Section 7.1, provides the required domain expertise for establishing a well-matched *ATR-RM*. Additionally, a system-atic approach for obtaining and maintaining the *ATR-RM* needs to be specified. Only with those two aspects, high quality of the resulting *ATR-RM* with respect to appropriate model quality criteria is ensured [16].

7.2.1 *Generic model-based systems engineering approach*

The generally established planning approach for new and complex products in the related system is shown in Figure 7.3 [17]. The focus is on a coherent and continu-ous approach according to *Model-based Systems Engineering (MBSE)* over several

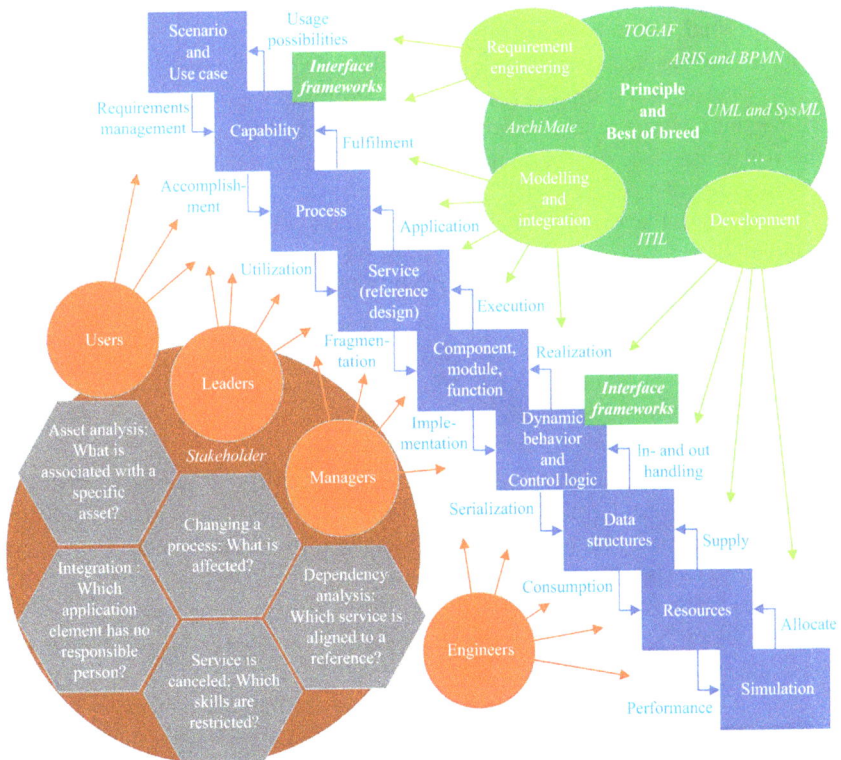

Figure 7.3   *Reference method for end-to-end model-based system development with a focus on architecture reuse of the various building blocks at the different levels through adequate interfaces and a solution continuum*

design levels [18]. Taking into account the domain of a recurring building block, we strive for comprehensive modeling from the strategic to the resource level. This provides us with a solid basis for subsequently ensuring a suitable implementation. The focus here is on the instantiation of recurring modules to enable transferability to other domains and generate synergy effects. Interfaces must be taken into account to facilitate interchangeability, particularly at lower levels. By mapping building blocks via architecture and solution continua, a kind of inheritance concept is realized, enabling reusability and interchangeability.

The approach is technology- and tool-neutral. According to the principle *Best of Breed*, the freedom to use the individual specialized and domain-specific tools is given [19]. This means that the most suitable tools and methods can be used for the respective area of application, enabling the most promising solution for each case. These range from enterprise architecture and requirement engineering to modeling and integration through development and testing. The necessary compatibility of the heterogeneous IT landscape must be ensured via appropriate associations and interfaces, as well as the intermediate levels. Various accruing artifacts are loosely connected via linked data. Therefore, a *Service-Oriented Architecture (SOA)* design realizes the interplay [20,21].

In addition, each stakeholder is provided with needs-oriented models and views that correspond to their respective perspectives. These serve to provide targeted answers to various questions at different levels and specialist areas. Specific guidelines per level and view support the creation of artifacts with specific instructions on detailing and maturity level. In particular, interfaces and transitions must be defined in detail. The interface descriptions are to be duplicated when used at the connection points of different models.

At the highest level, the USE CASE is considered as part of a SCENARIO – or mission threat – as in *NATO Architecture Framework, Version 4 (NAFv4)* and *Federated Mission Networking, Spiral 6 (FMN6)* [22,23]. The necessary CAPABILITIES to successfully implement the USE CASE can be derived from the resulting requirements as well as in *Concept of Operations (CONOPS)* [24–27]. Initial requirements for the interfaces can also be defined here. In addition, further possible uses of the CAPABILITIES emerge in retrospect, which in turn enable new applications SCENARIOS. This cycle is typically the driver for new innovations.

The CAPABILITIES are integrated, linked, and used within the framework of PROCESSES. This creates corresponding added value for the user. Appropriate connections between the elements must be taken into account to ensure a smooth process. Any gaps that arise can be identified and closed using additional skills among other things. Here, digitization for automation is a key driver for increasing performance. Furthermore, the discipline of *Product Lifecycle Management (PLM)* serves to ensure long-term sustainability [28–33].

Individual PROCESS steps must then be mapped and implemented using SERVICES and SUBSERVICES in accordance with the principle of *SOA*. Attention must be paid to the use of standardized reference designs. The goal is the reusability of self-contained elements through configuration options instead of the implementation of specific individual solutions. The SERVICES must be geared toward the user with the objective of generating added value. Interfaces are the key design criterion for the flexible connection and linking of SERVICES. These should be kept stable, comprehensive, and generic, with the changes possible. Genericity and extensibility can be realized via optional components.

The individual (partial) SERVICES are in turn implemented using various COMPONENTS, MODULES, and FUNCTIONS. For long-term success and maintainability, the modular structure is an essential design criterion. The following illustration shows a universal solution design for this approach. A further subdivision of individual solution units can be undertaken here. This can be understood as a kind of structure tree for a SERVICE and can extend over several levels.

The COMPONENTS, MODULES, and FUNCTIONS can be described both in terms of their more static composition as well as their DYNAMIC BEHAVIOR. This CONTROL LOGIC is used to implement the respective function, taking into account secondary conditions by means of input and output parameters. Consistent and safe states must always be ensured during the process.

The associated DATA STRUCTURES must be described in connection with the input and output parameters of the COMPONENTS, MODULES, and FUNCTIONS, as well as the CONTROL LOGIC. This includes both serialization and data management.

Various RESOURCES are required to provide and develop everything from the SERVICES to the DATA STRUCTURES.

Together with the implementation, these form the lowest level for executing the system – via SIMULATION.

The *Universal Solution Unit* is described by a configurable reference design, see Figure 7.4. Based on this template, solution units consist of both a BEHAVIOR ELEMENT that abstractly describes functionality and an ACTIVE ELEMENT with a corresponding external INTERFACE and internal structure.

The external INTERFACE is used to describe the INPUT and OUTPUT information in detail to enable interaction with the solution unit. This information can take the form of events or continuous data streams, and its type and structure are regulated via typing. Solution units can be chained by connecting the OUTPUT of the INTERFACE of one solution unit to the INPUT of the INTERFACE of another one. Moreover, recursive chaining can be achieved by connecting the OUTPUT of the INTERFACE of a solution unit to the INPUT of the same INTERFACE. Information exchanged via such a connection is represented by one or more PASSIVE ELEMENTS.

The internal structure is described via the subparts. These include a persistent and temporary MEMORY, a status register and the functions offered as BEHAVIOR. These must be individually designed and implemented for each solution unit. The complexity of the structural design is kept manageable by means of cascading. The elements and functional options mapped at a higher level are essential for the description, which are provided in this way and thus enable the higher-level overall context.

There are several ways of realizing the MEMORY and BEHAVIOR of the ACTIVE ELEMENT as part of the *Universal Solution Unit*. For example, the former could be realized using a list of attributes, while the latter could be realized using a list of operation heads with corresponding bodies. In Figure 7.5, a combined realization employing both an Information Model and a State Machine is exemplified.

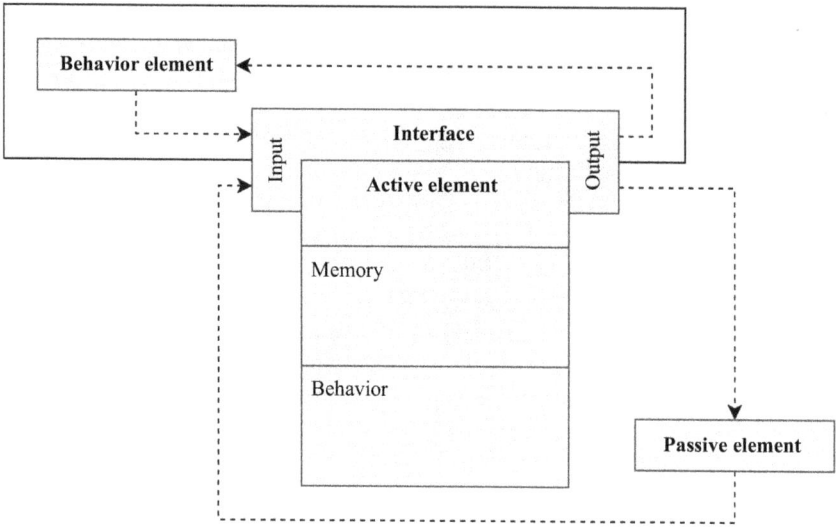

Figure 7.4   Universal solution unit

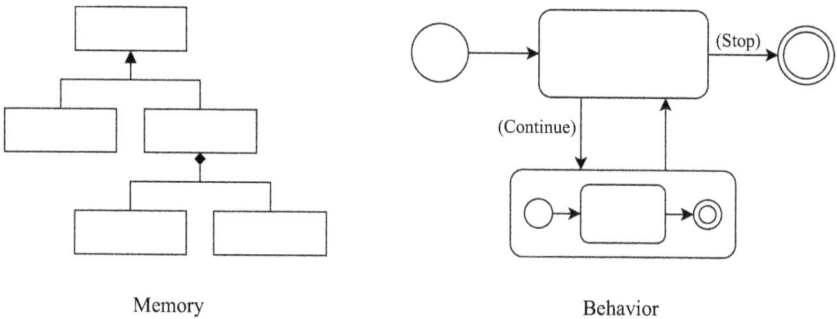

Memory   Behavior

Figure 7.5   Exemplary realization of memory and behavior of active elements as part of the universal solution unit

Altogether, this approach facilitates alignment between the *ATR-RM* and solution models derived from it, as well as simulations of these models [34]. This is fundamental for establishing the process "Maintenance of the *ATR-RM*" in the way described below, in Section 7.2.4.

7.2.2   Application to the Unified Architecture Framework

When applying our generic approach to a real-world problem, choosing an enterprise architecture framework for reference is highly beneficial as the specification of viewpoints ensures the clear delivery and thereby the understandability of information communicated through corresponding views. Here, we opted for the *Unified Architecture Framework (UAF)* due to its focus on holistically consistent modeling in the context of a *System of Systems*, but *ArchiMate* would have been an equally appropriate choice [25,26,35–37].

In the domain of *ATR*, *Systems of Systems* – i.e., large systems that are composed of independently useful subsystems – frequently occur, as such complex solutions are necessary to handle equally complex challenges [25,26]. For example, in the subdomain of *Underwater ATR*, an ATR SYSTEM might consist of a CARRIER and several DRONES, with the latter again being extended in its functionality by several modules. Such a case is illustrated in Figure 7.6.

Here, the complexity of an unknown number of targets to be classified, which are located within a large region, is handled by a CARRIER simultaneously deploying several DRONES as autonomous units. This enables the coverage of large regions in parallel and thus in a fraction of the time required when performing the task sequentially by only employing a single drone.

Accordingly, a *UAF-specific Solution Unit* is derived from the *Universal Solution Unit* – see Figure 7.4 – from Section 7.2.1 as depicted in Figure 7.7. The BEHAVIOR ELEMENT provides an abstract description of some behavior. This behavior is performed by the ACTIVE ELEMENT. To this end, it contains both ATTRIBUTES and OPERATIONS. The former encompass both PARAMETERS and MEMBER VARIABLES, while the latter can take the form of INTERNAL CODE, STATE MACHINES, or other manners of description. While a single OPERATION can only be specified through one of these manners, another OPERATION within the same ACTIVE ELEMENT may employ a different manner. Being modeled as *Systems Modeling Language (SysML) Blocks*, ACTIVE ELEMENTS may be connected using ports, which are again connected by BLOCK CONNECTORS [38]. Should such a connection require the exchange of some information, this information can be modeled as a PASSIVE ELEMENT that is transferred between the ACTIVE ELEMENTS via an INFORMATION EXCHANGE.

In Figure 7.8, an overview of the employed excerpt of the *UAF Domain Metamodel (UAF DMM)* is given [35]. For our purpose, the layers "Strategic," "Operational," "Services," and "Resources" are sufficient. Hence, we focus on them. Table 7.1 lists the integral element types for each of these layers. It can be seen that, for each layer, the corresponding element types match the *UAF-specific Solution Unit* – see Figure 7.7. In the case of "Strategic" and "Services" layers, only parts of the *UAF-specific Solution Unit* are taken into account, as the missing parts express

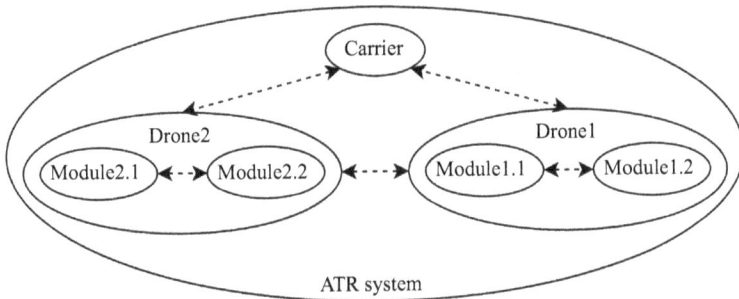

Figure 7.6   Exemplary system of systems in the domain of underwater ATR

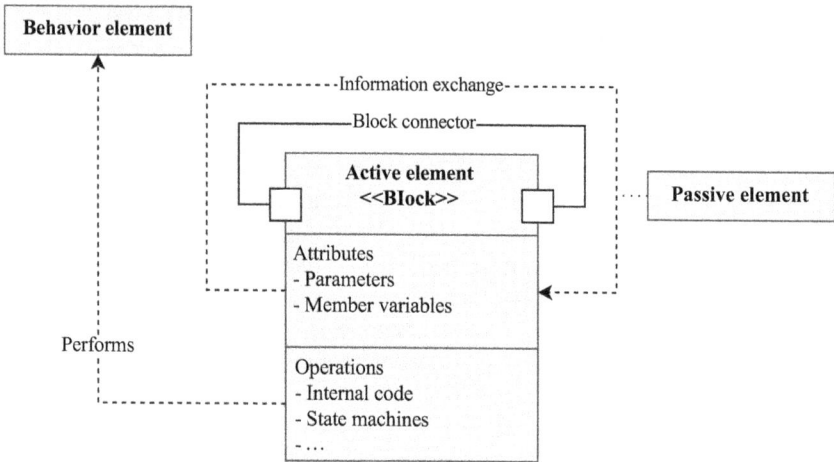

Figure 7.7   UAF-specific solution unit

a level of detail that, considering the respective intended information implementation, is detrimental to these layers.

In accordance with ArchiMate and the *Universal Solution Unit* – see Figure 7.4 – from Section 7.2.1, we distinguish between elements describing behavior (behavior elements), elements of active nature (active elements), and elements of passive nature (passive elements) [37]. The central element type of the layer "Strategic" represents behavior elements, while there is a central element type representing active elements and one representing behavior elements for each of the other layers. Additionally, there is a secondary element type for both layers "Operational" and "Resources" that represents passive elements.

Furthermore, we define elements as corresponding if their element type and thereby they themselves belong to the same layer. Corresponding active elements can be connected via ports. Additionally, active elements can be connected to corresponding behavior elements they are capable of performing. Finally, on both "Operational" and "Resources" layers, passive elements can be exchanged between corresponding active elements through the relations OPERATIONAL EXCHANGE and RESOURCE EXCHANGE, respectively.

The four layers are primarily connected through abstractions between behavior and active elements of adjacent layers: RESOURCE ARTIFACT can perform FUNCTION which can implement SERVICE FUNCTION. SERVICE SPECIFICATION can perform SERVICE FUNCTION. OPERATIONAL ACTIVITY can consume SERVICE SPECIFICATION and map to CAPABILITY. OPERATIONAL PERFORMER can perform OPERATIONAL ACTIVITY and exhibit CAPABILITY.

Secondarily, there are additional abstractions between behavior and active elements of layers that are not adjacent: FUNCTION can implement OPERATIONAL ACTIVITY and SERVICE SPECIFICATION can exhibit CAPABILITY.

Altogether, the excerpt constitutes a simplified meta model similar to ones like the views "IT4IT Meta Model - Solution Pattern" in the *Reference Model of IT4IT,*

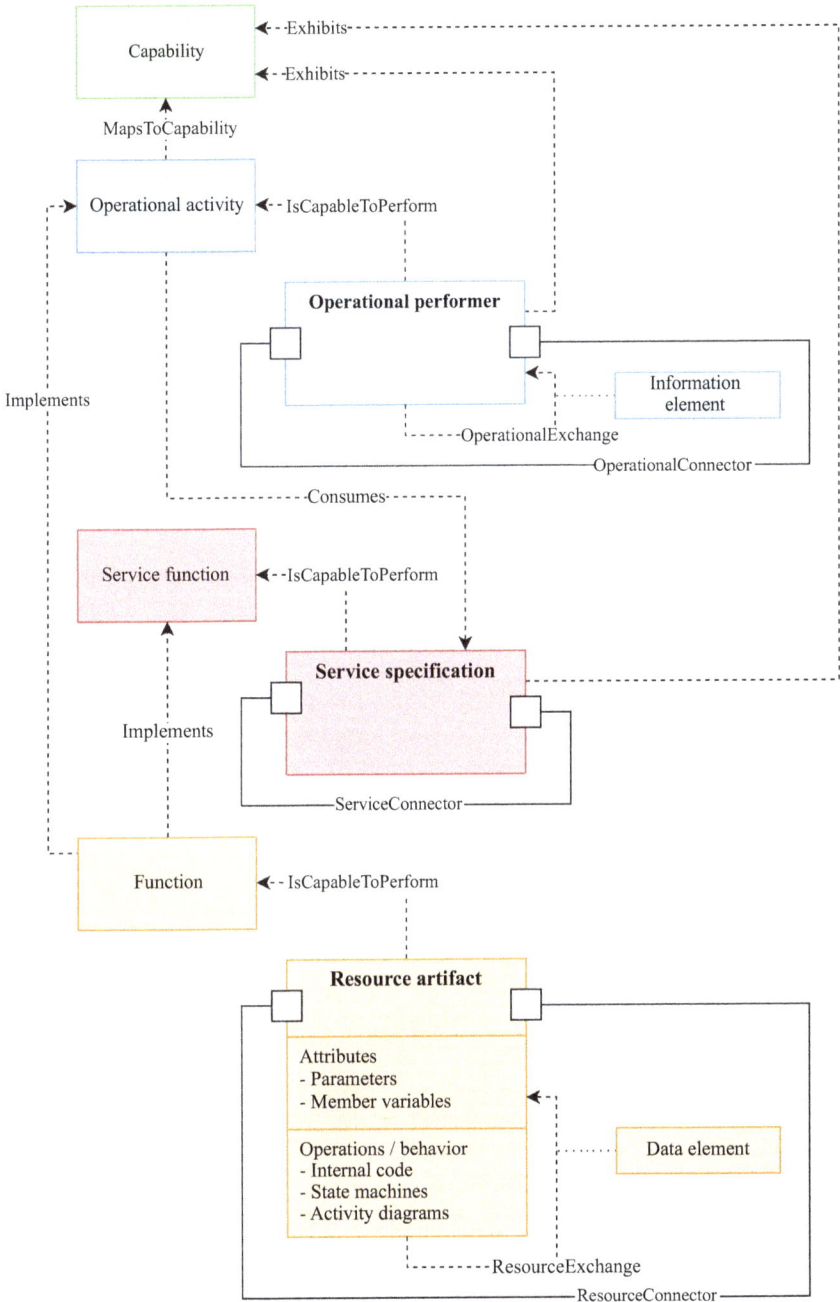

Figure 7.8   Application of the generic model-based systems engineering approach to UAF

Table 7.1   UAF layers and their corresponding element types

Layer	Element type(s)
"Strategic"	CAPABILITY
"Operational"	OPERATIONAL ACTIVITY, OPERATIONAL PERFORMER, and INFORMATION ELEMENT
"Services"	SERVICE FUNCTION and SERVICE SPECIFICATION
"Resources"	FUNCTION, RESOURCE ARTIFACT, and DATA ELEMENT

Version 3.0 (IT4ITv3-RM) as well as "NAFv4 Simplified Meta-Model" and "NAFv4 SMM – One Pager" in the *NATO Architecture Framework, Version 4 – Modeling Guidelines for Use of the UAF DMM* [39,40].

7.2.3   Automatic target recognition reference model template

From our generic approach (Section 7.2.1) and its application to *UAF* (Section 7.2.2), a template for the *ATR-RM* emerges. This template is the foundation for generating the *ATR-RM*. To this end, it relies on two premises and four concepts:

ATR taxonomy premise: As the *ATR-RM* models, the domain of *ATR*, it primarily addresses TARGET RECOGNITION and its automated specialization, AUTOMATIC TARGET RECOGNITION. Furthermore, based on their respective spatial environment, applications of AUTOMATIC TARGET RECOGNITION can be classified as either AERIAL AUTOMATIC TARGET RECOGNITION, TERRESTRIAL AUTOMATIC TARGET RECOGNITION, or MARITIME AUTOMATIC TARGET RECOGNITION, with UNDERWATER AUTOMATIC TARGET RECOGNITION being a specialization of MARITIME AUTOMATIC TARGET RECOGNITION. Hence, the *ATR-RM* secondarily addresses these subtypes of AUTOMATIC TARGET RECOGNITION. Altogether, the corresponding *ATR Taxonomy* that is illustrated in Figure 7.9, reappears in some manifestation on every modeled layer – i.e., from "Strategic" down to "Resources" – of the *ATR-RM*.

ATR process premise: The overarching *ATR Process* – see Figure 7.1 – from Section 7.2.4 reappears in some manifestation on every modeled layer – i.e., from "Strategic" down to "Resources" – of the *ATR-RM*, as well. To this end, PREPROCESSING, DETECTION, FEATURE EXTRACTION, CLASSIFICATION, IDENTIFICATION, DECISION-MAKING, and OPTIMIZATION are aggregated to realize TARGET RECOGNITION, as depicted in Figure 7.10.

Triad of model types concept: We distinguish between three types of models, the first one naturally being the *ATR-RM*. The other ones are solution models derived from the *ATR-RM* that represent certain subdomains of *ATR*, and simulation models that are employed to facilitate simulations of either the *ATR-RM* or solution models derived from it (base model). To this end, simulation models extend their corresponding base models. Consulting the *ATR Taxonomy*

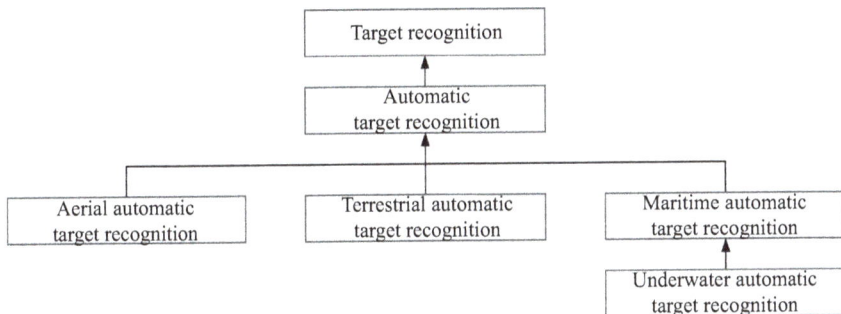

Figure 7.9   Target recognition taxonomy

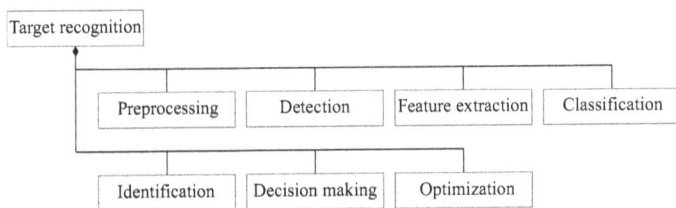

Figure 7.10   Target recognition structure

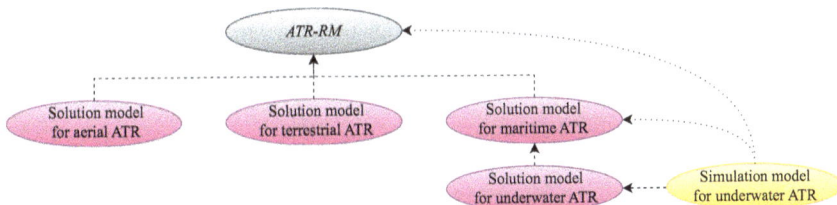

Figure 7.11   Triad of model types

Premise, the relationships between these three model types are exemplarily illustrated in Figure 7.11.

The example of the SOLUTION MODEL FOR MARITIME ATR reveals that although solution models are already derived from the *ATR-RM*, they may nevertheless be utilized as reference models. Hence, other solution models can be derived from them. Here, the SOLUTION MODEL FOR UNDERWATER ATR is derived from the aforementioned solution model. By transitivity, we treat the set of "solution models derived from solution models derived from the *ATR-RM*," the set of "solution models derived from solution models derived from solutions models derived from the *ATR-RM*," and so on as parts of the set of "solution models derived from the *ATR-RM*." Accordingly, we abbreviate "solution models derived from the *ATR-RM*" with "solution models" and use the term "solution model" to denote an element of this set.

Furthermore, the two dotted lines going out from the SIMULATION MODEL FOR UNDERWATER ATR highlight that although simulation models extend exactly one base model, contents of models from which this base model is derived from can be part of simulations. Hence, these contents may be referenced by simulation models. Here, contents of the SOLUTION MODEL FOR MARITIME ATR and the *ATR-RM* may additionally be referenced by the aforementioned model.

Package/folder structure concept: We establish a package/folder structure for storing both ELEMENTS and VIEWS within models [41]. The root level of this structure is based on naming conventions for *ArchiMate* models as outlined in Figure 7.12 [37]. It can be seen that the structure has two parts, one for storing ELEMENTS (element subtree) and one for storing VIEWS (view subtree).

The element subtree's first level is based on the layers – i.e., STRATEGIC, OPERATIONAL, SERVICES, and RESOURCES – of the *UAF DMM*, while its second level is based on the corresponding element types – i.e., CAPABILITY, OPERATIONAL PERFORMER, OPERATIONAL ACTIVITY, INFORMATION ELEMENT, SERVICE SPECIFICATION, SERVICE FUNCTION, RESOURCE ARTIFACT, FUNCTION, and DATA ELEMENT – as illustrated in Figure 7.8 and listed in Table 7.1, respectively [35]. The element subtree's structure is outlined in Figure 7.13.

Similarly, the view subtree's first level is based on the viewpoint groups – i.e., STRATEGIC, OPERATIONAL, SERVICES, RESOURCES, and INFORMATION – of the *UAF DMM*, while its second level is based on the concerns – i.e., TAXONOMY, STRUCTURE, CONNECTIVITY, PROCESSES, and TRACEABILITY – of the *UAF DMM* [35]. In our case, only for the viewpoint group SERVICES, the concern PROCESSES is employed for secondary partitioning. Diverging from all the other

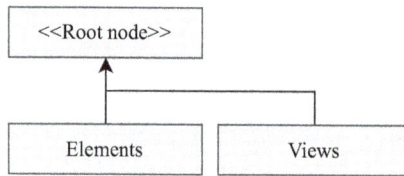

Figure 7.12   Package/folder structure – root level

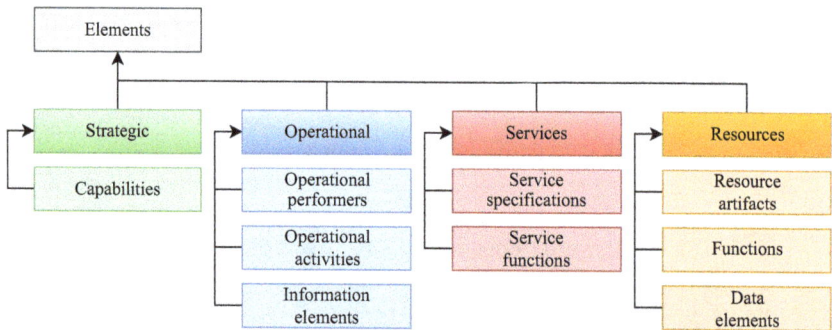

Figure 7.13   Package/folder structure – elements

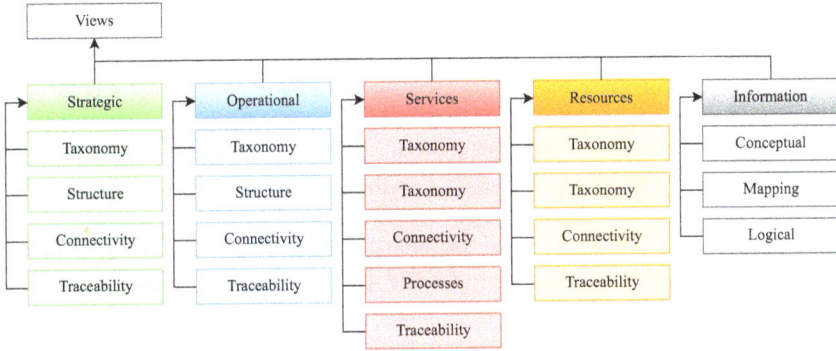

Figure 7.14   Package/folder structure – views

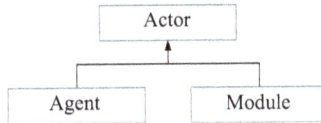

Figure 7.15   Actor taxonomy

viewpoint groups, for the viewpoint group INFORMATION, the degrees of detail CONCEPTUAL and LOGICAL as well as a MAPPING between them are employed for secondary partitioning. The result is outlined in Figure 7.14.

Agent module concept: With respect to planned *Agent-based Simulations* of the *ATR-RM* and solution models derived from it, we distinguish two types of ACTORS within such simulations as illustrated in Figure 7.15 [42,43]. While AGENTS denote top-level ACTORS, MODULES denote ACTORS extending the functionality of other ACTORS. Accordingly, we establish three top-level elements within the layer "Resources" of the *ATR-RM* – AGENT, MODULE, and CHARACTERISTIC. The latter represents complex data types – such as vectors – which are necessary to model certain attributes – such as positions in three-dimensional space.

Replaceable alternatives concept: Extending the *Agent Module Concept*, we account for the possibility of modules being replaced with different ones that fulfill the same purpose. That is, we expect cases in which there are several suitable approaches for performing some task. In such a case, the best approach shall be identified by determining the performance of each alternative with respect to some metrics through testing by way of the aforementioned *Agent-based Simulations*. For example, in a given scenario, there may be several algorithms – "Detection Algorithm 1," "Detection Algorithm 2," etc. – that are suitable for implementing "Detection" as part of the overarching *ATR Process* – see Figure 7.1 – from Section 7.2.4. Consequently, corresponding modules – DETECTION UNIT 1, DETECTION UNIT 2, etc. – are stored in a taxonomic structure as illustrated in Figure 7.16.

7.2.4 Automatic target recognition reference model maintenance

Based on the template from Section 7.2.3, a coherent initial state of the *ATR-RM* is obtained. Consequently, to preserve the coherence of the *ATR-RM*, it needs to be maintained consistently. We ensure this by conceptualizing and implementing an appropriate process "Maintenance of the *ATR-RM*" that includes all maintenance tasks regarding the *ATR-RM*. For this purpose, we once again employ *BPMN* [3].

As depicted in Figure 7.17, we distinguish between three maintenance tasks – FURTHER DEVELOPMENT OF THE *ATR-RM* OR A DERIVED SOLUTION MODEL, DERIVATION OF A SOLUTION MODEL, and SIMULATION OF THE *ATR-RM* OR A DERIVED SOLUTION MODEL. For each of these tasks, a separate corresponding subprocess is specified. Consequently, these subprocesses can be seen in Figures 7.18, 7.19, and 7.20.

The subprocess "Further Development of the *ATR-RM* or a Derived Solution Model" always begins with a distinction between an expansion and an adaptation of the respective model. While the former denotes the addition of new contents, the latter denotes the adjustment of existing contents.

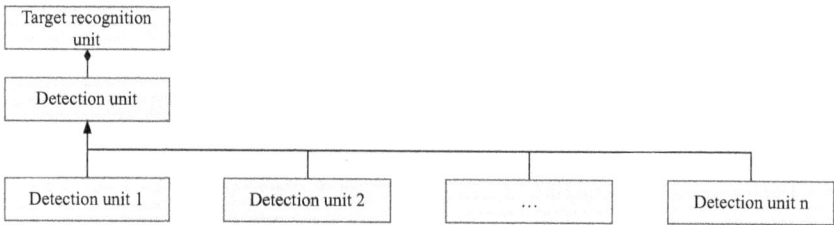

Figure 7.16   Replaceable alternatives concept

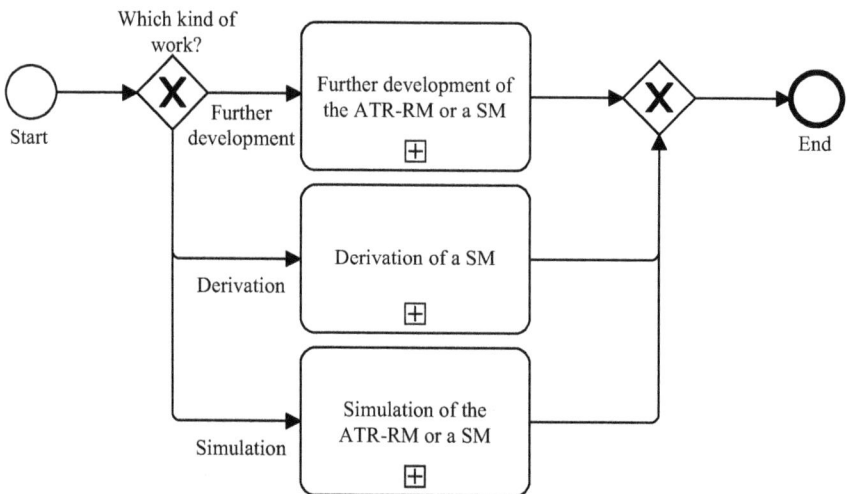

Figure 7.17   Maintenance of the ATR-RM – overview

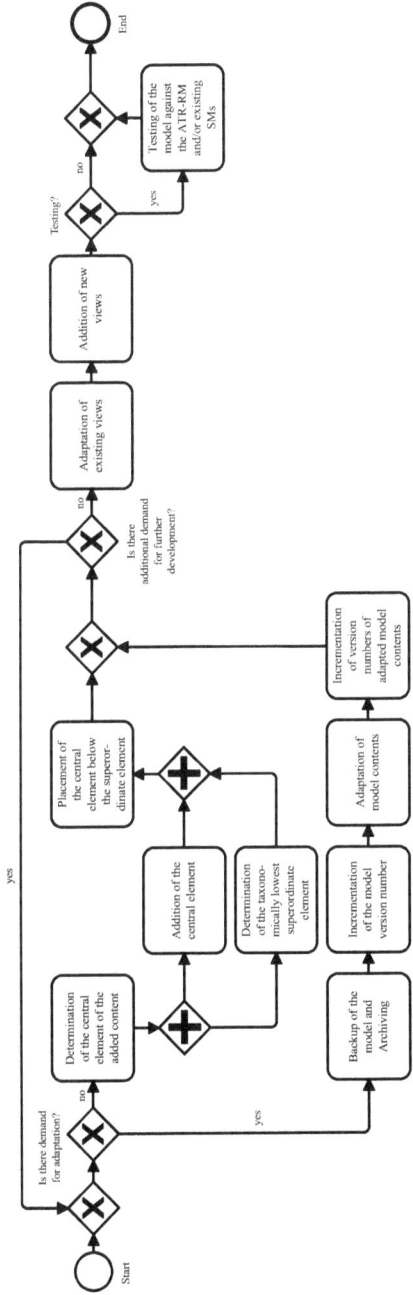

Figure 7.18   Maintenance of the ATR-RM – further development of the ATR-RM or a derived solution model

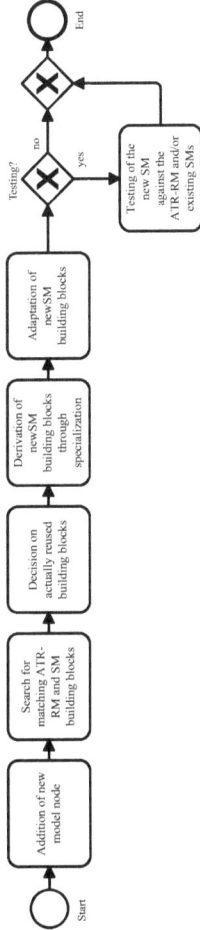

Figure 7.19   Maintenance of the ATR-RM – derivation of a solution model

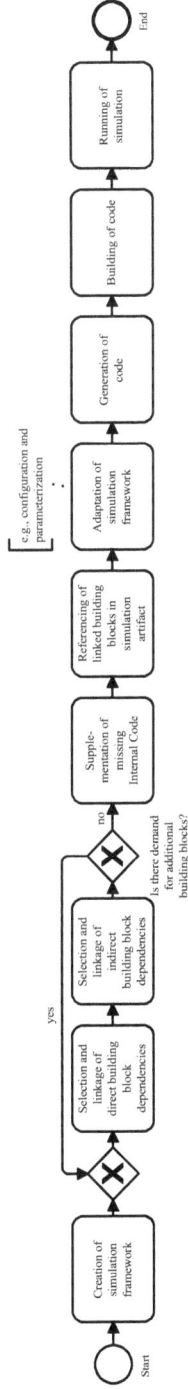

Figure 7.20   Maintenance of the ATR-RM – simulation of the ATR-RM or a derived solution model

In the case of an expansion, first, the central element of the added content (child) is determined. Second, the taxonomically lowest superordinate element (parent) is determined and the child is added to the model. Finally, in the taxonomy, the child is placed below the parent by adding one or more corresponding generalizations [44,45].

Contrarily, in case of an adaptation, first, a backup of the existing model is generated and archived before any adjustments to its contents are made. Second, the version number of the model is incremented. Third, the actual adaptation of model contents is performed. This can take several forms – e.g., adding, changing, or removing attributes or relations. Finally, for all model contents that were adjusted as part of the adaptation, their version number is incremented, as well. The versioning allows for reproducing changes of the *ATR-RM* and solution models derived from it. For instance, this is useful when some solution model is derived from the *ATR-RM* and afterward the *ATR-RM* is further developed. In such a case, the changes that were made to the *ATR-RM* need to be known so that they can later on be transferred to the solution model.

In both cases, the previous steps are repeated if additional further development of the model is required. Otherwise, first, existing views are adjusted in accordance with previously made changes and second, new views are added if necessary.

Optionally, this subprocess can be extended by testing the further developed model against the *ATR-RM* and/or solution models derived from it. Thereby, possible conflicts between contents of these models can be detected and resolved in subsequent executions of the subprocess "Further Development of the *ATR-RM* or a Derived Solution Model." While this task constitutes an optional step, it is highly recommended as some modeling tools – such as *Sparx Systems Enterprise Architect (Sparx EA)* – do not provide automatic functionality for detecting effects of changes to one model on other models [46]. The testing can be done by executing the subprocess "Simulation of the *ATR-RM* or a Derived Solution Model." In comparison with previous simulations on the same model, inconsistencies can be revealed.

The subprocess "Derivation of a Solution Model" starts with the addition of a model node for the new solution model. Second, both the *ATR-RM* and already existing solution models are examined to collect building blocks matching the subdomain of *ATR* represented by the new solution model (represented subdomain). Third, out of the collected building blocks, these are selected that shall actually be reused for the new solution model. Fourth, for the new solution model, new building blocks are derived from the selected existing ones through specialization [44,45]. Finally, the new building blocks are adjusted to better match the represented subdomain if necessary.

Optionally, this subprocess can again be extended by testing the newly derived solution model against both the *ATR-RM* and previously existing solution models. Thereby, possible conflicts between contents of these models can be detected. These conflicts can then be resolved by executing the subprocess "Further Development of the *ATR-RM* or a Derived Solution Model." This task is an optional step as adhering to the subprocess "Derivation of a Solution Model" already reduces the risk of conflicts since new solution models are built upon both the *ATR-RM* and previously existing solution models. Executing the subprocess "Simulation of the *ATR-RM* or a Derived Solution Model" for the newly derived solution model is one way of conducting such a test. In comparison with previous simulations on other solution models, inconsistencies can be revealed.

The subprocess "Simulation of the *ATR-RM* or a Derived Solution Model" begins with the creation of a simulation framework. To this end, first, a model node is added for the simulation (simulation model). Second, a diagram of the type "SysML Block Diagram" (simulation diagram) and an artifact of the type "Executable StateMachine" (simulation artifact) is added to the simulation model [38,47]. Finally, the simulation artifact is added to the simulation diagram. Optionally, additional elements may be added to the simulation model and simulation diagram if doing so is necessary to facilitate the simulation. As exemplified below, in Section 7.3.2, element types not listed in Table 7.1 from Section 7.2.2 may be assigned to these elements. In such cases, these other element types must be subtypes of element types listed in that table. It may be necessary to link these additional elements within the simulation artifact.

Afterward, from the contents of the to be simulated solution model, building blocks are selected and added to the simulation diagram (simulation elements). This includes direct dependencies of the intended simulation, primarily, and indirect dependencies – i.e., dependencies of dependencies – of the intended simulation, secondarily. Moreover, the indirect dependencies include transitive dependencies – e.g., indirect dependency *A.1.1* of indirect dependency *A.1* of direct dependency *A*. The tasks of adding some direct dependency and subsequently adding its corresponding indirect dependencies are repeated until no direct dependency is missing from the simulation diagram.

At this point, if any simulation element is missing *Internal Code*, it is supplemented. All the simulation elements are then linked within the simulation artifacts. The obtainment of the simulation diagram concludes with adapting the simulation framework to the simulation scenario – i.e., some missions as mentioned in Section 7.2.1 – through configuration and parameterization.

Based on the simulation diagram, the simulation code is first generated and then built. Afterward, the subprocess "Simulation of the *ATR-RM* or a Derived Solution Model" concludes with the simulation being run.

7.3   Resulting automatic target recognition reference model

Applying the methodology from Section 7.2 results in an *ATR-RM* abstracting the domain introduced in Section 7.1. In the following, we demonstrate its utilization for a given Scenario – see Section 7.3.1 – by presenting a corresponding derived solution model – see Section 7.3.2 – and simulation run – see Section 7.3.3.

7.3.1   Scenario

For our scenario, we adopt our exemplary case from "Aligning Models with Their Realization through Model-based Systems Engineering" [34]. In the following, a short summary is provided.

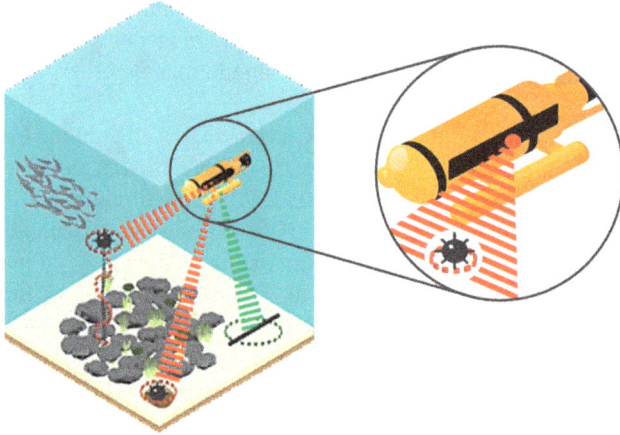

Figure 7.21   Exemplary underwater scenario

As illustrated in Figure 7.21, an *Autonomous Underwater Vehicle (AUV)* traverses an underwater environment to catch signals from targets located in this environment. Some of these targets are wanted, while the remaining ones are unwanted. They shall be classified accordingly.

This classification of targets into "Wanted Targets" and "Unwanted Targets" is performed by a *Target Classification Unit (TCU)*. The decision is based on the perceived strength of the caught signals – which depends on the actual strength of these signals and the intensity of background noise – in proportion to a threshold chosen in advance and not adapted during the simulation. A current induces a passive movement, because of which the *AUVs'* actual movement deviates from its desired movement. A *Movement Control Unit (MCU)* corrects this deviation by adjusting the *AUVs'* active movement such that its actual movement once again equals its desired movement. While the *MCU* ensures that the *AUV* reaches its intended final position, it also negatively impacts the *TCU* as it falsifies the perceived strength of caught signals by increasing the intensity of background noise. Hence, the *TCU* produces false positive decisions.

To more precisely specify these circumstances, we utilize mathematical formulae. In those formulae, we employ the symbols listed in Table 7.2.

These symbols refer to times, positions, velocities, and sound intensities of the *AUV*, the *TCU*, the *MCU*, the targets, and their environment. Among these symbols, t represents a parameter, while all symbols referring to positions or velocities as well as $N(t)$ represent variables. The remaining symbols represent constants. Some of these constants are fixed, while the remaining ones are chosen by the user to parameterize the simulation. The latter free variables of the simulation encompass t_i and t_n, $p_{\text{actual}}(t)$, v_{desired}, and v_{passive}, as well as $s_j, \forall j \in [0, m]$, N_0, and δN.

Based on their perceived signal strength, the *TCU* decides on the classification of targets as indicated in Table 7.3. The negative impact of the *MCU* finds its way into the formula via $N(t)$ as $N(t) = N_0 + \delta N, \forall t \in [t_i, t_n] \subseteq T$.

Table 7.2   Formula symbols used in the exemplary case

Symbol	Meaning
t	Point in time within the simulation
t_0	Starting time of the simulation/first t
t_i	Waiting time of the *MCU* until its activation
t_n	Duration of the simulation/last t
T	Set of all t
δt	Time step/distance between two points in time
$p_{\text{desired}}(t)$	Desired position of the *AUV* at t
$p_{\text{actual}}(t)$	Actual position of the *AUV* at t
$p_{\text{deviation}}(t)$	Position deviation of the *AUV* at t
v_{desired}	Desired velocity of the *AUV*
$v_{\text{actual}}(t)$	Actual velocity of the *AUV* at t
$v_{\text{active}}(t)$	Active velocity of the *AUV* at t
v_{passive}	Passive velocity of the *AUV*
h	Threshold of the *TCU*
s_j	Strength of signal $j \in [0, m]$
$N(t)$	Background noise at t
N_0	Background noise with inactive *MCU*
δN	Background noise increase due to active *MCU*

Table 7.3   Decision-making of the TCU

Condition	Decision
$s_j + N(t) \geq h$	Target is a "Wanted Target"
$s_j + N(t) < h$	Target is an "Unwanted Target"

7.3.2   *Derived solution and simulation models*

In accordance with the subprocess "Derivation of a Solution Model" – see Figure 7.19 – from Section 7.2.4, we create a solution model. Thereafter, we create a simulation model matching the scenario from Section 7.3.1 in accordance with the subprocess "Simulation of the *ATR-RM* or a Derived Solution Model" – see Figure 7.20 – from Section 7.2.4.

The corresponding simulation diagram is outlined in Figure 7.22. In this simplified illustration, neither attributes nor operations are included to improve readability. Within the simulation diagram, there are three elements of type SOFTWARE, which have been added to facilitate the simulation [35,36]. While SOFTWARE is a subtype of RESOURCE ARTIFACT and the elements could therefore be typed more generically, we emphasize their special status by assigning them a more specific type.

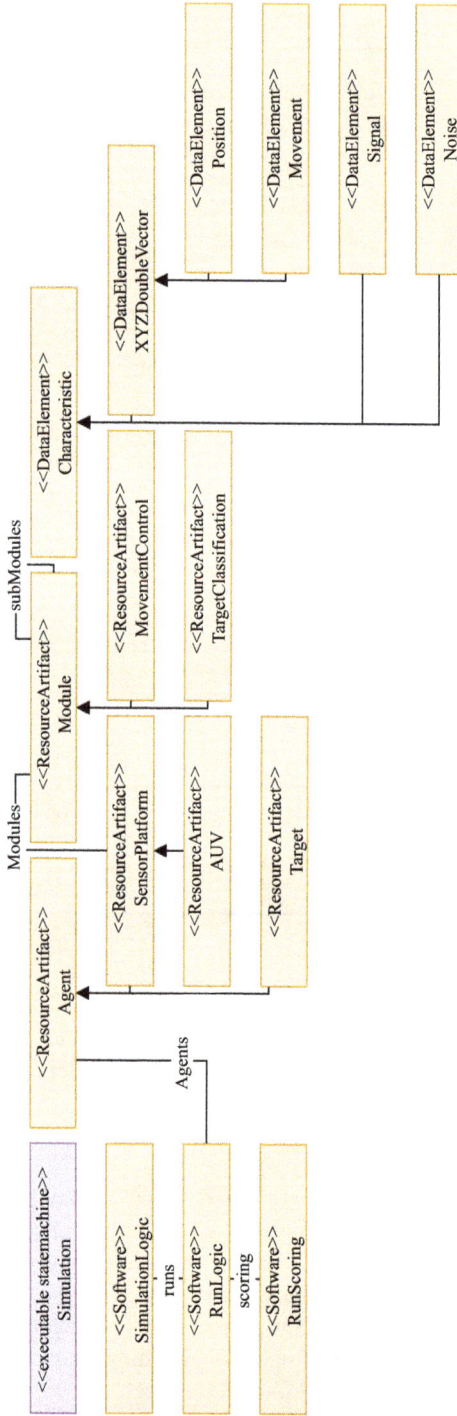

Figure 7.22   Simulation diagram of the simulation model

This allows for distinction between elements originating from the solution model and those originating from the simulation model. This distinction is important, as the latter elements would not be relevant for any other applications of the solution model.

The remaining elements of element types RESOURCE ARTIFACT and DATA ELEMENT are all part of the solution model – more specifically, its layer RESOURCES.

Furthermore, three associations – such as AGENTS, MODULES, and SUBMODULES – are contained in the simulation diagram [44,45]. In actuality, there are three attributes "agents," "modules," and "subModules" of the three elements RUNLOGIC, SENSORPLATFORM, and MODULE, respectively. As mentioned above, our simplified representation of the simulation diagram does not include any attributes. Hence, these three especially important ones are instead depicted as associations.

Altogether, the simulation diagram contains all the information required to successfully generate, build, and run the corresponding simulation. This information consists of the involved elements with their attributes and operations, the *Internal Code* specifying these operations, and the relations connecting these elements.

7.3.3   Simulation run

In accordance with the subprocess "Simulation of the *ATR-RM* or a Derived Solution Model" – see Figure 7.20 – from Section 7.2.4, we generate, build, and run the simulation [48].

As mentioned above, with the mathematical formalization of our scenario from Section 7.3.1, there are user-defined variables that allow for some dynamism of the simulation. Analogous to our scenario, we adopt our assignments of these variables from "Aligning Models with Their Realization through Model-based Systems Engineering" when running the simulation [34]. They are listed in Table 7.4.

With these assignments, we expect three false positives – i.e., one each per step beginning with the third step – in total. As illustrated in Figure 7.23, the simulation correctly computes the false positives as the scoring results match our prediction.

Table 7.4   Variable assignments for the simulation run

Reference point	Parameter	Value
AUV	t_n	5
	$p_{\text{actual}}(t_0)$	(0,0,0)
	$v_{\text{desired}}(t_0)$	(2,0,0)
	v_{passive}	(0,1,0)
MCU	t_i	2
	δN	1
TCU	h	3
	N_0	0
"Wanted Target"	s_0	3
"Unwanted Target"	s_1	2

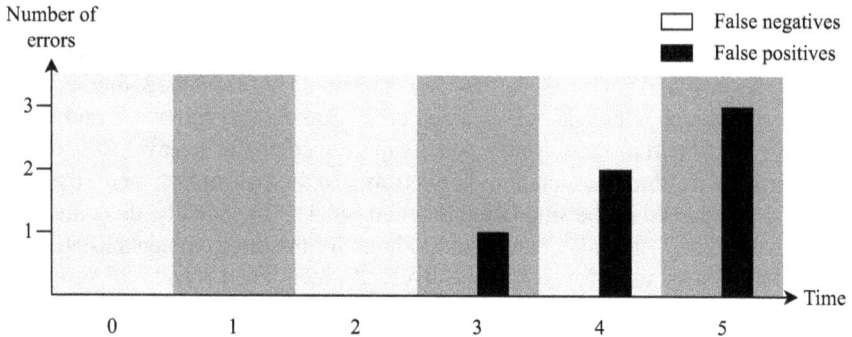

Figure 7.23   Scoring results of the simulation run

7.4   Conclusion

In the preceding, we have summarized which core concepts of *Automatic Target Recognition* are relevant for a *Reference Model for Automatic Target Recognition* – see Section 7.1. Furthermore, we have conceptualized an approach for consequently obtaining and maintaining such a reference model – see Section 7.2. Finally, we have exemplified how a *Reference Model for Automatic Target Recognition* can be generated consistently and validated through agent-based simulation according to our approach. Altogether, we have presented an approach for performing domain-oriented reference modeling in a sustainable and validatable manner.

By stepwise derivation of new models from existing models, where each derived model has exactly one base model it is derived from, a hierarchical structure of models is established. Within this hierarchical structure of knowledge, for two subdomains *A.1* and *A.2* of some domain *A*, information applicable for both *A.1* and *A.2* can be stored in the model of *A* (base model) and linked within the models of *A.1* and *A.2* (derived models). Hence, the approach facilitates the intended fostering of cross-domain sharing of knowledge through models. Thereby, it furthermore allows for *Data Deduplication* [49].

As our approach is based on well-established concepts and frameworks – e.g., *Model-based Systems Engineering* and the *Unified Architecture Framework* – and not necessarily dependent on the domain of *Automatic Target Recognition*, it can also easily be applied to other domains [18,35,36]. Moreover, it is especially well suited for any *System of Systems* domain [25,26].

References

[1]   Blacknell D. and Griffiths H. (eds.): *Radar Automatic Target Recognition (ATR) and Non-Cooperative Target Recognition (NCTR)*. Radar, Sonar and Navigation Series, Vol 33. IET Digital Library; 2013.

[2]   Melvin W.L. and Scheer J.A. (eds.): "Radar applications." *Principles of Modern Radar*. Vol III. Edison, NJ: SciTech Publishing Inc.; 2014.

[3] BPMN Team: *Information Technology – Object Management Group Business Model and Notation (OMG BPMN)*. Milford, MA: Object Management Group; 2013. ISO/IEC 19510:2013.

[4] Braga-Neto U.: *Fundamentals of Pattern Recognition and Machine Learning*. Berlin: Springer; 2020.

[5] Majumder U.K., Blasch E., and Garren D.A.: *Deep Learning for Radar and Communications Automatic Target Recognition*. Artech House Radar Series. Boston: Artech House; 2020.

[6] Fieguth P.: *An Introduction to Pattern Recognition and Machine Learning*. Berlin: Springer; 2022.

[7] vom Brocke J., and Buddendick C.: "Konstruktionstechniken für die Referenzmodellierung – Systematisierung, Sprachgestaltung und Werkzeugunterstützung" in Becker J., Delfmann P. (eds.). *Referenzmodellierung – Grundlagen, Techniken und domänenbezogene Anwendung – Mit 56 Abbildungen und 6 Tabellen*. Berlin, Heidelberg: Physica; 2004. pp. 19–49.

[8] Ascher D., Heiland E., Schnell D., *et al.*: "Methodology for holistic reference modeling in systems engineering." in Koç H., Stirna J., Sandkuhl K., *et al.* (eds.), *Joint Proceedings of the BIR 2022 Workshops and Doctoral Consortium (BIR-WS 2022) – co-located with 21st International Conference on Perspectives in Business Informatics Research (BIR 2022)*; Rostock/Germany, 2022. Vol 3223. pp. 140–151. Available from: https://ceur-ws.org/Vol-3223/paper13.pdf.

[9] Pellegrini P.F., Cuomo S., and Piazza E.: "ATC primary radar signal analysis for target characterisation – A model validation." in *The Record of the IEEE 1995 International Radar Conference*. Piscataway, NJ: IEEE; 1995. pp. 516–520.

[10] Bishel R.A.: "Short-range radar for use in vehicle lateral guidance and control." in *2023 IEEE International Automated Vehicle Validation Conference (IAVVC)*. Piscataway, NJ: IEEE; 2023. pp. 157–163.

[11] Zhang F., Liu S.Q., Wang D.B., *et al.* "Aircraft recognition in infrared image using wavelet moment invariants." *Image and Vision Computing*. 2009;27(4):313–318.

[12] Xu A., Gao J., Sui X., *et al.* "LiDAR dynamic target detection based on multidimensional features." *Sensors*. 2024;24(5):1369.

[13] Steiniger Y., Kraus D., and Meisen T.: "Survey on deep learning based computer vision for sonar imagery." *Engineering Applications of Artificial Intelligence*. 2022;114:105157.

[14] Yin F., Li C., Wang H., *et al.*: "Automatic acoustic target detecting and tracking on the azimuth recording diagram with image processing methods." *Sensors*. 2019;19(24):5391.

[15] Fahmy A., and Moustafa K.: "A survey of IFF systems." in *Proceedings of the 5th International Conference on Electrical Engineering*. 2006. Vol 5. pp. 1–11.

[16] Becker J., Probandt W., and Vering O.: Grundsätze ordnungsmäßiger Modellierung – Konzeption und Praxisbeispiel für ein effizientes Prozessmanagement. Berlin, Heidelberg: Springer; 2012.

[17] Hillmann P. and Karcher A.: *Are the Major Weapon Platforms Obsolete? – Session 8: The Case against Major Weapon Platforms - Redundancy – Looking into the Crystal Ball.* Harstad: NATO SAS-174; 2023. Available from: https://www.unibw.de/ia/publikationen-2/2023-nato-sas-174-major-platforms-obsolete-presentation_v4_reduced-1.pdf.

[18] Singam C.: "Model-Based Systems Engineering (MBSE)." in SEBoK Editorial Board (ed.) *The Guide to the Systems Engineering Body of Knowledge (SEBoK) – Version 2.7.* NJ: The Trustees of the Stevens Institute of Technology; 2022. pp. 258–264. Available from: https://sebokwiki.org/wiki/Model-Based_Systems_Engineering_(MBSE).

[19] Light B., Holland C.P., Kelly S., *et al.*: "Best of breed IT strategy – An alternative to enterprise resource planning systems." in Hansen H.R., Bichler M., Mahrer H. (eds.). *Proceedings of the 8th European Conference on Information Systems – Trends in Information and Communication Systems for the 21st Century – ECIS 2000*, Vienna, Austria, July 3–5, 2000; 2000. pp. 652–659. Available from: http://aisel.aisnet.org/ecis2000/180.

[20] SOA Work Group: *SOA Governance Framework.* UK: The Open Group; 2009. p. C093.

[21] SOA Work Group: *SOA Reference Architecture.* UK: The Open Group; 2011. p. C119.

[22] Architecture Capability Team: *NATO Architecture Framework – Version 4.* NATO C3 Board; 2020. AC/322-D(2018)0002-REV1. Available from: https://www.nato.int/cps/en/natohq/topics_157575.htm.

[23] Capability Planning Working Group: *Federated Mission Networking (FMN) – Spiral 6 Specification.* Norfolk, VA: NATO Military Committee; 2024.

[24] IEEE Guide for a Concept of Operations Document Working Group: *IEEE Guide for Information Technology – System Definition – Concept of Operations (ConOps) Document.* New York: IEEE-SA Standards Board; 2007. IEEE 1362-1998 (R2007). Available from: https://ieeexplore.ieee.org/stamp/stamp.jsp?tp=&arnumber=761853.

[25] ISO/IEC JTC 1/SC7, IEEE-CS S2ESC: *Systems and Software Engineering – Vocabulary.* Geneva, New York: ISO/IEC/IEEE; 2017. ISO/IEC/IEEE 24765:2017.

[26] ISO/IEC JTC 1/SC7, IEEE-CS S2ESC: *Systems and Software Engineering – System Life Cycle Processes.* Geneva, New York: ISO/IEC/IEEE; 2023. ISO/IEC/IEEE 15288:2023.

[27] Heiland E., Hillmann P., and Karcher A.: "Constraint based modeling according to reference design." in *Joint Proceedings of the BIR 2023 Workshops and Doctoral Consortium (BIR-WS 2023) – co-located with 22nd International Conference on Perspectives in Business Informatics Research (BIR 2023)*; Ascoli Piceno, Italy. Vol 3514; 2023. pp. 108–121. Available from: https://ceur-ws.org/Vol-3514/paper80.pdf.

[28] Stark J.: "Product Lifecycle Management (Volume 1) – 21st Century Paradigm for Product Realisation." in *Decision Engineering.* 5th edn. Cham: Springer; 2022.

[29] Stark J.: "Product Lifecycle Management (Volume 2) – The Devil is in the Details." in *Decision Engineering*. 4th edn. Cham: Springer; 2024.

[30] Stark J.: "Product Lifecycle Management (Volume 3) – The Executive Summary." in *Decision Engineering*. 1st edn. Cham: Springer; 2018.

[31] Stark J.: "Product Lifecycle Management (Volume 4) – The Case Studies." in *Decision Engineering*. 1st edn. Cham: Springer; 2019.

[32] Stark J.: "Product Lifecycle Management (Volume 5) – What Happens Across the Product Lifecycle?" in *Decision Engineering*. 1st edn. Cham: Springer; 2024.

[33] Arnold V., Dettmering H., Engel T., *et al.*: *Product Lifecycle Management beherrschen – Ein Anwenderhandbuch für den Mittelstand*. Berlin: Springer; 2011.

[34] Zenz L.J.I., Heiland E., Hillmann P., *et al.*: Aligning models with their realization through model-based systems engineering in 2023 International Conference on Advanced Enterprise Information System (AEIS); London, UK, 2023. pp. 63–69.

[35] UAF Team: *Information Technology – Object Management Group Unified Architecture Framework (OMG UAF) – Part 1: Domain Metamodel (DMM)*. Milford, MA: Object Management Group. ISO/IEC 19540-1:2022; 2022 35.020. Available from: https://www.omg.org/spec/UAF.

[36] UAF Team: *Information Technology – Object Management Group Unified Architecture Framework (OMG UAF) – Part 2: Unified Architecture Framework Profile (UAFP)*. Milford, MA: Object Management Group. ISO/IEC 19540-2:2022; 2022 35.020. Available from: https://www.omg.org/spec/UAF.

[37] ArchiMate Forum: *ArchiMate Specification, Version 3.1*. UK: The Open Group; 2019. p. C197.

[38] SysML Team: Information Technology – Object Management Group Systems Modeling Language (OMG SysML). Milford, MA: Object Management Group. ISO/IEC 19514:2017; 2017.

[39] IT4IT Forum: *IT4IT – A Reference Architecture for Managing Digital – Version 3.0*. United Kingdom: The Open Group; 2022. p. C221.

[40] Architecture Capability Team: *NATO Architecture Framework – Version 4 – Modelling Guidelines for Use of the UAF DMM*. Brussels: NATO C3 Board; 2019. Available from: https://www.nato.int/cps/en/natohq/topics_157575.htm.

[41] Hillmann P., Schnell D., Hagel H., *et al.*: "Enterprise model library for business-IT-alignment." in Signal, Image Processing and Embedded Systems Trends (SIGEM 2022). Vol abs/2211.11369. Zürich: AIRCC; 2022. pp. 157–170.

[42] Niazi M., and Hussain A.: "Agent-based computing from multi-agent systems to agent-based models: A visual survey." *Scientometrics*. 2011;89(2): 479–499.

[43] Hillmann P., Uhlig T., Rodosek G.D., *et al.*: "A novel multi-agent system for complex scheduling problems." in Tolk A., Diallo S.Y., Ryzhov I.O., *et al.*

(eds.). *Proceedings of the 2014 Winter Simulation Conference*. Piscataway, NJ: IEEE; 2014. pp. 231–241.

[44]   UML Team: *Information Technology – Object Management Group Unified Modelling Language (OMG UML) – Part 1: Infrastructure*. Milford, MA: Object Management Group. ISO/IEC 19505-1:2012; 2012.

[45]   UML Team: *Information Technology – Object Management Group Unified Modelling Language (OMG UML) – Part 2: Superstructure*. Milford, MA: Object Management Group. ISO/IEC 19505-2:2012; 2012.

[46]   Sparx Systems: *Enterprise Architect, Version 16.0 – User Guide Series*. Australia: Sparx Systems; 2022.

[47]   Sparx Systems: *Enterprise Architect, Version 16.0 – User Guide Series – Executable StateMachines*. Australia: Sparx Systems; 2022. Available from: https://sparxsystems.com/resources/user-guides/16.0/simulation/executable -state-machines.pdf.

[48]   Golling M., Koch R., Hillmann P., *et al.*: "On the evaluation of military simulations." in *2015 Military Communications and Information Systems Conference (MilCIS)*. Australia: IEEE; 2015. pp. 16–22.

[49]   Kim D., Song S., Choi B.Y.: *Data Deduplication for Data Optimization for Storage and Network Systems*. 1st edn. Cham: Springer; 2017.

Chapter 8

Surrogate model framework for explainable autonomous behaviour in maritime robotic systems

*Konstantinos Gavriilidis[1], Andrea Munafo[2], Wei Pang[1]
and Helen Hastie[3]*

With the surge in the application and deployment of robotic and autonomous systems across various sectors, there is an ever-growing demand for transparency and explainability. The maritime domain, which has embraced these technologies for tasks ranging from disaster site inspections to pipeline maintenance, is no exception. This chapter will discuss a novel surrogate model framework that aims to bring explainability to autonomous behaviours in maritime robotic systems, bridging the existing gap between complex robotic decisions and the human understanding of their actions with the desired mission intent. The chapter will provide an overview of why transparency and explainability are important for robotic and autonomous systems, especially in the maritime sector, exploring the complexity of conveying complex autonomous decisions in understandable terms for various stakeholders with potentially conflicting needs (e.g., in-field operators, remote controllers, and trained/untrained operators). It will introduce the framework of surrogate models for explainable decision-making, focusing on their utility in simplifying and approximating deterministic agent policies and offering explanations that are independent of the underlying autonomy model. Finally, the chapter will discuss the journey from simulated experiments to actual trials with maritime robots, emphasising the changes, challenges, and learnings experienced during this transition. The chapter will conclude by highlighting the potential growth areas, applications, and refinements for the surrogate model framework in the future.

8.1 Introduction

With the surge in the application and deployment of robotic and autonomous systems across various sectors, there is an ever-growing demand for transparency and

[1]School of Mathematical and Computer Sciences, Heriot-Watt University, UK
[2]Department of Information Engineering, University of Pisa, Italy
[3]School of Informatics, University of Edinburgh, UK

explainability. The maritime domain, which has embraced these technologies for tasks ranging from offshore site inspections [1] to defence and security [2], is no exception. The rapid advancement of Autonomous Underwater Vehicle (AUV) technology has significantly expanded the scope of tasks these systems can undertake and the duration of their operational capabilities. AUVs are increasingly deployed to complex and dynamic environments, such as disaster site inspections, underwater pipeline maintenance and environmental monitoring. These missions often involve unanticipated events that cannot be pre-planned, necessitating autonomous systems capable of adaptive decision-making. Traditional pre-planned waypoint-based missions are insufficient for these scenarios, as they cannot account for unforeseen obstacles and changes in the environment [3].

In behavioural autonomy [4], multiple behaviours are implemented to enable AUVs to adapt to varying circumstances and complete missions autonomously. For example, a robot may switch from a surveying behaviour to a transit behaviour to navigate to a new waypoint and may temporarily interrupt its current task to perform a GPS fix for position updates. However, these autonomous decisions can create a gap between operator expectations and the actual behaviour of the AUVs, leading to decreased trust and potential mission interruptions [4].

Explainability in autonomous systems is critical to bridging this gap. Operators must understand the rationale behind the actions taken by AUVs to ensure mission success and calibrate their trust in the system, for example, to rely on its decisions or to identify when the system is not working and intervention is needed. This need for transparency is heightened in the maritime domain, where communication bandwidth is limited, and delays are common [5]. In such environments, unexpected behaviours need to be explained clearly and promptly to operators, who might be remotely monitoring the mission [6].

To address this challenge, a framework for explaining autonomous decisions and actions becomes important. This framework should be autonomy-agnostic, allowing it to be integrated with various autonomy architectures without requiring extensive modifications. One promising approach is the use of surrogate models [7], which simplify and approximate the decision-making processes of complex autonomous systems. By distilling the decision logic into an interpretable form, such as a decision tree, and combining it with natural language explanations, operators can receive clear and understandable explanations of the AUVs' behaviour in real time [8].

This chapter introduces a novel surrogate model framework that aims to bridge the existing gap between complex robotic decisions and human understanding of their actions with the desired mission intent. The framework simplifies and approximates deterministic agent policies, offering explanations that are independent of the underlying autonomy model. The journey from simulated experiments to actual trials with maritime robots is discussed, emphasising the changes, challenges, and learnings experienced during this transition. The chapter concludes by highlighting potential growth areas, applications, and refinements for the surrogate model framework in the future. The remainder of this chapter is organised as follows: Section 8.1.1 discusses the importance of explainability in automated systems, while Section 8.2 defines key terminology and identifies the stakeholders relevant to transparency and explainability in maritime robotic systems. Section 8.3 introduces surrogate models and categorises them, whereas Section 8.4 presents an abstract overview of the framework for explainable autonomous behaviour. Finally, Section 8.5 examines the proposed approach in greater detail, presenting results from both simulated and real-world trials, and Section 8.6 concludes the chapter by reflecting on the lessons learned and the challenges encountered.

8.1.1   Importance of explainability

The significance of explainability in autonomous systems has become increasingly important with the growing deployment of these technologies across various domains, often in safety-critical applications.

Explainability refers to the ability of an autonomous system to provide clear, understandable and transparent explanations of its actions and decisions to human users [9]. This is critical for several reasons. First and foremost, explainability helps operators develop the appropriate amount of trust and reliability in autonomous systems [10]. This means not only understanding when the system is making correct decisions but also recognising when its behaviour might be unreliable or require intervention. Operators and stakeholders need to trust that the system will perform as expected under varying conditions. When autonomous systems can explain their actions, it reassures users that the decisions are based on sound reasoning, and in this way, it provides the appropriate amount of trust. This is especially important in the maritime domain, where unexpected behaviours can lead to significant operational risks and financial costs [11].

Moreover, explainability supports accountability and safety; autonomous systems often operate in complex environments where they must make real-time decisions. In these contexts, understanding the rationale behind decisions is important for diagnosing and mitigating potential issues. For instance, if an AUV deviates from its planned path to avoid an obstacle, an explanation of its behaviour makes it possible for operators to verify that the system is functioning correctly and taking the correct decisions [4].

Additionally, explainability can also be important for regulatory compliance and ethical considerations. As autonomous systems become more prevalent, there is an increasing demand from regulatory bodies for transparency in their decision-making processes [11]. One of the major barriers to a broader uptake of autonomous system technology lies in the lack of clear guidelines and regulations for its usage. Explainable systems can help meet these regulatory requirements by providing detailed logs and rationales for their actions [12].

The concept of explainability also extends to improving human–robot interaction. In operational settings, operators often have to make quick decisions based on the system's actions [13]. Clear explanations enable better collaboration between humans and robots, as operators can anticipate and understand the system's behaviour, leading to more efficient and effective mission outcomes. For example, if an AUV changes its course due to detected anomalies, operators can quickly adapt their strategies based on the provided explanations [14]. At the mission planning level, the ability of the system to provide explanations can speed up planning and reduce errors [15]. An important part of explainability is to ensure that operators neither over-trust nor under-trust the system but instead develop an informed understanding of its capabilities and limitations.

Within this context, the use of surrogate models in providing explainability can be particularly beneficial. Surrogate models can approximate the decision-making process of complex autonomous systems and present them in an interpretable manner, translating complex and often opaque autonomy models into more understandable forms. Figure 8.1 illustrates a simplified decision tree used as a surrogate model to explain the actions of an AUV during a mission and describes the sequence of actions that can be performed by the vehicle during a survey mission. The mission begins with the AUV navigating to a predetermined survey area. Upon arrival, the AUV assesses

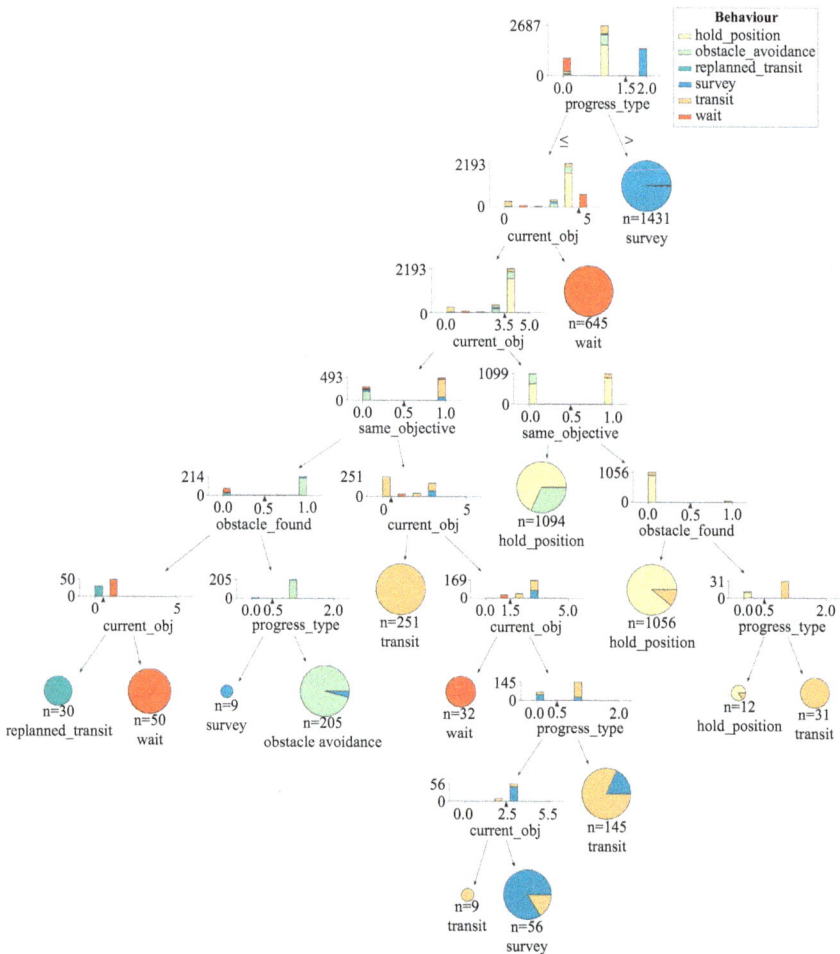

Figure 8.1   Decision tree used as a surrogate model to explain the actions of an AUV during an oceanographic survey mission

whether the survey area has been reached. If the area is not reached, the AUV returns to base; otherwise, it proceeds to conduct the survey. During the survey, data collection occurs, after which the AUV evaluates whether the survey is complete. If not, the AUV navigates to the next survey area and repeats the process. Once all survey areas are covered, the mission concludes. This decision tree provides a clear and interpretable explanation of the AUV's behaviour during the mission, serving as an effective surrogate model for the underlying autonomous system. This decision tree, which is characterised by a clear and sequential structure of conditional decisions, provides an interpretable explanation of the AUV's behaviour and is able to capture the main state and actions of the underlying autonomy. Understanding how surrogate models can help provide a layer of explainability for autonomous systems will be the focus of the remainder of this chapter.

It is also worth highlighting that recent research has demonstrated the effectiveness of knowledge distillation and natural language generation in producing verbalised explanations that are both accurate and user-friendly [14,16] and are able to ensure that these explanations are technically correct and understandable to users across varying levels of expertise.

To conclude this section, it is important to recognise that in the maritime domain, the deployment and operation of autonomous systems are subjected to unique challenges due to the harsh and unpredictable environment [17]. In this respect, the ability of autonomous systems to generate clear and accessible explanations becomes even more relevant. Communication constraints further complicate maritime operations. Underwater communication primarily relies on acoustic signals, which have limited bandwidth and are susceptible to delays and signal degradation [18]. This limitation makes real-time data transmission and remote control of autonomous maritime vehicles, particularly underwater vehicles, challenging, requiring minimal or no human supervision.

The complexity of underwater tasks also poses significant challenges. Maritime missions often involve complicated tasks such as pipeline inspection, seabed mapping, and inspection of infrastructures [19]. These tasks require precise navigation and control, as well as the ability to interact with the environment using various sensors and manipulators. The vehicle's autonomous system must be able to process and interpret sensor data accurately, make informed decisions based on this data [3,20] and communicate the results of these decisions to operators.

Moreover, AUVs must be prepared to handle unexpected obstacles, equipment malfunctions and mission deviations. Explainable autonomous systems play an important role in this context by providing operators with clear information on the AUV's decision-making process, enabling them to understand and address any issues that arise during the mission [8].

Another challenge is the integration of multiple autonomous systems. Maritime operations can include multiple AUVs, Remotely Operated Vehicles (ROVs) and Unmanned Surface Vehicles (USVs), each with distinct capabilities and roles. This requires a high level of interoperability and standardised communication protocols, as well as the ability to provide explainable decision-making processes that can be understood across different platforms [11].

The rest of the chapter will tackle these aspects, focusing on how surrogate models can be used to support the explainability of autonomous decision-making in the maritime environment, independent of the specific autonomy implementations.

8.2   Transparency and explainability in maritime robotic systems

Transparency in autonomous systems refers to the extent to which the internal processes of a system are made visible and understandable to users. It includes the availability of information about the system's decision-making processes, its actions and the rationale behind them. Explainability, which is a subset of transparency, focuses specifically on providing clear and understandable explanations for the decisions and actions taken by autonomous systems [9,21]. In the context of maritime robotic systems, these concepts can ensure that operators, stakeholders and regulators build an informed understanding of the system's behaviour, allowing them to oversee autonomous operations effectively and identify potential failures when necessary. An explainable system provides insight into its behaviour through various means, including visualisations, verbal explanations and interactive interfaces. As operators, stakeholders, regulators and robots have their own models of the scenario, it is important that the system can facilitate their communication. As a consequence, it is possible to develop approaches able to map natural language queries onto the machine's problem representation and mission concepts, as well as providing visual and textual methods of communicating the results back to the operator [15].

Each of these components can reinforce specific elements of explainability:

- Interpretability: The degree to which a human can understand the cause of a decision.
- Comprehensibility: The extent to which a system's operations can be described in understandable terms.
- Traceability: The ability to trace a system's decisions back to the underlying data and algorithms used to make them.

To design explainable systems, it is hence important to understand the specific stakeholders for which the system is being developed, because different stakeholders have varying needs for transparency and explainability. A possible stakeholder list includes:

- Operators: Require real-time, detailed explanations of autonomous system behaviours to monitor and intervene when necessary. Clear explanations help operators understand and predict system actions, enhancing situational awareness [16,22] and efficiency of the missions under execution [23].
- Engineers and developers: Need detailed insights into the system's decision-making processes for debugging, validation and improvement of algorithms. Explainability aids in identifying errors during and post-mission and understanding the impact of different parameters on system behaviour [24].
- Regulators: Demand transparency to ensure compliance with safety standards and regulatory requirements. Explainable systems can provide the necessary

documentation and rationale to demonstrate that the system operates within legal and safety guidelines [11].
- General public and other stakeholders: Require high-level explanations that assure them of the safety and reliability of autonomous systems. Public trust is crucial for the acceptance and adoption of these technologies [11].

Example use cases are given here.

- Pipeline inspection: During an underwater pipeline inspection mission, an AUV encountered unexpected obstacles. The operators need clear explanations of the AUV's decision to deviate from its planned path. An explainable system provides real-time information during the obstacle avoidance manoeuvres, ensuring that the mission can continue without manual intervention [4,8].
- Environmental monitoring: During environmental monitoring, multiple AUVs can be deployed to collect data. The complexity of coordinating these vehicles requires a high level of transparency. Explainable systems make vehicle coordination easier providing rationales for the AUVs' actions, such as changes in sampling locations due to detected anomalies [25]. Note that this can be done both in-mission and post-mission. In the latter case, the system can propose a new mission plan based on the collected data, providing explanation on why the new plan is required [15].
- Disaster response: Consider a scenario where an AUV is deployed to inspect underwater damage after an earthquake. The system prioritises areas with a high likelihood of structural damage. By explaining this prioritisation based on metrics of interest (e.g., risk assessment), the operators can focus their efforts on the most critical regions, speeding up the response time and improving the effectiveness of the recovery operations [26].

8.3   Surrogate models for explainable decision-making

Surrogate models are simplified representations of complex systems that are used to approximate the behaviour of the original models while providing interpretable information on their functioning. In the context of explainable decision-making, surrogate models serve as interpretable proxies for more complex, often opaque, machine learning or autonomy models. These models facilitate the understanding of how decisions are made by providing a more transparent and comprehensible overview of the decision-making process [7].

A surrogate model typically mimics the input–output behaviour of a more complex model by learning an approximate mapping from inputs to outputs [7]. This approximation allows for the extraction of interpretable rules that can be easily communicated to stakeholders. This concept is summarised in Figure 8.2. A surrogate model is trained on data coming from a complex decision-making system and is able to provide a more interpretable approximation of the decision-making process. The usage of the approximation makes it possible to communicate information on the actions and decisions taken by the more complex decision-making system.

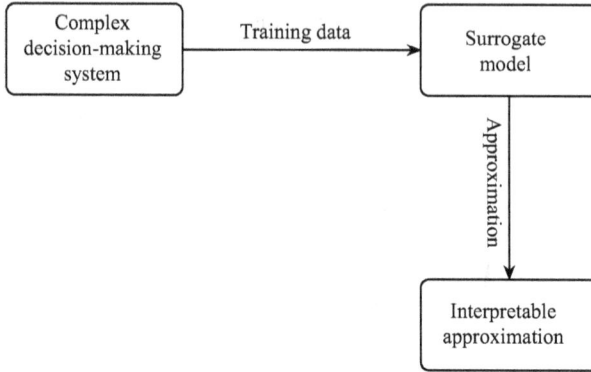

Figure 8.2   Surrogate models, once properly trained, can provide an interpretable approximation of a complex decision-making system

8.3.1   *Types of surrogate models*

There are several types of surrogate models, each with its strengths and applications. The choice of surrogate model depends on the specific requirements for interpretability, accuracy and the nature of the underlying complex model.

Surrogate models include:

8.3.1.1   Linear models

Linear models, such as linear regression, are among the simplest forms of surrogate models. They approximate the relationship between inputs and outputs using a linear equation, making them highly interpretable. However, their simplicity can limit their accuracy in capturing complex nonlinear relationships [27].

8.3.1.2   Decision trees

Decision trees are widely used as surrogate models due to their intuitive and visual nature. They represent decisions as a series of binary choices, making them easy to interpret and understand. Decision trees can capture nonlinear relationships, but their accuracy can be limited by overfitting or underfitting [8,28].

8.3.1.3   Rule-based models

Rule-based models, such as decision rules and association rules, use a set of if–then rules to represent the decision-making process. These models are highly interpretable and can be directly translated into human-understandable explanations. However, they may struggle to capture complex interactions between variables [29].

8.3.1.4   Neural network distillation

Neural network distillation involves training a simpler neural network to mimic the behaviour of a more complex one. This technique can reduce the complexity of the model while retaining the majority of its performance, making it more interpretable [30].

Table 8.1   Classification performance metrics for five models derived using nested cross-validation on simulation data. Transparent models are grouped at the top, followed by opaque models. Training and score times (in seconds) are also provided to indicate computational effort

Models	Accuracy	Precision	Recall	F1-Score	Fit Time	Score Time
Transparent models						
Decision Tree	**0.8981**	**0.9464**	0.8498	**0.8712**	25.0936	0.0025
CategoricalNB	0.8247	0.8701	**0.8616**	0.8379	**0.1127**	**0.0019**
KNN	0.6655	0.7806	0.8291	0.6953	4.8554	0.0721
Opaque models						
SVM	0.8846	0.9163	0.8378	0.8535	14.9298	0.0547
Multilayer Perceptron (MLP)	**0.8987**	0.9459	0.8496	0.8707	147.8816	0.0075

8.3.1.5   Model-agnostic methods

Model-agnostic methods, such as LIME (Local Interpretable Model-agnostic Explanations) [31] and SHAP (SHapley Additive exPlanations) [32], do not rely on a specific type of model. Instead, they approximate the complex model locally around a specific prediction to provide interpretable information. These methods offer flexibility and can be applied to a wide range of models [31,32].

A deeper discussion on these models goes beyond the objectives of this chapter. While we initially evaluated multiple surrogate models, we ultimately focused on decision trees because they offer a balance between interpretability and predictive performance. Their hierarchical structure provides intuitive, rule-based explanations that are easy for operators to understand. Moreover, as shown in our cross-validation experiments (Table 8.1), decision trees achieved high accuracy while significantly reducing training and inference time compared to more complex models. This makes them practical for real-time onboard execution, where computational resources are limited.

8.4   Framework for explainable autonomous behaviour

This section goes into more detail on the proposed framework for explainable autonomous behaviours, with a focus on maritime robotic systems. The underlying objective is to ensure transparency and interpretability of the system's actions. At its core, the framework leverages surrogate models to provide clear and understandable explanations of autonomous decisions. In doing so, it integrates three main elements: data collection, model interpretation and explanation generation.

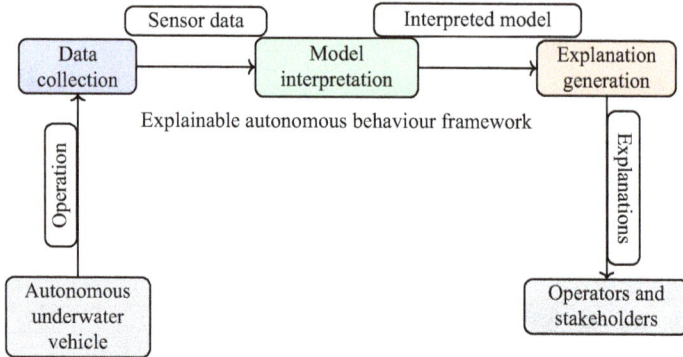

Figure 8.3   Conceptual framework for explainable autonomous behaviour in maritime robotic systems

Data collection involves gathering relevant sensor data and contextual information during the AUV's operation. This data is used to build and refine the surrogate models, ensuring that they accurately reflect the system's decision-making process. Model interpretation focuses on translating the complex decision-making processes of the autonomous system into an interpretable form, such as a decision tree. Finally, explanation generation involves producing human-readable explanations based on the interpreted model, which can be communicated to operators and stakeholders. Figure 8.3 illustrates the conceptual framework for explainable autonomous behaviour.

Note that the described framework makes the explanation layer independent of the specific autonomy that one is aiming to explain, i.e., the model is able to provide explanations that do not rely on the specifics of the underlying autonomy architecture. This is important for several reasons. First, model-independent explanations ensure that the interpretability of the system's behaviour is preserved regardless of the complexity or nature of the underlying model. This improves the robustness of the explanations, making them applicable across different models and contexts [31]. Second, it facilitates the adoption of explainable AI in regulatory and safety-critical applications, as the explanations are consistent and reliable irrespective of the underlying model changes [33]. Third, model-independent explanations support transparency and trust, as they provide a consistent framework for understanding the system's decisions, which is important for operator training and acceptance [7].

8.4.1   Integration with existing maritime robotic systems

Integrating the explainable autonomy framework with existing maritime robotic systems requires careful consideration of both hardware and software components. The framework is designed to work seamlessly with the AUV's onboard systems, including its sensors, actuators, navigation, control and communication modules.

One approach to integration is through middleware that acts as an intermediary layer between the AUV's control system and the explainability framework [34,35].

This middleware collects data from the AUV's sensors and control algorithms, processes it using the surrogate models and generates explanations in real time. This allows the framework to operate independently of the specific autonomy architecture of the AUV, making it adaptable to different systems and configurations.

The conceptual integration of the explainable autonomy module with an existing system is represented in Figure 8.4. The framework leverages the presence of a middleware layer able to gather relevant information coming from the various components composing the system (e.g., sensors, actuators and controllers). Raw sensor data from the vehicle is preprocessed to remove noise and irrelevant information. This step is important for ensuring the accuracy and reliability of the surrogate models. This step is also possible to identify information that needs to be added to properly explain the system.

The middleware makes this information available to the surrogate model that can use it to update its approximation of the underlying autonomous behaviours. This is done in two ways: during training, the data is used to train the model and approximate the decision-making process of the AUV's control system [36]. During deployment, the data is used by the model to update the approximation of the underlying autonomy engine (e.g., update the model state to match the received data) and make real-time inferences about the system's actions. This involves mapping the current inputs to predicted outputs based on the surrogate models. The

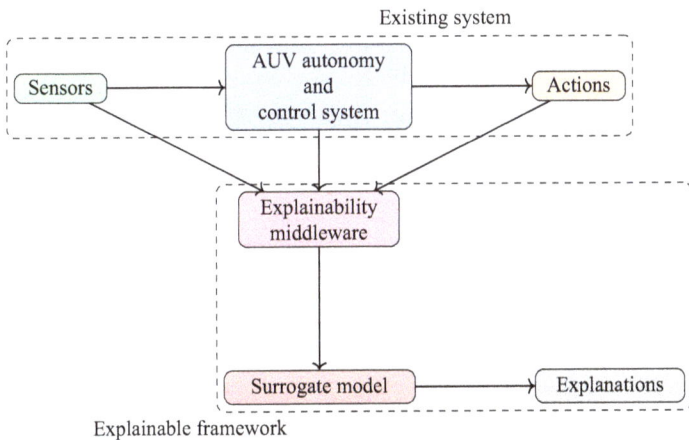

Explainable framework

Figure 8.4   *Integration of the explainable autonomy framework with an existing maritime robotic system. The framework leverages the presence of a middleware layer able to gather relevant information coming from the various components composing the system (e.g., sensors, actuators and controllers). The middleware makes this information available to the surrogate model that can use it to update its approximation of the underlying autonomous behaviour. Explanations are generated based on the surrogate model.*

inferred outputs are then translated into human-readable explanations using natural language generation techniques. These explanations can be tailored to the needs of different stakeholders, ultimately providing information on the rationale behind the AUV's actions [14,31,37]. Finally, the framework includes a feedback loop that allows operators to provide input and corrections to the surrogate models. This continuous feedback helps improve the accuracy and relevance of the explanations over time [24]. Note that explanations are generated based only on the surrogate model, thereby decoupling the complexity of the autonomy from the explanations provided to operators and stakeholders.

8.4.2   *Simplifying and approximating autonomy systems*

When the goal is to approximate a complex autonomous system with a more transparent model, such as a surrogate model, it becomes important to carefully decide how this simplification should be done. Using surrogate models to simplify autonomous systems involves reducing the complexity of decision-making processes while preserving their core behaviours, thereby improving transparency without compromising essential functionality. Various methodologies can be employed to achieve this, including the following:

8.4.2.1   Policy compression

Policy compression techniques aim to reduce the size and complexity of the policy representation. This can be done using methods such as pruning, where less critical branches of a decision tree are removed, or by merging similar states and actions to simplify the overall policy structure [38].

8.4.2.2   Hierarchical decomposition

Hierarchical decomposition involves breaking down complex policies into simpler, more manageable sub-policies. This method exploits the hierarchical nature of many tasks, allowing for more straightforward and interpretable policy representations. It facilitates easier debugging and understanding of the policy's behaviour [39].

8.4.2.3   State abstraction

State abstraction reduces the complexity of the policy by grouping similar states together. This method involves creating higher-level representations of states that capture the essential features relevant to the decision-making process, reducing the number of unique states the policy must handle [40].

8.4.2.4   Action clustering

Action clustering simplifies policies by grouping similar actions together. By reducing the number of distinct actions the policy needs to consider, this method helps streamline decision-making and ultimately improves interpretability [41].

Figure 8.5 shows an example of an autonomous navigation scenario using a simplified decision tree policy.

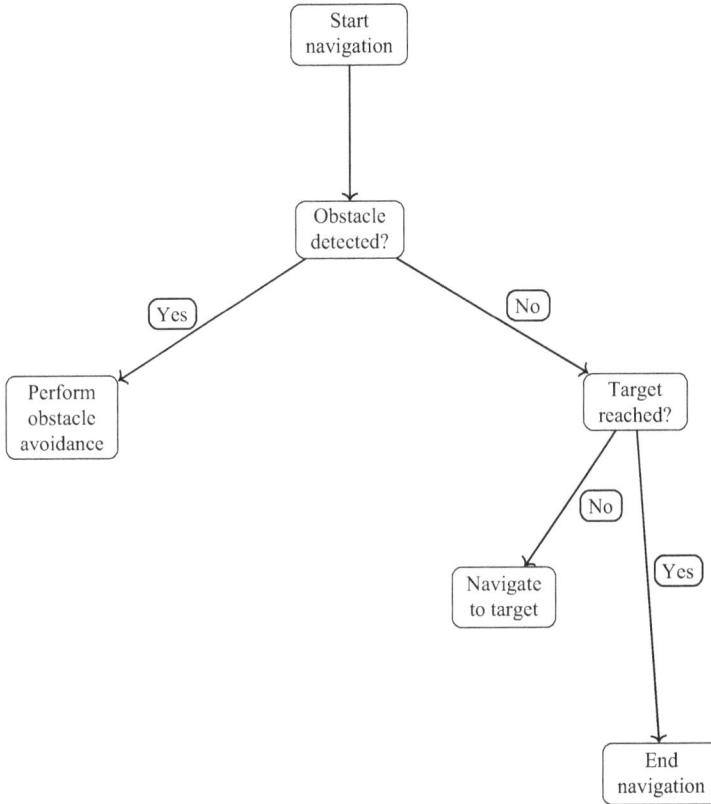

Figure 8.5   Autonomous navigation scenario represented as a decision tree

8.4.3   Evaluation of explanation effectiveness

Evaluation of explanations is a complex topic [42], with metrics including Interpretability, Trust and Reliability, Actionability and Consistency, Informativeness and clarity.

Interpretability refers to how easily a human can understand the explanation, and a detailed evaluation is out of the scope of this work. This can be evaluated through user studies where participants are asked to rate the clarity and comprehensibility of the explanations [33].

Trust and reliability are related but distinct concepts. Reliability refers to the system's consistency in performing as expected, while trust involves an operator's willingness to rely on the system based on its perceived reliability, predictability and transparency [43]. This can be assessed by observing user interactions with the system and measuring their willingness to rely on its decisions [23] or subjective measures, such as structured questionnaires, e.g., the Multi-Dimensional Measure of Trust (MDMT) [44,45]. These assessments gauge user perceptions of trustworthiness, which can be complemented by observational studies measuring operator reliance on system decisions. Actionability evaluates whether the explanations provide users with actionable insights. This involves determining if users can use the explanations to make informed decisions or take appropriate actions [46]. Consistency refers to the ability of the explanations to remain stable across similar scenarios. This can be tested by providing explanations for a set of similar inputs and checking for consistency in the reasoning [7].

8.5   A surrogate model to explain complex AUV behaviours

The increasing complexity of AUVs requires advanced methods for understanding and explaining their behaviours. This section introduces a surrogate model framework designed to explain complex AUV actions. Based on the work presented in [36], this section goes into the use case and explanation types, methodology for model development and evaluation and the results and discussions from both simulated and real-world trials. The following subsections provide an overview of the surrogate models' application, the approaches used to generate explanations, and the effectiveness of these models in real-time scenarios, demonstrating the robustness and applicability of surrogate models in providing explainability of AUV behaviours.

8.5.1   Use case and explanation types

The focus of our use case is on hybrid autonomy, which integrates a ROS-based deterministic agent with a reactive agent. These agents prioritise behaviours through multi-objective optimisation tailored for maritime applications, specifically targeting USVs and AUVs (see Figure 8.6). Our industry partner, SeeByte Ltd, has developed an autonomous agent capable of driving these vehicles for a variety of tasks, including inspection and monitoring. To gain insights into the decision-making process of the autonomous agent, we conducted an in-depth interview with a domain expert. This interview helped us abstract the agent's behaviour decision-making into an empirical decision tree, represented in Figure 8.7. The decision tree represents the various behaviours the agent may exhibit during a mission and the conditions under which these behaviours are activated. This step directly influences the data collection process before model interpretation, as well as the data streams monitored and processed by the explainability middleware, as described in Section 8.4.

The scenario designed to validate this decision tree involved two vehicles operating in a restricted coastal area of the River Charles in Boston. The mission required collaboration between two USVs, where one vehicle (USV Heron) followed a predefined plan to achieve various objectives, while the other vehicle (USV Philos) actively inspected the area for obstacles and provided real-time updates.

Each mission included several objectives, such as launching, recovering, surveying and holding position at specific target areas. Obstacles encountered during the mission were classified as either static (persisting throughout the mission) or dynamic (appearing intermittently). Both USVs were programmed with six behaviours: wait, transit, survey, hold position, replanned transit and avoid obstacle. These behaviours

*Figure 8.6   Unmanned Surface Vehicles (USVs) Heron (left) and Philos (right)
            used during the trials on the Charles River in Boston*

were managed by running two autonomy models simultaneously in a primary-secondary configuration [36].

The types of explanations generated by our framework were determined in consultation with the expert and are captured by the empirical decision tree in Figure 8.7.

The explanation types include:

- **Behaviour causality**: This explanation type clarifies why the robot selected its current behaviour or action. It helps operators understand how the robot perceives its environment, updates its world model and acts according to its goals. For example, if the robot switches to an obstacle avoidance behaviour, the explanation would specify the detected obstacle and the rationale behind this behaviour change. This addresses the question: *Why did you do that?*

- **Replanning clarification**: This explanation type deals with scenarios where the robot must deviate from its initial plan due to unforeseen circumstances, such as encountering an obstacle or facing a potential system failure. The explanation provides details on why the replanning was necessary, ensuring that operators are aware of safety-critical decisions. For instance, if the robot stops at a new location to avoid a collision, the explanation would highlight this reasoning. This answers the question: *Why do I need to replan at this point?*

- **Counterfactual explanation**: This type allows operators to explore alternative scenarios by asking how the robot would behave if certain conditions were different. It helps users understand the flexibility and robustness of the robot's decision-making process. For example, an operator might ask what the robot

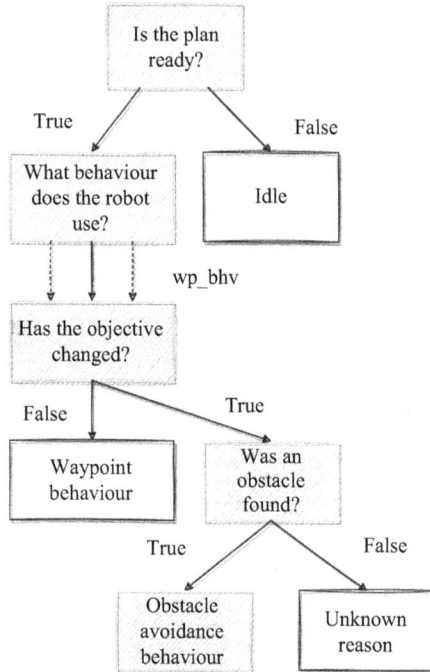

*Figure 8.7   Empirical decision tree for behaviour activation derived from a
domain expert*

would do if an obstacle appeared in a different location, and the explanation would describe the expected behaviour under those new conditions. This
addresses the question: *What if?*

These explanations are designed to be delivered in a natural language format that
is easily understandable by operators, enhancing their ability to monitor and intervene
in the robot's operations effectively.

8.5.2   *Methodology for developing and evaluating surrogate models*

To develop a robust framework for generating explanations of complex AUV
behaviours, we follow the paradigm described in Section 8.4 and have designed a
pipeline that integrates with the existing autonomous systems. This section outlines
the methodological steps taken to create and evaluate our surrogate models, ensuring
that they provide accurate and meaningful explanations.

8.5.2.1   Pipeline architecture

The pipeline architecture, illustrated in Figure 8.8, serves as a wrapper application
that interfaces with the AUVs' existing autonomy modules. It includes components
such as ROSListening for data collection, surrogate model training, feature contribution estimation and explanation generation. This architecture ensures that the

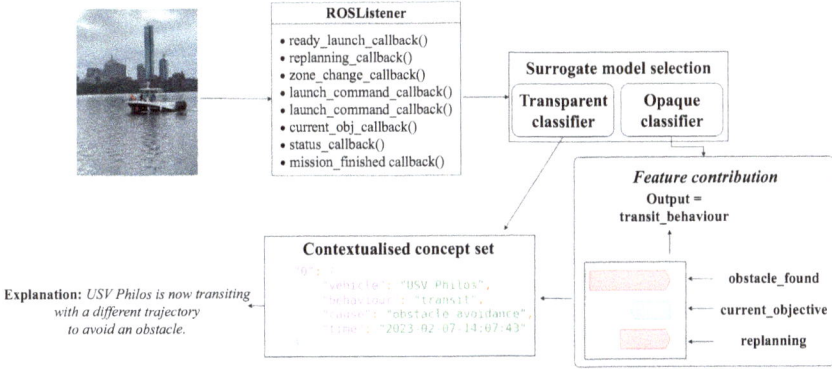

Figure 8.8   *Pipeline architecture where a surrogate model approximates agent policy, and feature contribution is estimated to detect behaviour causality*

explanation framework operates independently without disrupting the AUVs' core functionalities.

8.5.2.2   Data collection

Data collection involved monitoring ten simulated missions, each lasting approximately 22 min, resulting in a dataset of 5056 instances with five categorical values and one target value. This was obtained capturing data from eight ROS topics corresponding to five vehicle states: $S = \{$ *ready plan, current objective, progress type, same objective* and *obstacle found* $\}$, together with the associated behaviour.

8.5.2.3   Surrogate model training and selection

We trained several classifiers to serve as surrogate models [7], including K-Nearest Neighbors (KNN), Categorical Naive Bayes (CategoricalNB), Decision Tree, Support Vector Machine (SVM) and Multilayer Perceptron Neural Networks (MLP). Each model received five categorical features as the input to predict the vehicle's current behaviour from the set of six behaviours.

Nested cross-validation was used to optimise hyperparameters and evaluate model performance. The Decision Tree emerged as the most suitable surrogate model due to its high accuracy and reduced training and inference time when compared with more complex models, as shown in Table 8.1 and further discussed in Section 8.5.3.

Figure 8.9 provides a confusion matrix of the predictions of the decision tree per behaviour. For *transit, hold position* and *survey behaviours*, there are some false classifications due to some inconsistency between the progress type feature and the corresponding behaviour, which indicates that an internal autonomy state could be missing. As for false classifications between *hold position, survey* and *obs avoid* behaviours, we noticed that even though replanning is triggered and an obstacle is found, the vehicle finds a way to perform its objective; however, the explanation

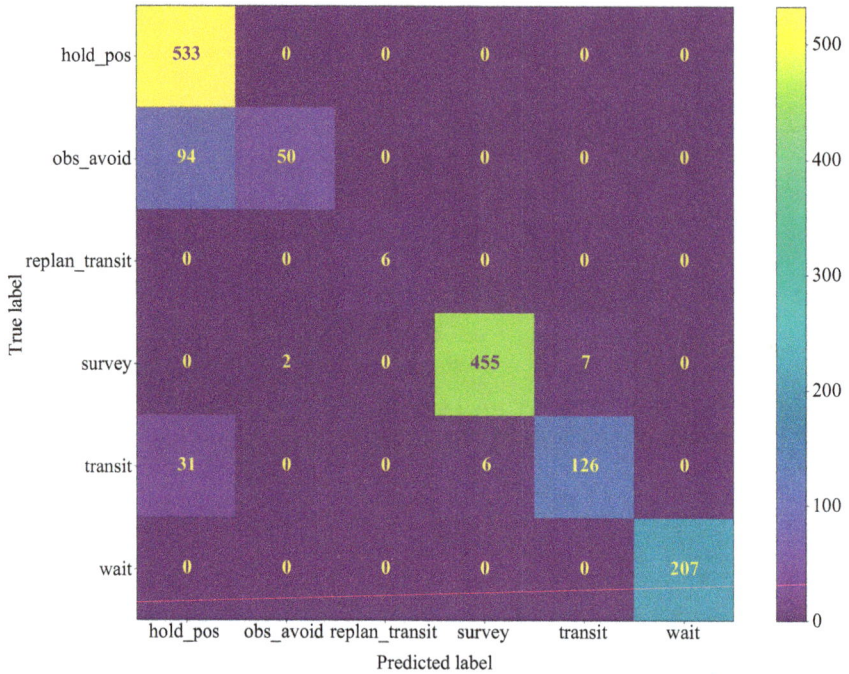

Figure 8.9   Confusion matrix indicating classification performance per behaviour with a decision tree during the simulated trial

framework misses this fact probably because an internal autonomy state is missing once again.

8.5.2.4   Feature contribution estimation

Once a trained surrogate model is in place to predict the corresponding behaviour of a vehicle state, the feature contribution for the classification of the behaviour is used as a basis for the causal reasoning explanation. For transparent models like the decision tree, feature contributions can be derived directly by analysing the model. For more complex models, such as Neural Networks, Shapley values [32] can instead be used to estimate feature contributions, providing a robust and descriptive approach. Shapley values quantify the contribution of each feature to the difference between the expected and actual prediction; each model initially has a prior belief about what the expected value will be, and a Shapley value describes how a specific feature creates the difference between the expected and actual values $(E(x) - f(x))$.

8.5.2.5   Explanation generation

The final step involved generating natural language explanations based on the surrogate model's predictions and feature contributions. Based on the importance of each feature towards a prediction, behaviour causality is inferred and this knowledge is represented with contextualised concept sets. Contextualisation is incorporated with the use of key-value pairs, as opposed to simple triplets to indicate the role of each value. The end result is a knowledge base with (vessel, behaviour, causality and time) sets, which describe the sequence of behaviours exhibited by the robot in JSON format.

These now structured explanations are finally converted into user-friendly formats using SimpleNLG [47], a rule-based natural language generator. An example of a contextualised concept set can be found in Figure 8.8, where the current behaviour (Transit) and its trigger (Obstacle) can be distinguished. This entry indicates that the current transit behaviour has a modified trajectory that goes around the obstacle to avoid collision. With regard to natural language generation, for each new entry in the Knowledge Base, the key-value pairs are passed to a Surface Realiser, which produces an explanation that has been syntactically checked with SimpleNLG.

8.5.3   Results and discussion

The surrogate models developed in this study were evaluated on their ability to accurately predict the behaviours of AUVs and generate clear, understandable explanations for these behaviours. The evaluation was conducted through a series of simulations followed by real-world trials.

8.5.3.1   Simulation results

The surrogate models have accuracies for behaviour prediction of around 90%, as shown in Table 8.1. This could further be improved by training on more data both in simulation and with real vehicles. To illustrate the model's effectiveness, we present four continuous scenarios from a single mission, each demonstrating the surrogate model's predictions and the corresponding explanations. These scenarios included typical mission activities such as transiting to a launch point, avoiding obstacles and performing surveys. For most scenarios, the surrogate model accurately predicted the AUV's behaviour and provided clear explanations based on the features identified as most influential by the model. However, some false predictions occurred, primarily due to inconsistencies in the progress type feature, highlighting areas for further refinement.

8.5.3.2   Real-world trial results

To verify how the performance of the explanation engine is affected when going from simulated data to real trials with real vehicles a real-world trial took place with two USVs, Heron and Philos, collaborating to complete a survey mission in a dynamic environment. As a result, the surrogate model was tested on a separate trial dataset comprising 1331 instances with five features and corresponding behaviours. The decision tree model achieved an impressive accuracy of 99% during these trials, as shown in Figure 8.11. One finding from the real-world trials was the model's ability to handle dynamic obstacles effectively, providing accurate behaviour predictions and timely explanations. The few false classifications observed were primarily between the survey and transit behaviours, likely due to a missing vehicle state that could better indicate transitions between these behaviours. Finally, the hold position behaviour is missing from Figure 8.11, because this objective was not used during the trial for practical reasons. The high accuracy in real-world conditions validated the robustness of the surrogate model and its capability to generalise from simulations to real-world applications.

8.5.3.3 Explanation quality

While various methods exist for evaluating the effectiveness of explanations, including interpretability, trust and actionability, this study primarily relied on an ad hoc assessment by domain experts, as described in [36]. Future work should include a structured evaluation using both subjective (e.g., user feedback) and objective (e.g., task performance) measures. The assessment focused on clarity, relevance and completeness, particularly in terms of how well the explanations supported operators' understanding of the AUVs' decision-making processes. The results suggested that the decision tree model, with its inherently high interpretability, contributed to improved understandability.

To further assess explanation quality, we compared decision tree-generated explanations with those produced by SHAP, as illustrated in Figure 8.10. This figure presents four continuous scenarios from a single mission, along with the corresponding explanations.

*   **Scenario 1:** The vessel moves to the launch point to retrieve the relative positions of the objective areas and begins working on each objective. The surrogate model correctly predicts the current behaviour using the *progress_type*, *current_objective* and *same_objective* features. Shapley values confirm this, identifying the *current_objective* as the primary contributor.

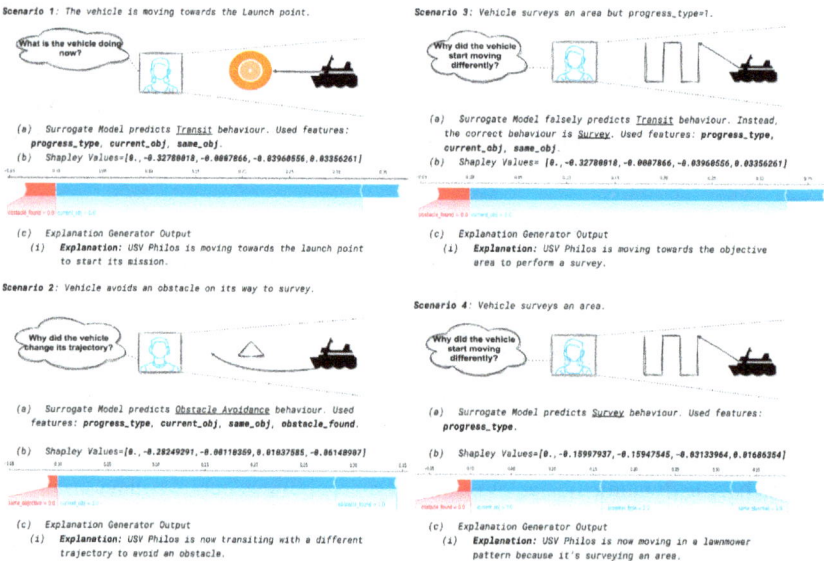

Figure 8.10 Four continuous events from a single mission along with their behaviour predictions and the corresponding explanations

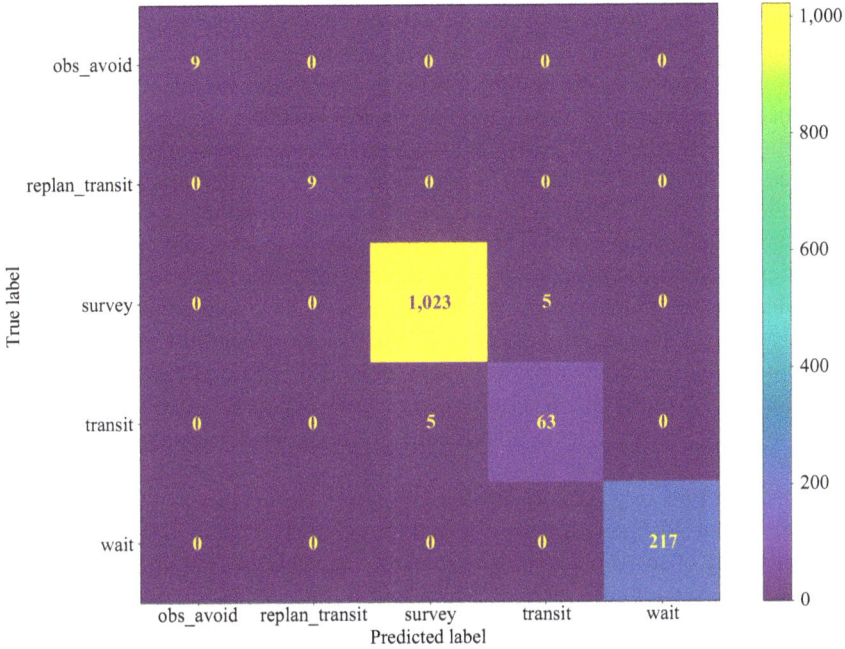

Figure 8.11   Confusion matrix indicating classification performance per behaviour with a decision tree during the real trial

- **Scenario 2:** While moving from the launch point to the survey area, the vehicle encounters an obstacle and adjusts its trajectory to avoid it. The behaviour prediction is again accurate, with the model utilising *progress_type*, *current_obj*, *same_objective* and *obstacle_found* features.
- **Scenario 3:** A false explanation is generated due to the *progress_type* value, despite the survey having already started. Here, the surrogate model relies on *progress_type*, *current_objective* and *same_objective*, whereas SHAP attributes the prediction solely to the *current_objective*.
- **Scenario 4:** During the survey, the surrogate model correctly detects the new behaviour based on its distinct *progress_type*. SHAP, meanwhile, attributes causality to both the *current_objective* and the *same_objective*, which appear to be reasonable contributing factors.

This evaluation method proved effective in providing detailed insights into the factors influencing the AUVs' behaviours, enhancing the model's transparency and reliability.

8.6   Conclusions: discussions, challenges and lessons learned

The results from both simulations and real-world trials demonstrate the effectiveness of using surrogate models to explain complex AUV behaviours. The high accuracy of the decision tree model and the quality of the generated explanations highlight the potential of this approach for improving transparency and trust in autonomous maritime systems. However, the findings also underscore the importance of continuous refinement, particularly in addressing inconsistencies in feature representation and enhancing the robustness of explanations across different mission scenarios.

One of the primary challenges was ensuring the quality and consistency of data collected from simulations and real-world trials. Variations in sensor readings, environmental conditions, and vehicle states required robust data preprocessing and validation techniques to maintain the accuracy and reliability of the surrogate models.

Selecting and training surrogate models that balance interpretability and performance was another significant challenge. While transparent models like decision trees provide clear explanations, achieving high accuracy often requires the use of more complex models, which in turn require more sophisticated explanation techniques such as Shapley values. Additionally, deciding the level of approximation is often application-dependent, left to the interpretation of the model developers and in turn, this might not fully capture the expectations of the operators.

Generating real-time explanations requires efficient computational methods, and the availability of communication streams with the vehicles existing autonomy systems. Communication interruptions might have a direct impact on the ability of the system to relay explanations to the users.

Additionally, while the framework has been designed to be autonomy-agnostic, adapting it to various platforms and mission scenarios requires a good understanding of different autonomy architectures, the implementation of specific interfaces, sensor configurations and operational requirements, all of which add complexity to the framework's development and deployment.

Finally, engaging with domain experts proved invaluable for understanding the nuances of AUV behaviours and developing meaningful explanations. Their insights helped shape the empirical decision tree and validate the surrogate models, ensuring that the explanations were both accurate and operationally relevant.

Future work will focus on extending the framework to incorporate more complex behaviours and scenarios, as well as integrating feedback mechanisms to continuously improve explanation quality based on operator input. A structured evaluation of the explanation framework is an important element of future work. This should involve both subjective measures (e.g., user comprehension and levels of trust) and objective performance metrics (e.g., mission success rate and operator intervention frequency) to systematically assess the impact of explainability. Additionally, incorporating advanced natural language generation techniques, including large language models, can improve the clarity and accessibility of explanations. This would enable more nuanced and context-aware communication with operators, improving their ability to understand and respond to AUV behaviours.

Acknowledgements

This work was supported in part by the Engineering and Physical Sciences Research Council (EPSRC), the Massachusetts Institute of Technology (MIT), and SeeByte

Ltd. We would like to thank our collaborators at MIT, Mike Benjamin, Conlan Cesar and Mike DeFilippo, and SeeByte for their valuable insights and technical support throughout this work.

References

[1] Willners J.S., Xu S., Luczynski T., *et al.*: 'Shared autonomy for robotic inspection' *UKRAS22 Conference "Robotics for Unconstrained Environments" Proceedings*. 2022, pp 48–49. doi:10.31256/Nc9Mp9N

[2] Ferri G., Munafò A. and LePage K.D.: 'An autonomous underwater vehicle data-driven control strategy for target tracking'. *IEEE Journal of Oceanic Engineering*. 2018;**43**(2):323–343.

[3] Cesar C., DeFilippo M., Benjamin M.R., *et al.*: 'Coordinating multiple autonomies to improve mission performance' *OCEANS 2021: San Diego–Porto*. Piscataway: IEEE; 2021.

[4] Benjamin M.R., Schmidt H., Newman P.M. and Leonard J.J.: 'Nested autonomy for unmanned marine vehicles with MOOS-IvP'. *Journal of Field Robotics*. 2010;**27**(6):834–875.

[5] Hamilton A., Holdcroft S., Fenucci D., *et al.*: 'Adaptable underwater networks: The relation between autonomy and communications'. *Remote Sensing*. 2020;**12**(20):3290.

[6] Garcia F.J.C., Robb D.A., Liu X., *et al.*: 'Explainable autonomy: A study of explanation styles for building clear mental models'. *Proceedings of the 11th International Conference on Natural Language Generation*. Pennsylvania: Association for Computational Linguistics; 2018. pp. 99–108.

[7] Molnar C.: *Interpretable Machine Learning*. Leanpub, 3rd edition; 2019. Available at: https://leanpub.com/interpretable-machine-learning and https://christophm.github.io/interpretable-ml-book/.

[8] Gavriilidis K., Munafo A., Hastie H., Cesar C., DeFilippo M. and Benjamin M.R.: *Towards Explaining Autonomy with Verbalised Decision Tree States* arXiv:2209.13985; 2022.

[9] Arrieta A.B., Díaz-Rodríguez N., Del Ser J., *et al.* 'Explainable artificial intelligence (XAI): Concepts, taxonomies, opportunities and challenges toward responsible AI'. *Information Fusion*. 2020;**58**:82–115.

[10] Lee J.D. and See K.A.: 'Trust in automation: Designing for appropriate reliance'. *Human Factors*. 2004;**46**(1):50–80.

[11] Winfield A.F., Booth S., Dennis L.A., *et al.*: 'IEEE p7001: A proposed standard on transparency'. *Frontiers in Robotics and AI*. 2021;**8**:665729.

[12] Omeiza D., Webb H., Jirotka M. and Kunze L.: 'Explanations in autonomous driving: A survey'. *IEEE Transactions on Intelligent Transportation Systems*. 2021;**23**(8):10142–10162.

[13] Jackson S., Golsteijn C., Johnson R. and Munafo A. 'Scheduling and tasking of autonomous systems for collaborative missions'. *OCEANS 2022* Hampton Roads, VA; 2022. pp 1–5. doi:10.1109/OCEANS47191.2022.9976994

[14]   Garcia F.J.C., Robb D.A., Liu X., Laskov A., Patron P. and Hastie H.: 'Explainable autonomy: A study of explanation styles for building clear mental models' *Proceedings of the 11th International Conference on Natural Language Generation*. Association for Computing Machinery, New York, NY; 2018. pp 99–108.

[15]   Lindsay A., Ramírez-Duque A.A., Craenen B., *et al.*: 'Supporting explainable planning and human-aware mission specification for underwater robots'. In: Huda M.N., Wang, M., and Kalganova, T. (eds.), *Towards Autonomous Robotic Systems*. TAROS 2024. Cham: Springer; 2025. https://doi.org/10.1 007/978-3-031-72059-8_3.

[16]   Gavriilidis K., Konstas I., Hastie H. and Pang W.: 'Enhancing situation awareness through model-based explanation generation'. *Proceedings of the 2nd Workshop on Practical LLM-assisted Data-to-Text Generation*. Association for Computational Linguistics, Tokyo, Japan; 2024. pp 7–16.

[17]   Ferri G., Munafò A., Tesei A., *et al.* 'Cooperative robotic networks for underwater surveillance: an overview'. *IET Radar, Sonar & Navigation*. 2017;**11**(12):1740–1761.

[18]   Caiti A., Casalino G., Munafo A. and Turetta A. 'Linking acoustic communications and network performance: Integration and experimentation of an underwater acoustic network'. *IEEE Journal of Oceanic Engineering*. 2013;**38**(4):758–771.

[19]   Rumson A.G.: 'The application of fully unmanned robotic systems for inspection of subsea pipelines'. *Ocean Engineering* 2021;**235**:109214.

[20]   Bharti V., Koskinopoulou M., Munafo A. and Petillot Y.: 'Enhancing image quality assessment using CNN-based edge detection'. *OCEANS 2024 – Singapore*. Singapore: School of Physical and Engineering Sciences I Dept. Information Engineering; 2024.

[21]   Clinciu M.A. and Hastie H.F.: 'A survey of explainable AI terminology'. *1st Workshop on Interactive Natural Language Technology for Explainable Artificial Intelligence 2019*; 2019. London: Association for Computational Linguistics. pp. 8–13.

[22]   Robb D.A., Garcia F.J.C., Laskov A., Liu X., Patron P. and Hastie H.: 'Keep me in the loop: Increasing operator situation awareness through a conversational multimodal interface'. *Proceedings of the 20th ACM International Conference on Multimodal Interaction*. Association for Computing Machinery, New York, NY; 2018. pp. 384–392.

[23]   Gunning D., Stefik M., Choi J., Miller T., Stumpf S., and Yang G.Z.: 'XAI—explainable artificial intelligence'. *AAAI Science Robotics*; 2019;**4**(37): eaay7120.

[24]   Samek W., Wiegand T. and Müller K.-R.: 'Towards explainable artificial intelligence'. *Lecture Notes in Computer Science* 2019;**11700**:1–16.

[25]   Ferri G., Jakuba M.V. and Yoerger D.R.: A novel trigger-based method for hydrothermal vents prospecting using an autonomous underwater robot *Autonomous Robots*. 2010;**29**(1):67–83.

[26] Sobrín-Hidalgo D., Gonzalez-Santamarta M.A., Guerrero-Higueras A.M., Rodríguez-Lera F.J. and Matellán-Olivera V.: *Enhancing Robot Explanation Capabilities through Vision-language Models: A Preliminary Study by Interpreting Visual Inputs for Improved Human-Robot Interaction.* arXiv:2404.09705v1; 2023.

[27] Hastie T., Tibshirani R. and Friedman J.: *The Elements of Statistical Learning: Data Mining, Inference, and Prediction.* Berlin: Springer; 2009.

[28] Quinlan J.R.: 'Induction of decision trees' in *Machine Learning.* Berlin: Springer; 1986. pp. 81–106.

[29] Rudin C.: 'Stop explaining black box machine learning models for high stakes decisions and use interpretable models instead'. *Nature Machine Intelligence.* 2019;**1**(5):206–215.

[30] Hinton G., Vinyals O. and Dean J.: *Distilling the Knowledge in a Neural Network.* arXiv:1503.02531; 2015.

[31] Ribeiro M.T., Singh S. and Guestrin C.: '"Why should i trust you?" Explaining the predictions of any classifier'. *Proceedings of the 22nd ACM SIGKDD Internatinal Conference on Knowledge Discovery and Data Mining*; 2016. New York: ACM. pp 1135–1144.

[32] Lundberg S.M. and Lee S.-I.: 'A unified approach to interpreting model predictions'. *Advances in Neural Information Processing Systems.* 2017;**30**:4765–4774.

[33] Doshi-Velez F. and Kim B.: *Towards a Rigorous Science of Interpretable Machine Learning.* arXiv:1702.08608, 2017.

[34] Brugali D. and Scandurra P.: 'Component-based robotic engineering (Part I)' *IEEE Robotics & Automation Magazine.* 2009;**16**(4):84–96.

[35] Quigley M., Conley K., Gerkey B., *et al.* 'Ros: an open-source robot operating system' *Proceedings of the IEEE International Conference on Robotics and Automation (ICRA) Workshop on Open Source Software.* Kobe, Japan; 2009.

[36] Gavriilidis K., Munafo A., Pang W. and Hastie H.: 'A surrogate model framework for explainable autonomous behaviour'. *ICRA 2023 Workshop on Explainable Robotics.* London, United Kingdom; 2023.

[37] Gavriilidis K., Carreno Y., Munafo A., Pang W., Petrick R.P.A. and Hastie H. 'Plan verbalisation for robots acting in dynamic environments' *ICAPS 2021 Workshop on Knowledge Engineering for Planning and Scheduling.* Guangzhou, China; 2021.

[38] Van Hasselt H., Guez A. and Silver D.: 'Deep reinforcement learning with double q-learning'. *Proceedings of the Thirtieth AAAI Conf. on Artificial Intelligence.* PKP Publishing Services Network; 2016. pp 2094–100.

[39] Dietterich T.G.: 'Hierarchical reinforcement learning with the MAXQ value function decomposition'. *Journal of Artificial Intelligence Research.* 2000; **13**:227–303.

[40] Li L., Walsh T.J., and Littman M.L.: 'Towards a unified theory of state abstraction for MDPs'. *Proceedings of the Ninth International Symp. on Artificial Intelligence and Mathematics (ISAIM)*; 2006.

[41]   Pinto L., Davidson J., Sukthankar R. and Gupta A.: Asymmetric actor critic for image-based robot learning. *Proceedings of Robotics: Science and Systems XIII*; 2017.

[42]   Clinciu M. and Hastie H.: *Let's evaluate explanations!*; 2020.

[43]   Hancock P.A., Billings D.R., Schaefer K.E., Chen J.Y.C., De Visser E.J. and Parasuraman R. 'A meta-analysis of factors affecting trust in human-robot interaction'. *Human Factors*. 2011;**53**(5):517–27.

[44]   Jian J.-Y., Bisantz A. and Drury C.: 'Foundations for an empirically determined scale of trust in automated systems'. *International Journal of Cognitive Ergonomics*. 2000;**4**:53–71.

[45]   Ullman D. and Malle B.F. 'Measuring gains and losses in human-robot trust: Evidence for differentiable components of trust'. *2019 14th ACM/IEEE International Conference on Human-Robot Interaction (HRI)*. ACM/IEEE, Daegu, Korea; 2019; pp 618–19.

[46]   Wachter S., Mittelstadt B. and Russell C.: 'Counterfactual explanations without opening the black box: Automated decisions and the GDPR'. *Harvard Journal of Law & Technology*. 2017;**31**(2):841–87.

[47]   Gatt A. and Reiter E.: 'SimpleNLG: A realisation engine for practical applications', in Krahmer E and Theune M (ed.) *Proceedings of the 12th European Workshop on Natural Language Generation (ENLG 2009)*. Athens, Greece, March 2009. London: Association for Computational Linguistics; 2009, pp 90–93.

Chapter 9

Toward a structured framework for control performance and safety assessment for Maritime Autonomous Surface Ships

V. Garofano[1], Y. Pang[1] and R.R. Negenborn[1]

As the maritime industry moves toward fully autonomous operations, it is becoming increasingly important to assess the control performance and safety of Maritime Autonomous Surface Ships. This chapter presents a structured framework designed to facilitate testing and data collection using autonomous ship systems, thereby supporting the verification that autonomous operations align with International Maritime Organization standards. We discuss the integration of key hardware and software components for robust autonomous operation and evaluate these systems using analytical performance criteria in both simulated and real-world scenarios. The chapter concludes by proposing new key performance indicators necessary for the continued development of autonomous maritime systems. Through extensive datasets and collaborative research via Open Science-focused algorithms and designs, we aim to set the groundwork for future advancements in this field.

9.1 Introduction

The technological advancement in autonomous navigation technologies has initiated a significant shift across various industries, with the maritime sector exploring the advancement toward fully autonomous Maritime Autonomous Surface Ships (MASSs). This technology promises enhanced efficiency, safety, and sustainability in maritime operations, yet it simultaneously poses complex challenges in terms of control performance and safety assessment evaluation [2,26]. Traditionally, these aspects have been governed by standards set by the International Maritime Organization (IMO), which, while comprehensive, were not designed with autonomous technologies in mind [7]. Therefore, it starts to be important to rethink these standards to promote the integration of smart navigation systems within the regulatory frameworks that ensure maritime safety and can evaluate the performance of decision-making algorithms [1].

[1]Department Maritime and Transport Technology, Faculty Mechanical Engineering, Delft University of Technology, The Netherlands

MASSs operate under different circumstances than manned ships, requiring robust control algorithms and hardware systems that can perform reliably in changing and often unpredictable marine environments [4,24]. To address these needs, this chapter proposes a structured test framework that can perform tests following existing IMO standards with state-of-the-art autonomous navigation technologies [20,25]. This approach is aimed at redefining control performance and safety assessment methodologies to accommodate the autonomous capabilities of future vessels [6,8].

Different policies and guidelines are already in place to regulate the safety of ship navigation. The SOLAS (Safety of Life at Sea) Convention, for example, regulates specific equipment and procedures to ensure the safety of navigation, fire safety systems, life-saving appliances, and communication protocols. However, in the context of autonomous ships, these provisions necessitate reinterpretation [5,11]. Autonomous vessels must be equipped with advanced navigation systems that not only comply with these regulations but also enhance them through technology. For example, automated systems and life-saving protocols need redesigning to function effectively with less human intervention on board [10]. Under the COLREGs (International Regulations for Preventing Collisions at Sea), the ability of ships to avoid collisions autonomously and safely becomes paramount [9]. Traditional rules designed for human interpretation must now be encoded within the upcoming AI systems of autonomous ships. These vessels must be capable of interpreting complex maritime scenarios and making real-time decisions to maintain navigational safety [21,22]. This requires a sophisticated integration of sensor data and decision-making algorithms that can robustly interpret and act according to the COLREGs [3,21].

The MARPOL (International Convention for the Prevention of Pollution from Ships) convention focuses on minimizing environmental impact of maritime operations. Autonomous ships offer a promising path to enhanced compliance with these regulations. Through optimized route planning and feedback controls, these ships can reduce fuel consumption and emissions, significantly lowering their environmental footprint [19].

Although the STCW (Standards of Training, Certification, and Watchkeeping) convention primarily addresses crew qualifications, its adaptation for autonomous shipping is becoming important as well. Operators who remotely monitor and control autonomous fleets must receive specialized training that reflects the technical and operational scenarios of unmanned shipping. This ensures that even without a traditional crew, the vessels are operated safely and effectively under rigorous oversight standards [36].

Performance standards for electronic navigational aids, such as ECDIS, AIS, and radar, are outlined by the IMO to ensure that maritime traffic is managed safely and efficiently [25]. For autonomous ships, these systems must also integrate advanced technological solutions that enhance their functionality and reliability [31]. This involves the implementation of redundant systems and robust sensor fusion technologies that can deliver precise and reliable navigation data in real time [12].

This chapter will explain and analyze the Researchlab Autonomous Shipping (RAS) idea of a structured framework, describing the hardware and software components that constitute autonomous ship systems as a whole. From sensors and communication devices to complex algorithms responsible for decision-making and obstacle avoidance, the framework will cover the spectrum of technological requisites and assessment performance criteria for autonomous operations [35]. This discussion will highlight the technological support of autonomous ships and investigate the reliability, redundancy, and real-time processing needs that these systems demand [18,33].

The practical application of these technologies will be described as well, where various configurations and scenarios, including open water tests and towing tank experiments, are employed to refine and validate the control strategies of autonomous ships [16]. The chapter will also propose the development of new Key Performance Indicators (KPIs) necessary for assessing the efficacy and safety of autonomous navigation systems [14,19,29].

The main contribution of the RAS-structured framework is its innovative approach to integrating and evaluating autonomous navigation technologies within existing maritime safety and performance standards. This framework is designed to bridge the gap between traditional maritime regulations and the advanced capabilities of MASSs. It redefines control performance and safety assessment methodologies, ensuring that autonomous vessels can operate reliably and safely under various conditions [5,38]. The RAS framework systematically incorporates robust control algorithms, advanced navigation systems, and sophisticated sensor fusion technologies, ensuring that these systems not only comply with but also enhance current regulatory requirements [13,17,28]. Additionally, it emphasizes the importance of developing new KPIs and sharing extensive datasets to propel the advancement of autonomous maritime operations [24]. In this context, a structured framework refers to a comprehensive and methodical approach that investigates all critical aspects of autonomous ship systems, from hardware and software components to performance criteria and practical testing scenarios [23]. It ensures a cohesive integration of technologies, regulatory compliance, and performance evaluation, facilitating the reliable and safe development of autonomous maritime vessels [32,34].

9.2   The RAS framework

The RAS framework is designed to address these evolving needs by incorporating technologies and methodologies to enhance safety, performance, and compliance in autonomous maritime operations. By focusing on real-time data acquisition, advanced sensor capabilities, and robust communication systems, the RAS framework aims to provide a comprehensive test field that aligns with and potentially extends the test scenario in existing IMO standards.

Through the RAS framework, it is possible to design and validate collision-avoidance applications, enhance redundancy and reliability in navigation systems, and potentially increase environmental compliance, all while adapting safety equipment and procedures to fit the unmanned nature of autonomous ships. These advancements represent a step forward in the maritime research field, enabling possibilities for a safer, more efficient, and environmentally responsible future for autonomous maritime operations.

Central to the RAS framework is its focus on developing a reliable and robust communication layer, based on the Robotic Operating System (ROS) [30]. Traditional maritime operations, reliant on onboard human decision-making, face inherent delays in data processing and action implementation. On the contrary, the RAS framework emphasizes real-time, low-latency communication between autonomous ships and shore-based control centers. This strategy is designed to monitor decision-making algorithms' efficiency and provide a better insight into reaction times to navigational and operational changes.

To achieve this, the framework incorporates:

- **Real-time diagnostics**: Automated diagnostic tools continuously assess the health and functionality of onboard systems, with the possibility to flag any irregularities for immediate attention.
- **Redundancy management**: The framework can potentially operate multiple layers of redundancy for hardware and sensor systems, ensuring that a single point of failure does not compromise the vessel's operational capability or safety.
- **Dynamic data sharing**: Information from individual ships is shared into a centralized database accessible via a secure VPN, allowing for dynamic data sharing and enhanced collective decision-making algorithms.
- **Fleet-wide operational optimization**: Utilizing data from across the fleet, the system can identify patterns, predict potential issues, and optimize fleet operations through coordinated control strategies. This is especially important in complex scenarios such as navigating busy shipping lanes or conducting synchronized logistical operations.
- **Scalability and flexibility**: The networked approach is designed to be scalable, accommodating an increasing number of autonomous vessels without compromising performance or security. It also provides the flexibility needed to customize control strategies for different types of vessels and operational tasks.

These technologies rely on robust, high-speed communication infrastructures that enable the real-time transmission of vast amounts of data to and from the autonomous ships. By leveraging wireless data exchange and/or advanced satellite communications, the system ensures that data flows are not disrupted, which is important for maintaining operational integrity and safety. Figure 9.1 provides a schematic of the RAS framework.

To ensure reliable monitoring of the autonomous ship's dynamic parameters and navigation information, the vessel is equipped with specific sensors. These sensors are connected to a central CPU, which processes and forwards vessels' information to a shore monitoring and control center. The ships used in the RAS framework are equipped with the following sensors:

- **GNSS receiver:** This sensor receives feedback about the real-time positioning of the Autonomous Surface Vessel. To enhance accuracy, Differential GNSS technology is employed, which mitigates atmospheric errors with the help of a ground station. Another method to improve position feedback is the implementation of Real-Time Kinematic technology and mathematical-based filters such as the Kalman Filter, which help in refining the accuracy of the positioning data.

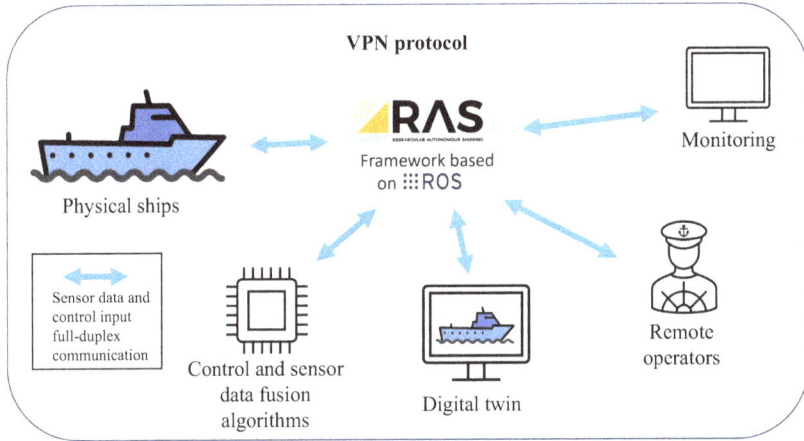

Figure 9.1   RAS framework overview

- **Stereo vision cameras:** Essential for autonomous ships to perceive their surroundings independently, these cameras are installed at various points around the ship to create a 360-degree view of the environment. The stereo vision technology not only helps in distance measurement but is also integrated with state-of-the-art AI computer vision technologies to autonomously detect and distinguish between different objects in the environment.
- **Inertial measurement unit (IMU):** Comprising a 6 Degrees of Freedom (DoF) accelerometer, gyro, and magnetometer, this sensor provides real-time data on various states of the ship. It helps in measuring forces applied to the ship's hull that can be used to extract hydrodynamic resistance values, angular velocity, and heading – crucial for understanding the internal state conditions without human onboard presence.
- **Encoders on propulsion shaft:** These measure the RPMs at which the ship's propellers are rotating, providing critical data for evaluating the thrust effort required by the control algorithm for precise navigation along a predefined path.
- **Energy measurement unit:** It evaluates the amount of mechanical and electrical energies required by the control algorithm under various navigational scenarios. This data is vital for developing control algorithms that optimize energy use efficiently.
- **Propeller azimuth angle feedback:** Whether using rudders or azimuth rotative propellers, this sensor provides feedback on the direction in which the propellers are pushing thrust, enhancing the capacity to assess control performance accurately.

All these sensors and their related data are then used in the assessment of control performance and safety of the autonomous navigation scenario. The combined data from these sensors are indeed important for determining whether the control algorithms are performing as expected.

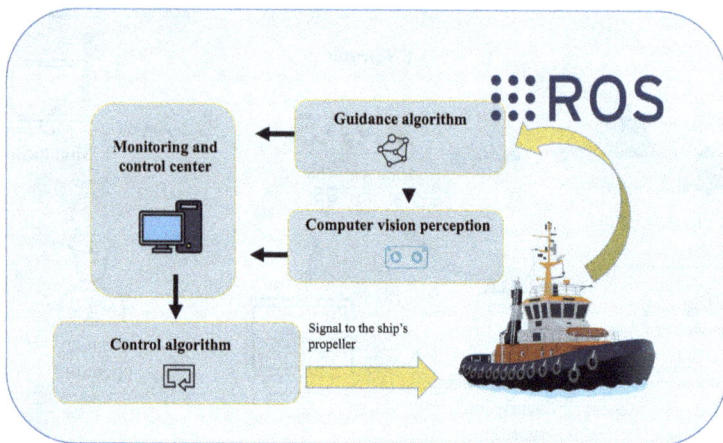

Figure 9.2   Overview of RAS framework control and feedback loop

The RAS framework provides control feedback closed-loop interaction between several hardware and software components, as shown in Figure 9.2 that can be used to check standard requirements in autonomous maritime operations, aligning with existing IMO standards and extending their scope to incorporate advanced technological capabilities [12].

Building on the general framework design described previously, we utilize this basis to gather data and conduct laboratory tests on autonomous ship systems. To ensure the reliability, safety, and effectiveness of control algorithms that manage the ship's automated systems, the testing involves a series of simulations and experiments in a controlled laboratory environment. This approach allows for methodical progress toward replicating more complex scenarios encountered in open water environments, thereby closely emulating real-world conditions, Figure 9.3.

The development of autonomous ship systems starts in simulation environments, where specialized tools try to recreate realistic maritime conditions. These digital testbeds must replicate the physical and operational dynamics accurately, ensuring that the simulations are as close to real-world scenarios as possible. This fidelity is achieved through state-of-the-art analytical models of aerodynamic and hydrodynamic interactions within the ship system, incorporating environmental disturbances such as currents, waves, and winds based on the latest research results.

Beyond virtual simulations, Hardware-in-the-Loop (HIL) testing incorporates physical components – such as sensors and actuators – into a simulated environment. This approach evaluates the hardware system in its entirety under controlled yet realistic conditions, bridging the gap between theoretical models and practical operations. Similarly, scale-model testing in facilities such as towing tanks or flume tanks (see Figure 9.4) provides critical insights into the physical dynamics of ships. These experimental setups yield valuable data on how well a vessel and its systems handle stability and maneuverability, thereby informing more robust design and validation processes.

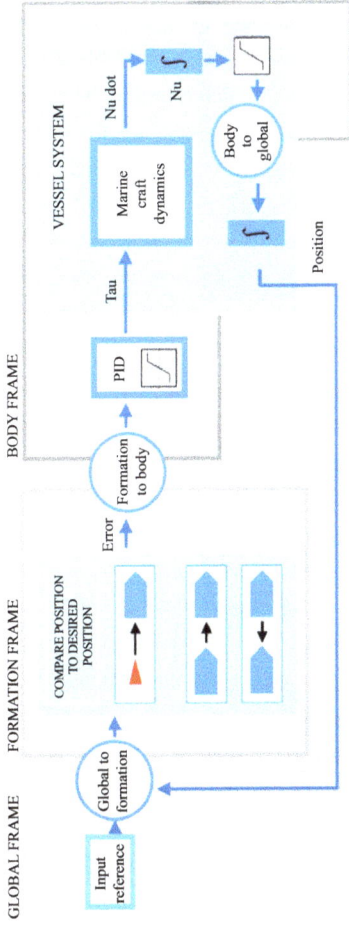

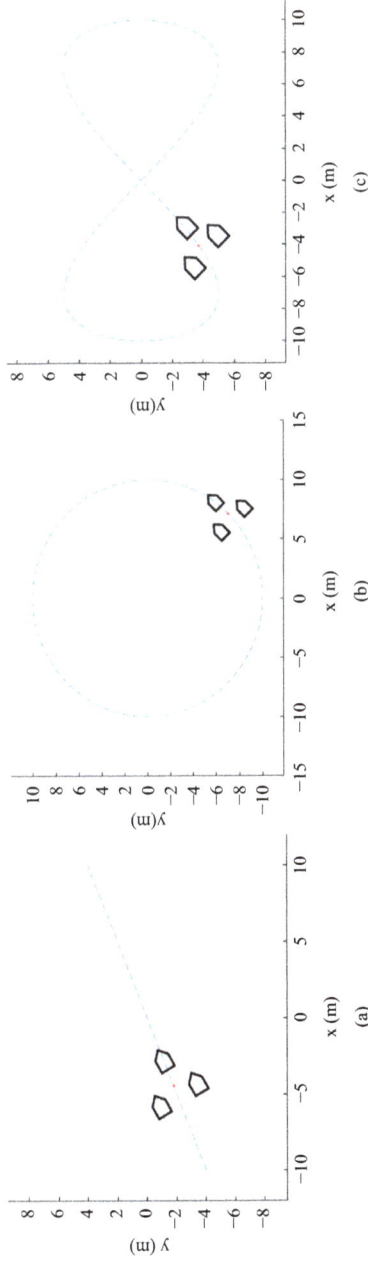

Figure 9.3 *Simulation environments of ResearchLab Autonomous Shipping. (a) Straight line path (b) Circular path and (c) Lemniscate path.*

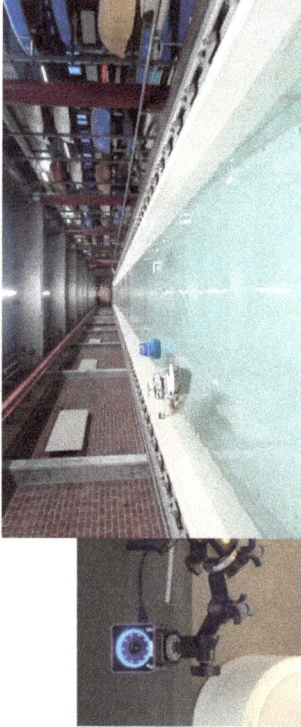

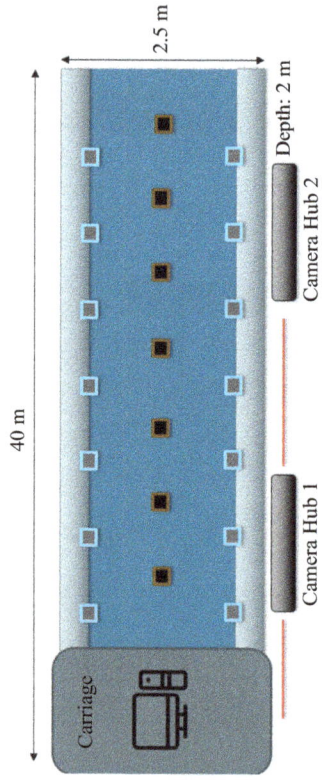

Figure 9.4 Experiments in water tank facilities at RAS

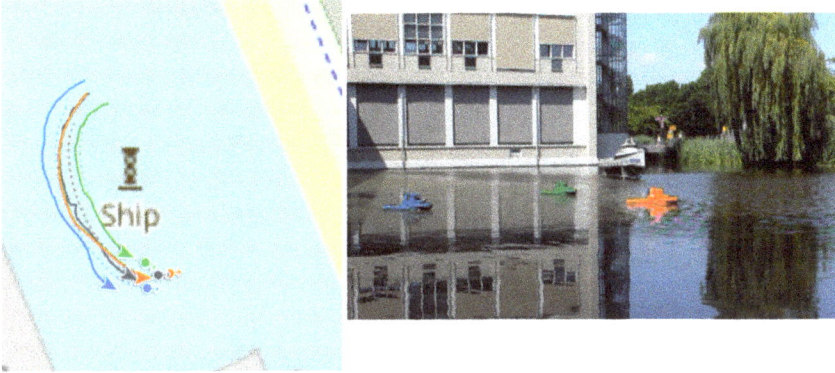

Figure 9.5   Experiments in Open Water Facilities at RAS

Throughout these tests, the real-time collection and analysis of data are essential. Monitoring the output from various sensors and systems during tests helps in fine-tuning the technology, enhancing the algorithms, and ensuring the systems' reliability across different scenarios. This iterative cycle of testing, data gathering, and analysis helps advance the systems toward real-world readiness (see Figure 9.5).

Moreover, ensuring that subsystems within an autonomous ship function in unison is another key aspect of experimental campaigns. Integration and interoperability testing in a laboratory setting can reveal potential communication and functionality issues, supporting robust operation across the vessel's navigation, communication, and operational systems.

Autonomous systems are then evaluated, Figures 9.6 and 9.7 provide examples to ensure compliance with international performance and safety standards, such as those outlined by the IMO.

Finally, the collaborative nature of research and development in this field drives innovation and refines autonomous systems. Partnerships among academia, industry leaders, and regulatory bodies are essential, as they consolidate expertise, resources, and perspectives. This collective effort not only fosters technological progress but also ensures alignment with global standards and practical maritime needs [15] This approach can support transitioning from theory to practice, turning innovative research outcomes into operational realities.

9.3   Assessment of the control and safety performance

Assessing the control and safety performance of the MASS is important for ensuring their reliable and safe autonomous operation. The methods for testing and assessing the performance of MASSs have been developed under the RAS framework [35], with relevant research projects contributing significantly to this field [29]. This section outlines the test scenario and the criteria that can be used for assessing the control

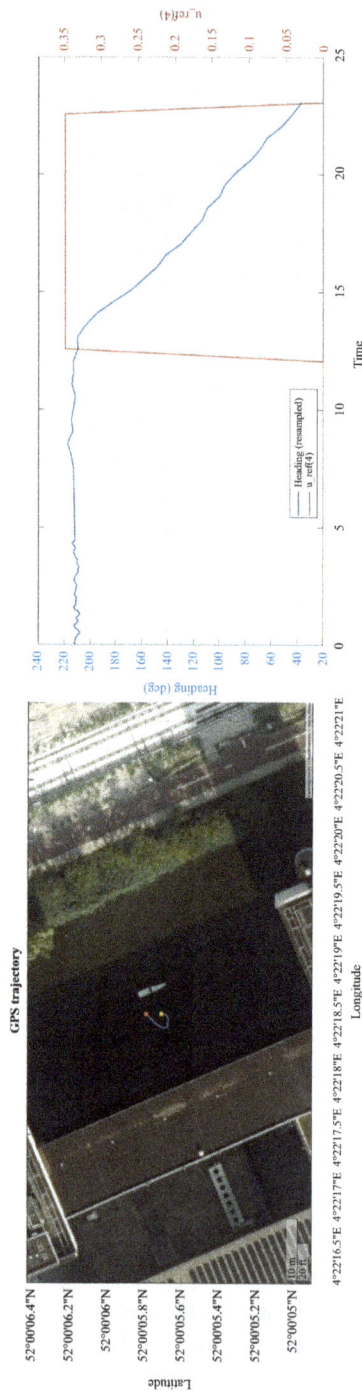

Figure 9.6   Example of maneuvering test and data collection for performance and operational safety assessment

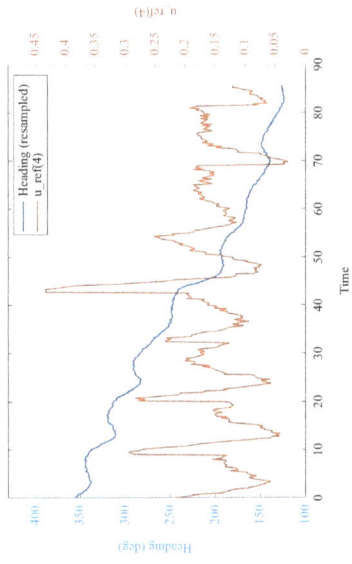

Figure 9.7   Autonomous path following dataquery

and safety performance of MASSs. Each scenario focuses on a different aspect of operation and performance, providing a comprehensive evaluation that covers safety, navigability, stability, and efficiency. The aim is to establish a set of criteria that can evaluate the vessel's ability to operate reliably and safely under various conditions.

By applying scenario-based testing to these main performance aspects, the idea is to have an objective evaluation of the MASS's performance. This methodical approach can evaluate whether an autonomous vessel meets required standards and performs effectively.

This section describes the primary testing scenarios, which include maneuvering, path following, collision avoidance, and multiple vessel coordination.

Maneuvering tests are designed to evaluate the MASS's ability to perform various movements such as turning, reversing, stopping, and changing speeds, focusing on the vessel's responsiveness and its capability in handling complex maneuvers safely and efficiently. Path-following tests examine the MASS precision in adhering to a predetermined course, testing its navigational systems under varying environmental disturbances to check accurate course maintenance and control. In collision avoidance tests, the MASS encounters both static and dynamic obstacles, testing its sensor systems and algorithms for effective detection and evasion, important for operational safety and reliability. Finally, multiple vessel coordination involves deploying several MASSs simultaneously to assess their collective operational capabilities, measuring their coordination capability, formation keeping, and cooperative behavior.

9.3.1   Maneuverability assessment aspects

9.3.1.1   Navigate in a straight line

Testing the autonomous surface ship for its ability to navigate straight is key for assessing navigation systems and stability. This is initially done in a controlled waterway to measure trajectory accuracy. This test, critical for ensuring efficient and safe operation, uses GNSS or camera systems to monitor deviations from a straight path at regular intervals Δt. Performance is measured by the average and maximum deviations from the intended path.

$$d(t) = \frac{\left|(x_{\text{end}} - x_{\text{start}})(y_{\text{start}} - y_{\text{mean}}(t)) - (x_{\text{start}} - x_{\text{mean}}(t))(y_{\text{end}} - y_{\text{start}})\right|}{\sqrt{(x_{\text{end}} - x_{\text{start}})^2 + (y_{\text{end}} - y_{\text{start}})^2}}$$

The MASS's performance is considered successful if the maximum and average deviations meet the defined acceptance thresholds, as described in Tables 9.1 and 9.2.

Table 9.1   Acceptance criteria for navigate in a straight line scenario

KPI	Acceptance criteria
Maximum deviation $\max\{d(t)\}$	Less than a_1 meters
Average deviation $\frac{\sum_{t=0}^{n} d(t)}{n}$	Less than a_2 meters

Table 9.2   Validation criteria for the mooring test

KPI	Acceptance criteria
Distance d	$d < a$ meters

9.3.1.2   The mooring test

The mooring test evaluates how precisely an autonomous surface vessel can use its stopping mechanisms to moor at a predetermined spot, crucial in crowded urban ports to avoid collisions and ensure safety. The test measures the MASS's ability to accurately stop at a specific berthing point, assessing navigational accuracy, control systems, and responsiveness. Effective docking reduces unnecessary maneuvers, saving time and energy, thus improving cost-efficiency.

A designated point, Ms(x_s, y_s), marks the target mooring location. The objective is for the ASV to halt as close to Ms as possible. The key performance indicator is the distance d between the actual stop point Mm(x_m, y_m) and Ms, calculated via GPS or cameras. The distance is given by:

$$d = \sqrt{(x_m - x_s)^2 + (y_m - y_s)^2}$$

To determine the success of the mooring test, an acceptance value a is predefined based on the operational requirements of the MASS. The test is considered successful if the calculated distance d is less than the acceptance value a.

9.3.1.3   Maximum speed

The maximum speed test assesses the performance and energy usage of an autonomous vessel at its speed limit. It requires a long waterway where the ship can accelerate to its maximum speed, testing its ability to reach and maintain this speed without further acceleration, signifying a steady state.

This test evaluates the vessel's top speed and its effects on maneuverability, including turning radius and stability. High-speed tests also identify potential control system issues that may emerge only at high speeds.

Speed is measured with accelerometers and the vessel's position via GNSS or cameras. The primary KPI is the maximum speed, V_m. The performance meets expectations if V_m is within an acceptable range of the designed maximum speed V_d, denoted as a.

The collected data includes time points, coordinates, and calculated speeds:
The scenario is considered successful if

$$V_d - \max\{V(t)\} < a$$

where $\max\{V(t)\}$ represents the highest speed recorded during the test, and a is the predefined acceptance value.

9.3.1.4 Stopping test

The stopping test evaluates the ability of a MASS to halt using a stop engine-full astern maneuver. This maneuver determines the MASS's maximum stopping distance, a key safety metric. The test applies to autonomous vessels that can reverse and requires a long waterway for execution.

The test procedure starts as the MASS cruises at a set speed and then shifts to full astern to decelerate. The key performance indicators are the head reach (HR) and the track reach (TR). HR is the distance the MASS travels forward from deceleration onset to a full stop. TR is the total distance traveled until it completely stops.

$$HR = |x(t_1) - x(t_0)|$$

$$TR = \sum_{i=1}^{ts} \sqrt{(x(t_i) - x(t_{i-1}))^2 + (y(t_i) - y(t_{i-1}))^2} < 10L$$

Both HR and TR can be measured using GNSS, cameras, and distance meters. The performance of the test is considered satisfactory if the TR is smaller than 10 times the vessel length (L), with decreasing HR values indicating better performance. The test should be repeated at different speeds to comprehensively assess the MASS's stopping ability across its entire operational speed range.

9.3.1.5 Turning test

The turning test assesses the autonomous vessel capability to perform a U-turn, crucial for route changes or dock departures in various maritime settings. This maneuver tests the MASS's minimum turning radius at different speeds, key for effective maneuvering in tight waterways. The procedure starts with the MASS on a straight path, then the rudder is set to its maximum design angle, usually up to 35 degrees, to initiate a full circle turn to both the starboard and the port.

Measurements begin at the maneuver command (t_0) and end once the MASS completes the turn. The KPIs are the advance (Ad) and the tactical diameter (TD). These are measured using GNSS to track the x–y coordinates of the ASV. The advance measures the distance along the initial course up to a 90-degree heading change (t_q), and the tactical diameter measures the perpendicular distance across the initial course until a 180-degree heading change is achieved (t_h).

$$Ad = |x(t_q) - x(t_0)|$$

$$TD = |y(t_h) - y(t_0)|$$

For the test conducted at the specified speed, the tactical diameter should be less than five times the ship length (L_{ship}), and the advance should be less than four and a half ship lengths. The test should be repeated at different speeds to assess the MASS's turning ability across its entire operational speed range. According to the IMO standard, if Ad $< 4.5L$ and TD $< 5L$, the requirements for turning ability are met. However, stricter requirements may be set based on the MASS's intended purpose.

9.3.1.6   Zigzag tests

The Zigzag tests involve the vessel performing maneuvers that mimic the pattern of saw teeth by alternating the rudder position between predetermined positive and negative angles once the vessel's heading matches these angles. This test is designed to provide insights into the vessel's dynamic response and maneuvering capabilities, assessing how quickly and effectively it can change its course without significant overshoot.

The 10°–10° zigzag test involves setting the rudder to 10 degrees to either side, assessing the vessel's responsiveness and agility in dynamic environments. The key performance indicators for this scenario include the first and second overshoot angles and the time to reach the initial 10-degree heading angle change on both sides. Measurements of the heading angle ($\Psi(t)$) commence at the start of the maneuver.

$$\alpha_1 = \max\{\Psi(t)\} - 10°, \quad \alpha_2 = \min\{\Psi(t)\} + 10°$$

The distance traveled during the first overshoot should not exceed 2.5 ship lengths (L), calculated using GNSS coordinates measured at regular intervals until the first overshoot angle reaches its maximum.

$$l_{10} = \sum_{i=t_1+\Delta t}^{t_{max}} \sqrt{(x(i) - x(i-1))^2 + (y(i) - y(i-1))^2} \leq 2.5L$$

The 20°–20° zigzag test increases the challenge by requiring the rudder to alternate at 20 degrees, testing the limits of the vessel's maneuvering system under more extreme conditions. This setup mirrors that of the 10°–10° zigzag, but the larger turning degrees place a greater demand on the vessel's maneuverability and the first overshoot angle becomes a more critical measure of the vessel's initial turning ability.

$$\alpha_1 = \max\{\Psi(t)\} - 20°$$

This scenario provides a rigorous test of the MASS's ability to execute tight maneuvers, ensuring its suitability for operation in environments requiring high maneuverability and precise handling.

9.3.1.7   Pull-out test

The pull-out test assesses the dynamic stability of the autonomous ship and its ability to return to a straight course after a turning maneuver. This test, critical for evaluating how quickly the vessel stabilizes post-turn, can be conducted immediately following the turning test. Initially, the rudder is set to its maximum allowable angle from the end of the turning test, setting the stage for the pull-out test.

During this test, the rudder is quickly adjusted back to a neutral position and held there until the vessel achieves a steady turning rate. Ideally, a perfect performance shows a zero steady turning rate, indicating dynamic stability. If there is dynamic instability, the steady turning rate will show a residual value, where a higher residual suggests poorer performance.

The rate of heading change $\dot{\Psi}(t)$ is monitored from the start of the test (t_0) until a steady rate is achieved (t_s), using a gyro compass or GNSS. Dynamic stability is confirmed if $\dot{\Psi}(t_s) = 0$; instability is indicated by $\dot{\Psi}(t_s) \neq 0$, with increasing $|\dot{\Psi}(t_s)|$ signifying reduced performance.

9.3.1.8   Path following

Path following tests evaluate the MASS's ability to adhere to a designated route composed of multiple way-points, assessing the navigation system's effectiveness. These tests involve trajectories that the MASS navigates, represented as longitude and latitude coordinates.

The MASS's precision in passing way-points is crucial and measured using GNSS or visual aids like cameras. The primary metric is the deviation from the optimal path length.

The accuracy per way-point, W_{pi}, is determined by tracking the MASS's path using GNSS coordinates sampled at intervals δt, with the distance from $P(t)$ to W_{pi} minimized. The overall path accuracy is the average deviation from the way-points, which should be below a predetermined threshold a.

KPI 1:

$$\sum_{i=1}^{n} A_{ci} < a_1 \text{ for } n \text{ way-points, where } A_{ci} = \min_{\forall t} \sqrt{(x(t) - x_{Wpi})^2 + (y(t) - y_{Wpi})^2}$$

KPI 2:

$$d_t - OR < a_2, \text{ where } d_t = \sum_{j=t_0+\Delta t}^{end} \sqrt{(x(j) - x(j-1))^2 + (y(j) - y(j-1))^2}$$

9.4   Collision avoidance

Collision avoidance capability is critical for safe operations of autonomous ships, particularly in busy urban waterways and port environments. A robust collision avoidance system not only increases safety but also optimizes cost in terms of time and energy required to alter the vessel's path. According to international regulations, such as those outlined by the IMO's COLREGs, MASSs as well must adhere to specific guidelines that regulate how vessels should maneuver around each other to prevent collisions.

9.4.1   Avoid obstacle

The avoid obstacle test evaluates the autonomous surface vessel's collision avoidance capabilities by introducing both stationary and moving obstacles like other vessels into its path. This test is crucial for assessing the MASS's ability to detect and efficiently navigate around obstructions, maintaining safety without unnecessary delays or excessive energy use.

The primary KPIs are the registration distance (RD) – the distance at which the MASS first identifies an obstacle – and the efficiency of the evasive maneuvers, which compares the additional distance traveled during the maneuver to the planned route.

- **KPI 1: Registration Distance (RD)** must exceed the minimum safe distance (MD) to effectively avoid the obstacle.
- **KPI 2: Efficiency of the Evasive Maneuver** quantifies the total extra distance traveled during the maneuver compared to the initial route:

$$\sum_{i=t1+\Delta t}^{end} \sqrt{(x(i) - x(i-1))^2 + (y(i) - y(i-1))^2} - IR < a$$

where *IR* is the length of the initial route and *end* marks the conclusion of the route.

9.4.2 *Overtaking*

In the overtaking scenario, an additional vessel moves at a lower speed along the same course as the MASS. The ship must overtake this vessel while maintaining a safe distance and optimal speed.

9.4.3 *Avoid crossing obstacles*

This test evaluates the MASS's response to a moving vessel on a crossing path, emphasizing timely and accurate decision-making to prevent collisions. The MASS must detect the crossing vessel's path, determine a collision course, and apply COLREG regulations to execute efficient evasive maneuvers.

Measurements track the additional distance traveled during maneuvers, the closest approach between vessels, and the distance at which the MASS first detects the crossing vessel. The successful application of COLREGs is crucial, assessing the MASS's ability to navigate safely with minimal impact on its route and energy use.

Key performance indicators include the following:

- **KPI 1: Registration Distance (RD)** must exceed the minimum safe distance (MD) for initiating evasive maneuvers.
- **KPI 2: Minimum Distance** ($\min\{d(t)\}$) between the MASS and the crossing vessel should surpass a set safety distance (SD).
- **KPI 3: Efficiency of the Evasive Maneuver** evaluates the extra distance traveled compared to the initial path:

$$\sum_{i=t1+\Delta t}^{end} \sqrt{(x(i) - x(i-1))^2 + (y(i) - y(i-1))^2} - IR < a$$

where *IR* is the initial route's length and *end* is the maneuver's conclusion.
- **KPI 4: COLREG Compliance** checks if the MASS adheres to giving way if the crossing vessel is from the starboard side.

9.4.4 *Avoid head-to-head collision*

This scenario tests the MASS ability to handle situations where an oncoming vessel is heading directly toward it. The key challenge is to determine whether the vessels can pass each other without altering course or if evasive maneuvers are necessary. According to COLREGs, both vessels should ideally move to starboard to pass each other safely. This test assesses the MASS capability to correctly register and react to potential collision paths.

9.4.5   Avoid shallow water

The avoid shallow scenario examines the MASS depth perception and decision-making capabilities when encountering shallow waters or underwater obstacles. The risk of grounding is a critical concern, and the MASS must demonstrate the ability to effectively avoid such hazards. This scenario can be simulated by setting a course through shallow parts of a waterway or creating underwater obstacles that the MASS must detect and evade using historical data, GNSS, depth measurement tools or sonar.

KPIs for this scenario include the following:

- **KPI 1: Registration Distance (RD)** should be greater than the minimum distance (MD) required to initiate an evasive maneuver safely.

  $$RD > MD$$

- **KPI 2: Efficiency of Navigation** is measured by the additional distance traveled compared to the initial route:

  $$\sum_{i=t_1+\Delta t}^{t_{end}} \sqrt{(x(i) - x(i-1))^2 + (y(i) - y(i-1))^2} - IR < a$$

  where IR is the initial route's length and t_{end} is the last time point of the route.

9.5   Vessel collaboration

The use of a fleet of autonomous vessels can significantly enhance efficiency and safety through coordinated operations. This section explores scenarios involving fleet formations, which are particularly beneficial in terms of energy consumption and control costs. The following vessels can utilize the wake of the leading vessel, reducing overall energy requirements, while simultaneously utilizing control outputs from the leader to maintain formation.

9.5.1   Keep constant distance

The "Keep Constant Distance" test examines an autonomous vessel's ability to maintain a set distance from another, a basic test for fleet formation capabilities. This test is essential for assessing the viability of operating MASS in a fleet, emphasizing safety and efficiency.

In this scenario, two MASS are used: one as the leader and the other as the follower. The objective is for the follower to autonomously adjust its speed and direction to keep a consistent distance from the leader as they navigate a predetermined course. This evaluates the fleet's navigational abilities and ensures that the maintained distance is safe to avoid collisions.

Distances between the MASS are continuously measured, with KPIs including the following:

- **KPI 1: Minimum Distance** ensures that the distance never falls below a safety threshold:

$$\min\{d(t)\} > D - a_1$$

- **KPI 2: Maximum Distance** ensures that the distance does not exceed a set limit:

$$\max\{d(t)\} < D + a_2$$

- **KPI 3: Average Distance** checks for consistent distance maintenance:

$$D - a_3 < \frac{\sum_{t=1}^{n} d(t)}{n} < D + a_3$$

with n being the number of measurement points.

9.5.2   Form a fleet and keep formation

This test evaluates the capability of at least three autonomous surface ships to form and maintain a specific formation. Initially, the MASSs are positioned within close proximity, with one designated as the leader. The leader sets a course while the others adjust their positions relative to the leader, optimizing fuel usage and computational load and leveraging aerodynamic or hydrodynamic benefits.

Key performance indicators include the average and extreme values of distances and angles between MASS, as well as the time taken to establish the formation. Precision in maintaining the formation is monitored using cameras, distance sensors, and GNSS.

Key performance indicators for maintaining the formation include the following:

- **KPI 1: Formation Time** measures the time to complete formation:

$$t_f - t_0 < a_1$$

- **KPI 2–4: Distance Maintenance** ensures that distances between vessels stay within predefined limits:

$$\min\{d_{ij}(t)\} > D_{ij} - a_2, \quad \max\{d_{ij}(t)\} < D_{ij} + a_3,$$

$$\frac{1}{n} \sum_{t=1}^{n} d_{ij}(t) > D_{ij} - a_4 < D_{ij} + a_5$$

- **KPI 5-7: Angle Consistency** checks that angles relative to the leader are maintained:

$$\min\{\theta_i(t)\} > \Theta_i - a_6, \quad \max\{\theta_i(t)\} < \Theta_i + a_7,$$

$$\frac{1}{n} \sum_{t=1}^{n} \theta_i(t) > \Theta_i - a_8 < \Theta_i + a_9$$

Table 9.3   Overview of testing scenarios and their objectives

Scenario	Objective	Assessment aspects
Maneuvering	Test basic maneuvering and responsiveness.	Control, safety, navigability, reliability
Path following	Assess the MASS ability to follow a designated path using way-points.	Control, cost efficiency, navigability, reliability
Collision avoidance	Evaluate the MASS systems for avoiding collisions under various conditions.	Control, safety, cost efficiency, reliability
Vessel collaboration	Test the operation of MASS in a fleet, focusing on formation maintenance and dynamic repositioning.	Control, safety, navigability, energy efficiency

9.5.3   Change position in formation

This scenario involves two vessels within a fleet switching positions while maintaining formation integrity and safety, which can be critical in adapting to environmental changes or operational needs.

These collaborative scenarios are essential for testing the operational dynamics of MASS fleets, ensuring that they can perform complex maneuvers and maintain formations under various maritime conditions. The results from these tests validate the MASS readiness for coordinated and efficient fleet operations.

9.6   Overview of the scenarios

This subsection provides a comprehensive overview of all testing scenarios included in the guideline. Below is a summary table that outlines the objectives of each scenario and the corresponding assessment aspects. This summary helps to contextualize the reasons each scenario has been included in the testing method and how it contributes to the comprehensive assessment of the autonomous vessels.

9.6.1   Key performance indicators (KPIs)

Table 9.3 summarizes the KPIs for assessing the control and safety performance of MASSs across various test scenarios.

9.7   Conclusion

Throughout this chapter, we have explored various test scenarios designed to evaluate the performance and safety of MASS systems in diverse navigational contexts and introduced the RAS framework as a comprehensive approach for these assessments. These tests including maneuvers such as navigating in straight lines, mooring, reaching maximum speeds, stopping, turning, and zigzag paths are critical for ensuring that MASS operates efficiently and safely within their operational environments.

The implementation of collision avoidance tests underscores the importance of MASS' ability to interact with other vessels and stationary obstacles dynamically, ensuring compliance with established maritime regulations such as COLREGs. These tests not only demonstrate an autonomous ship responsiveness to immediate obstructions but also its ability to adhere to legal and safety standards that govern maritime operations.

Moreover, scenarios focusing on fleet formation and maintaining formation with multiple MASS emphasize the technological advances in cooperative navigation. These tests, summarized in Tables 9.4 and 9.5, evaluate the collective operational effectiveness and intervehicle communication within a fleet, marking a significant step toward integrated maritime traffic management systems.

9.7.1   Future recommendations

As the development of MASS continues to accelerate, several areas require further research and development:

- **Enhanced sensor integration:** Future work should focus on integrating more advanced sensor technologies to improve the detection accuracy and operational range of MASS. This includes the development of sophisticated LiDAR, sonar, and radar systems that can provide more detailed environmental feedback and better obstacle detection capabilities.
- **Robust control/AI-based algorithms:** There is a pressing need to develop more robust control and artificial intelligence algorithms that can better assist human decision-making in navigation and collision avoidance scenarios. These algorithms should be capable of handling high-stakes situations with minimal human oversight.
- **Intervehicle communication protocols:** Developing standardized protocols for intervehicle communication will be essential for the effective operation of MASS fleets. This will facilitate smoother coordination and operation across various types of maritime vessels.
- **Regulatory and legal frameworks:** As MASSs become more prevalent, updating regulatory frameworks to include provisions for autonomous operations will be crucial. This includes revising maritime laws to accommodate the unique aspects of unmanned vessels, ensuring that they can operate legally and safely alongside manned vessels.
- **Public acceptance and trust:** Efforts should also be directed toward building public trust and acceptance of autonomous surface vessels. This involves conducting extensive safety trials and transparently sharing results with the public and stakeholders in the maritime industry.

In conclusion, while autonomous ship systems hold substantial promise in revolutionizing maritime transport by enhancing operational efficiency and safety,

Table 9.4   Experimental scenarios and associated SI units

Category	Scenario	KPIs	Unit
Manoeuvring	Navigate in straight line	Average and maximum deviation from straight line	m
	Mooring test	Deviation of stopping location to set location	m
	Reach maximum speed	Maximum speed	m/s
	Stopping test	Track reach, Head reach	m
	Turning test	Tactical diameter, Advance	m
	10-10 zigzag	First overshoot angle, Second overshoot angle, Traveled distance during first overshoot	deg, m
	20-20 zigzag	First overshoot angle	deg
	Pull-out test	Residual turning rate	deg/s
Path following	Path following	Cumulative absolute deviation from the way-points, Deviation from the optimal path length	m
Collision avoidance	Avoid stationary obstacle	Registration range, Extra distance covered by performing maneuver	m
	Overtaking	Registration range, Minimum distance between the vessels, Extra distance covered by performing maneuver	m
	Avoid crossing obstacle	Registration range, Minimum distance between the vessels, Extra distance covered by performing maneuver, Correct implementation of COLREG	m, yes/no
	Avoid head-to-head collision	Registration range, Minimum distance between the vessels, Extra distance covered by performing maneuver	m
	Avoid shallow	Registration range, Extra distance covered by performing maneuver	m
Vessel collaboration	Keep constant distance	Minimum, maximum and average distance to leading MASS	m
	Form a fleet and keep formation	Minimum, maximum and average distance between MASS, Minimum, maximum and average angle between MASS	m, deg
	Change position in formation	Minimum distance between MASS, Time to execute change, Deviation from optimal route	m, s, m

efforts are still required to refine their technological capabilities and integrate them into existing maritime regulations and operational frameworks. A consistent set of performance indicators covering key aspects such as navigation accuracy, reliability, and environmental impact will be essential to monitor progress, guide

Table 9.5 Overview of test methods and KPIs

Scenario	KPIs	Measurement parameters	Measurement equipment	Assessment
Navigate in straight line	Maximum deviation; Average deviation	x–y coordinates over time	GPS; Camera	$\max\{d(t)\} < a1$; $\Sigma_n t = 0(d(t)/n < a2$
Mooring test	Deviation from location	x–y coordinates when moored	Distance sensor; GPS; Camera	$d < a$
Reach maximum speed	Maximum speed	x–y coordinates over time	Camera; GPS	$V_d - \max\{V(t)\} < a$
Stopping test	Track reach; Head reach	x–y coordinates over time	Camera; Distance sensor; GPS	$TR < 10L$; $HR = |x(t_1) - x(t_0)|$
Turning test	Tactical diameter; Advance	x–y coordinates over time	Camera; GPS	$TD < 5L$; $Ad < 4.5L$
$10°$–$10°$ zigzag	First overshoot angle; Second overshoot angle	Heading over time	Camera; GPS; Gyro	$\alpha_1 < 10°$; $\alpha_2 > 10°$
$20°$–$20°$ zigzag	First overshoot angle	Heading over time	Camera; GPS; Gyro	$\alpha_1 < 10°$
Pull-out test	Residual turning rate	Heading over time	Camera; GPS; Gyro	$\dot{\Psi}(t_s) = 0$
Path following	Cumulative deviation from waypoints; Deviation from path length	x–y coordinates over time	Camera; GPS; LiDAR	$\Sigma_n i = 1Aci/n < a1$; $dt - OR < a2$
Avoid stationary obstacle	Registration range; Extra maneuver distance	x–y coordinates over time	Camera; GPS; LiDAR	$RD > MD$; $Extradist. < a$

technological improvements, and align standards across the sector. Finally, continued collaboration among researchers, industry stakeholders, and regulatory bodies will also be important in realizing the full potential of autonomous maritime navigation.

References

[1]   Abaei M.M. and Hekkenberg R.: "A method to assess the reliability of the machinery on autonomous ships." *19th Conf. on Computer Applications and Information Technology in the Maritime Industries*; 2020. pp 11–13.
[2]   Aslam S., Michaelides M.P., and Herodotou, H.: Internet of ships: A survey on architectures, emerging applications, and challenges. *IEEE Internet of Things Journal.* 2020;**7**(10):9714–27.
[3]   Benjamin M.R. and Curcio J.A.: "COLREGS-based navigation of autonomous marine vehicles." *2004 IEEE/OES Autonomous Underwater Vehicles (IEEE Cat. No.04CH37578)*; Sebasco, ME, 2004. Piscataway: IEEE. pp. 32–39.
[4]   Chaal M., Bahootoroody A., Basnet S., Banda O.A.V., and Goerlandt F.: "Towards system-theoretic risk assessment for future ships: A framework for selecting Risk Control Options." *Ocean Engineering.* 2022; **259**:111797.
[5]   Chaal M., Ren X., BahooToroody A., *et al.*: "Research on risk, safety, and reliability of autonomous ships: A bibliometric review." *Safety Science.* 2023;**167**:106256.
[6]   Chae C.-J., Kim M., and Kim H.-J.: "A study on identification of development status of MASS technologies and directions of improvement." *Applied Sciences.* 2020;**10**(13):4564.
[7]   Chang C.-H., Kontovas C., Yu Q., and Yang Z.: "Risk assessment of the operations of maritime autonomous surface ships." *Reliability Engineering & System Safety.* 2021;**207**:107324.
[8]   Choi J., Park J., Jung J., Lee Y., and Choi H.-T.: "Development of an autonomous surface vehicle and performance evaluation of autonomous navigation technologies." *International Journal of Control, Automation and Systems.* 2020; **18**(3):535–45.
[9]   Cockcroft A.N. and Lameijer J.N.F.: *Guide to the Collision Avoidance Rules.* Amsterdam: Elsevier; 2003.
[10] de Vos J., Hekkenberg R.G., and Banda O.A.V.: "The impact of autonomous ships on safety at sea–a statistical analysis." *Reliability Engineering & System Safety.* 2021;**210**:107558.
[11] Ellefsen A.L., Æsøy V., Ushakov S., and Zhang H.: "A comprehensive survey of prognostics and health management based on deep learning for autonomous ships." *IEEE Transactions on Reliability.* 2019;**68**(2):720–40.
[12] Garofano V., Hepworth M., and Shahin R.: Obstacle avoidance and trajectory optimization for an Autonomous Vessel Utilizing MILP Path Planning, Computer Vision based Perception and Feedback Control [Internet]. Institute of Marine Engineering, Science and Technology (IMarEST); 2022. Available from: http://dx.doi.org/10.24868/10713

[13]   Goerlandt F.: "Maritime autonomous surface ships from a risk governance perspective: Interpretation and implications." *Safety Science.* 2020;**128**:104758.

[14]   Hannaford E., Maes P., and Van Hassel E.: "Autonomous ships and the collision avoidance regulations: a licensed deck officer survey." *WMU Journal of Maritime Affairs*, 2022;**21**(2):233–66.

[15]   Haseltalab A., Garofano V., and Afzal M.: The Collaborative Autonomous Shipping Experiment (CASE): Motivations, theory, infrastructure, and experimental challenges [Internet]. *Proceedings of the International Ship Control Systems Symposium (iSCSS)*. IMarEST; 2020. Available from: http://dx.doi.org/10.24868/issn.2631-8741.2020.014

[16]   Hinostroza M.A., Xu H., and Soares C.G.: "Experimental and numerical simulations of zig-zag manoeuvres of a self-running ship model." In: Soares C.G. and Teixeira A.P. (eds), *Maritime Transportation and Harvesting of Sea Resources*. London: Taylor & Francis Group; 2018, pp. 563–70.

[17]   Hirst H.: *COLREGS: Still Fit for Purpose?*; 2020.

[18]   Hoem Å.S., Fjørtoft K.E., and Rødseth Ø.J.: *Addressing the Accidental Risks of Maritime Transportation: Could Autonomous Shipping Technology Improve the Statistics?* Poland: Gdynia Maritime University; 2019.

[19]   Huang Y., Chen L., Chen P., Negenborn R.R., and Van Gelder, P.: "Ship collision avoidance methods: State-of-the-art." *Safety Science.* 2020;**121**:451–73.

[20]   Organization International Maritime: *COLREG: Convention on the International Regulations for Preventing Collisions at Sea, 1972*. London: International Maritime Organization; 2003.

[21]   Johansen T.A., Perez T., and Cristofaro A.: "Ship collision avoidance and COLREGS compliance using simulation-based control behavior selection with predictive hazard assessment." *IEEE Transactions on Intelligent Transportation Systems*. 2016; **17**(12):3407–22.

[22]   Kim M., Joung T.-H., Jeong B., and Park H.-S.: "Autonomous shipping and its impact on regulations, technologies, and industries." *Journal of International Maritime Safety, Environmental Affairs, and Shipping.* 2020;**4**(2):17–25.

[23]   Madsen A.N., Aarset M.V., and Alsos O.A.: "Safe and efficient maneuvering of a Maritime Autonomous Surface Ship (MASS) during encounters at sea: A novel approach." *Maritime Transport Research.* 2022;**3**:100077.

[24]   Mahacek P., Kitts C.A., and Mas I.: "Dynamic guarding of marine assets through cluster control of automated surface vessel fleets." *IEEE/ASME Transactions on Mechatronics.* 2012;**17**(1):65–75.

[25]   International Maritime: *Standard for Ship Manoeuvrability, 1972*. London: International Maritime Organization; 2002.

[26]   Negenborn R.R., Goerlandt F., Johansen T.A., *et al.*: "Autonomous ships are on the horizon: here's what we need to know." *Nature.* 2023;**615**(7950):30–3.

[27]   International Maritime Organization: *COLREG: Convention on the International Regulations for Preventing Collisions at Sea, 1972*. London: International Maritime Organization; 2003.

[28]   Orzechowski S.C.: "Regulatory scoping exercise for the future adoption of autonomous inland ships in Europe." *Journal of Shipping and Trade.* 2024;**9**(1):2.

[29]   Pang Y., Changyuan C., Garofano V., and Negenborn R.: *Assessment of Performance of Urban Freight Vessels.* Project co-funded by the INTER-REG North Sea Region Programme 2014–2020 (ERDF). Publication in the Framework of AVATAR; 2023.

[30]   Quigley M., Gerkey B., Conley K., *et al.* "Ros: an open-source robot operating system." *Proceedings of the IEEE International Conference on Robotics and Automation (ICRA) Workshop on Open Source Robotics*, Kobe, Japan; 2009.

[31]   Ramos M.A., Utne I.B., and Mosleh A.: "Collision avoidance on maritime autonomous surface ships: Operators' tasks and human failure events." *Safety Science.* 2019;**116**:33–44.

[32]   Rødseth Ø.J. and Burmeister H.-C.: "Developments toward the unmanned ship" *Proceedings of International Symposium Information on Ships–ISIS.* Vol. 201, German Institute of Navigation (DGON), Hamburg; 2012. pp. 30–3.

[33]   Statheros T., Howells G., and Maier K.M.: "Autonomous ship collision avoidance navigation concepts, technologies and techniques." *Journal of Navigation.* 2008;**61**(1):129–42.

[34]   Størkersen K.V.: "Safety management in remotely controlled vessel operations." *Marine Policy.* 2021;**130**:104349.

[35]   Tas J.: *Method for Testing and Assessing the Performance of Autonomous Surface Vessels.* Technical Report #2022.MME, 2022.

[36]   Utne I.B., Rokseth B., Sørensen A.J., and Vinnem J.E.: "Towards supervisory risk control of autonomous ships." *Reliability Engineering & System Safety.* 2020;**196**:106757.

[37]   Åström K.J. and Murray R.M.: *Feedback Systems: An Introduction for Scientists and Engineers.* Princeton: Princeton University Press; 2010.

[38]   Öztürk Ü., Akdağ M., and Ayabakan T.: "A review of path planning algorithms in maritime autonomous surface ships: Navigation safety perspective." *Ocean Engineering.* 2022;**251**:111010.

Chapter 10
Summary, conclusions, and future work

Frank Ehlers[1]

In this final chapter, after reading nine chapters that deal with various aspects of the Methods and Measurements, it is worthwhile to summarize the findings, draw conclusions, and look into the future. Thereby, we continue the discussion started in the Introduction (Chapter 1). In Section 10.1, summaries, conclusions, and recommendations for future work from the authors of the chapters are listed. In Section 10.2, the conclusion from the (holistic) viewpoint on autonomy described in Section 1.2 is given: a list of the new Methods presented in this book.

10.1 Chapters' summary, conclusion, and future work

Wanderlingh *et al.* described that future research in maritime autonomy is expected to enhance the robustness, adaptability, and intelligence of autonomous systems operating in complex marine environments. A significant focus is on the advancement of machine learning techniques to improve decision-making and real-time adaptability of autonomous maritime vehicles. Deep learning methods are being explored to enable these systems to effectively respond to uncertain oceanic conditions and sensor noise. Another critical research direction involves the coordinated control of multiple autonomous surface vehicles, addressing challenges in communication, navigation, and control to facilitate cooperative missions such as environmental monitoring and search-and-rescue operations. Multi-agent ASV-AUV networks could coordinate large-scale searches, autonomously dividing tasks and optimizing search patterns to cover vast areas efficiently. Furthermore, energy-efficient and solar-powered autonomous surface vehicles (ASVs) will enable continuous monitoring of high-risk areas, ensuring rapid response with the need for minimal human intervention. By reducing response times, increasing operational reach, and minimizing risk to human rescuers, autonomous maritime systems are set to play an even greater role in saving lives, mitigating disaster impacts, and supporting global humanitarian efforts in the coming years. Advancements in underwater robotics are also notable, with the development of autonomous underwater vehicles capable of deep-sea exploration and infrastructure inspection, addressing challenges such as underwater communication

[1]Bundeswehr Technical Center for Ships and Naval Weapons, Maritime Technology and Research (WTD 71), Kiel, Germany

and harsh environmental conditions. Furthermore, the rise of autonomous surface vessels is revolutionizing naval capabilities, offering cost-effective and efficient alternatives to traditional manned vessels for various military and commercial applications. Collectively, these research directions determine a transformative period in maritime autonomy, driven by interdisciplinary advancements that bridge robotics, artificial intelligence (AI), and marine engineering.

Rego *et al.* proposed a quantized observer and controller with adaptive quantization for limited bandwidth environments, under strong emitter–receiver communication bandwidth constraints, that is applicable to systems operating under state-feedback control. The proposed method is shown to yield a bounded estimation error. The authors derived a set of conditions on the quantizer design parameters that guarantee ultimate boundedness of the estimation error. Future work will focus on refining the quantizer parameters, including the "zoom-in" and "zoom-out" coefficients, to further enhance system performance under various noise and disturbance conditions. Another key direction involves applying the proposed methods to real-world networked marine systems, such as autonomous underwater and surface vehicles. This will allow for the evaluation of their effectiveness in environments characterized by severe communication bandwidth constraints. Additionally, the focus will be on leveraging distributed navigation and control strategies to enable coordinated operations in challenging settings. Furthermore, efforts will be directed toward improving robustness in the presence of packet loss, a critical challenge in acoustic communication networks.

Heshmati-Alamdari *et al.* defined and addressed the problem of cooperative object transportation by a team of underwater vehicle–manipulator systems (UVMSs) operating within a constrained workspace containing static obstacles. Two control strategies are given where the coordination is achieved solely through implicit communication, which emerges from each robot's onboard sensor measurements and the physical interactions between the robots and the commonly grasped object. Consequently, no explicit online data exchange occurs among the robots. Although the control strategies discussed in this chapter do not entirely eliminate the necessity for communication in underwater intervention tasks (especially in terms of safety, adaptability, and efficiency), they significantly reduce the need for continuous inter-robot communication during task execution. This reduction enhances the robustness of the cooperative framework while also mitigating the constraints imposed by the limited bandwidth of acoustic communication channels, such as restrictions on the number of UVMSs that can effectively participate.

Gur *et al.* presented URSULA: a robotic squid designed specifically for dexterous underwater manipulation and seabed intervention tests. Due to potential risks such as entanglement that propeller-based thrusters possess, URSULA deploys two novel locomotion systems without a propeller. The fin system is presented in this chapter. A fin consists of elastic membranes stretched between multiple rigid rods (termed beams) protruding from the body of the robot. The fins are actuated through servomotors attached to the base of the beams and can swivel about the longitudinal axis of the robot to any offset angle between $-90°$ (pointing down vertically) and $90°$ (pointing up vertically). Analytical results and numerical computations compare well, except for the flapping mode, where, contrary to expectations, the analytical model predicts no net force in the sway direction. Following the simulation results, an actual prototype of the fin is built and evaluated in a test tank. Several improvements to the fin model and the actual fin design are planned for the future. The analytic model will be modified to incorporate the elasticity of the membranes and the flexibility of the beams. A more accurate model for the fin hydrodynamic force generation mechanism that can predict the sway forces in flapping mode is necessary. Smaller, rigid fins that are actuated by a single servomotor may represent a viable alternative to the elastic designs discussed in this chapter. Such rigid fins can also help improve reliability, consistency, and durability, which appear to be problematic with elastic fins.

Maehle *et al.* presented an approach and first results for monitoring vegetation and water quality with MONSUN micro-AUVs. The micro-AUV hovers above the lake ground at a constant distance of approximately 1–1.5 m. Various experiments have shown that MONSUN can do this with an accuracy of a few centimeters with a low-cost ping sonar and a pressure sensor for water depth measurements. While hovering, the robot takes videos of the macrophyte vegetation using a downward-looking video camera with LED lights. The quality of these videos proved to be sufficient for the intended automatic recognition of the macrophyte images with AI methods. For mapping the macrophyte vegetation, MONSUN has to dive along a transect from the shore perpendicularly until the vegetation boundary is reached. The images taken on its way must be geo-referenced with an accuracy of a few meters. For the necessary localization, several methods were considered. One is cooperative localization using three MONSUNs in a V-formation. Alternatively, a Point and Shoot strategy or the use of a GPS buoy, both requiring only a single AUV, is considered. Various test drives have been made at Lake Ratzeburg, which showed that all three methods are suitable and allow for the required georeferencing accuracy. For the offline evaluation of the macrophyte videos, an AI image recognition pipeline was developed, including an automatic frame extraction from the videos and subsequent preprocessing. The frame extraction was used successfully for building a macrophyte dataset for training a neural network. However, the required annotation of the images still had to be done by human experts, which was very time-consuming. The current data set comprises about 30,000 images, from which a total of 2105 images for 46 different species, as well as other structural elements have been annotated. The recognition of single macrophytes as required by the standard Phylib method turned out to be too challenging. The mapping of vegetation zones using a classification network and post-processing was, however, possible. Also, the vegetation boundary could be detected online by simpler machine learning techniques with good results. Using an optional sensor ring, chemical–physical values and the underwater light spectrum could also be recorded.

Zenz *et al.* summarized which core concepts of Automatic Target Recognition are relevant for a Reference Model for Automatic Target Recognition. Furthermore, they have conceptualized an approach for consequently obtaining and maintaining such a reference model. Finally, they have exemplified how a Reference Model for Automatic Target Recognition can be generated in a consistent manner and validated through agent-based simulation according to their approach. Altogether, they have presented an approach for performing domain-oriented reference modeling in a sustainable and validatable manner. By stepwise derivation of new models from existing models, whereby each derived model has exactly one base model, a hierarchical structure of models is established. Within this hierarchical structure of knowledge, for two subdomains A.1 and A.2 of some domain A, information applicable to both A.1 and A.2 can be stored in the model of A (base model) and linked within the models of A.1 and A.2 (derived models). Hence, the approach facilitates the intended fostering of cross-domain sharing of knowledge through models. Thereby, it furthermore allows for Data Deduplication. As the approach is based on well-established concepts and frameworks – e.g. Model-based Systems Engineering and the Unified Architecture Framework – and not necessarily dependent on the domain of Automatic Target Recognition, it can also be easily applied to other domains and is especially well suited for any System of Systems domain.

Gavriilidis *et al.* state that the results from both simulations and real-world trials demonstrate the effectiveness of using surrogate models to explain complex AUV behaviors. The high accuracy of the Decision Tree model and the quality of the generated explanations highlight the potential of this approach for improving transparency and trust in autonomous maritime systems. However, the findings also underscore the importance of continuous refinement, particularly in addressing inconsistencies in feature representation and enhancing the robustness of explanations across different mission scenarios. One of the primary challenges was ensuring the quality and consistency of data collected from simulations and real-world trials. Variations in sensor readings, environmental conditions, and vehicle states required robust data preprocessing and validation techniques to maintain the accuracy and reliability of the surrogate models. Selecting and training surrogate models that balance interpretability and performance was another significant challenge. While transparent models like decision trees provided clear explanations, achieving high accuracy often required the use of more complex models, which in turn required more sophisticated explanation techniques such as Shapley values. Additionally, deciding the level of approximation is often application-dependent and left to the interpretation of the model developers, and, in turn, this might not fully capture the expectations of the operators. Generating real-time explanations requires efficient computational methods and the availability of communication streams with the vehicles' existing autonomy systems. Communication interruptions might have a direct impact on the ability of the system to relay explanations to the users. Additionally, while the framework has been designed to be autonomy-agnostic, adapting it to various platforms and mission scenarios requires a good understanding of different autonomy architectures, the implementation of specific interfaces, sensor configurations, and operational requirements, all of which add complexity to the framework's development and deployment. Finally, engaging with domain experts proved invaluable for understanding the nuances of AUV behaviors and developing meaningful explanations. Their insights helped shape the empirical decision tree and validate the surrogate models, ensuring that the explanations were both accurate and operationally relevant. Future work will focus on extending the framework to incorporate more complex behaviors and scenarios, as well as integrating feedback mechanisms to continuously improve explanation quality based on operator input. A structured evaluation of the explanation framework is an important element of future work. This should involve both subjective measures (e.g. user comprehension and levels of trust) and objective performance metrics (e.g. mission success rate and operator intervention frequency) to systematically assess the impact of explainability. Additionally, incorporating advanced natural language generation techniques, including large language models, can improve the clarity and accessibility of explanations. This would enable more nuanced and context-aware communication with operators, improving their ability to understand and respond to AUV behaviors.

Garafano *et al.* explored various test scenarios designed to evaluate the performance and safety of Maritime Autonomous Surface Ship (MASS) systems in diverse navigational contexts and introduced the Researchlab Autonomous Shipping (RAS) framework as a comprehensive approach for these assessments. These tests including maneuvers such as navigating in straight lines, mooring, reaching maximum speeds, stopping, turning, and zigzag paths are critical for ensuring that MASS operates efficiently and safely within their operational environments. The implementation of collision avoidance tests underscores the importance of MASS's ability to interact with other vessels and stationary obstacles dynamically, ensuring compliance with established maritime regulations such as COLREGs. These tests not only demonstrate an autonomous ship's responsiveness to immediate obstructions but also its ability to adhere to legal and safety standards that govern maritime operations. Moreover, scenarios focusing on fleet formation and maintaining formation with multiple MASS emphasize the technological advances in cooperative navigation. These tests evaluate the collective operational effectiveness and inter-vehicle communication within a fleet, marking a significant step toward integrated maritime traffic management systems. As the development of MASS continues to accelerate, several areas require further research and development: Enhanced Sensor Integration, Robust Control/AI-based Algorithms, Inter-Vehicle Communication Protocols, Regulatory and Legal Frameworks, and Public Acceptance and Trust. In conclusion, while autonomous ship systems hold substantial promise in revolutionizing maritime transport by enhancing operational efficiency and safety, efforts are still required to refine their technological capabilities and integrate them into existing maritime regulations and operational frameworks. A consistent set of performance indicators covering key aspects such as navigation accuracy, reliability, and environmental impact will be essential to monitor progress, guide technological improvements, and align standards across the sector. Finally, continued collaboration among researchers, industry stakeholders, and regulatory bodies will also be important in realizing the full potential of autonomous maritime navigation.

10.2   New methods

Putting together what we have seen in this book's guided tour on how Maritime Autonomous Vehicles can be embedded into the currently ongoing Digitalization Process, utilizing model-based engineering paradigms and digital twin concepts, we see a scenario in which

- collaborating Maritime Autonomous Vehicles perform Manipulation Task with the help of AI,

- the collaboration is based on common and explained Reference Models, leading to surrogate models, which balance the information-to-knowledge workload, and providing a certified and trusted fleet of Maritime Autonomous Vehicles.

Certainly, a lot of effort is necessary to make such a scenario become a reality. Echoing Garafano *et al.*, we all should invest in "continued collaboration among researchers, industry stakeholders, and regulatory bodies" to generate sustainable progress from our efforts. One summarizing recommendation from all chapters of this book seems to be that model-based engineering paradigms and digital twin concepts should be used as methods for this collaboration, and AI should be used to gather and exploit the measurements from new experiments with Maritime Autonomous Vehicles.

Index